Structural Vibration

Guoyong Jin · Tiangui Ye
Zhu Su

Structural Vibration

A Uniform Accurate Solution for Laminated
Beams, Plates and Shells with General
Boundary Conditions

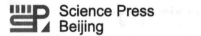 Science Press
Beijing

 Springer

Guoyong Jin
College of Power and Energy Engineering
Harbin Engineering University
Harbin
China

Zhu Su
College of Power and Energy Engineering
Harbin Engineering University
Harbin
China

Tiangui Ye
College of Power and Energy Engineering
Harbin Engineering University
Harbin
China

ISBN 978-3-662-51620-1   ISBN 978-3-662-46364-2   (eBook)
DOI 10.1007/978-3-662-46364-2

Jointly published with Science Press, Beijing
ISBN: 978-7-03-043594-1 Science Press, Beijing

Springer Heidelberg New York Dordrecht London

Printed on acid-free paper

Springer-Verlag GmbH Berlin Heidelberg is part of Springer Science+Business Media
(www.springer.com)

Preface

In practical applications that range from outer space to the deep oceans, engineering structures such as aircraft, rockets, automobiles, turbines, architectures, vessels, and submarines often work in complex environments and can be subjected to various dynamic loads, which can lead to the vibratory behaviors of the structures. In all these applications, the engineering structures may fail and collapse because of material fatigue resulting from vibrations. Many calamitous incidents have shown the destructive nature of vibrations. For instance, the main span of the famous Tacoma Narrows Bridge suffered severe forced resonance and collapsed in 1940 due to the fact that the wind provided an external periodic frequency that matched one of the natural structural frequencies of the bridge. Furthermore, noise generated by vibrations always causes annoyance, discomfort, and loss of efficiency to human beings. Therefore, it is of particular importance to understand the structural vibrations and reduce them through proper design to ensure a reliable, safe, and lasting structural performance. An important step in the vibration design of an engineering structure is the evaluation of its vibration modal characteristics, such as natural frequencies and mode shapes. This modal information plays a key role in the design and vibration suppression of the structure when subjected to dynamics excitations. In engineering applications, a variety of possible boundary restraining cases may be encountered for a structure. In recent decades, the ability of predicting the vibration characteristics of structures with general boundary conditions is of prime interest to engineers and designers and is the mutual concern of researchers in this field as well.

Beams, plates, and shells are basic structural elements of most engineering structures and machines. A thorough understanding of their vibration characteristics is of great significance for engineers to predict the vibrations of the whole structures and design suitable structures with low vibration and noise radiation characteristics. There exists many books, papers, and research reports on the vibration analysis of beams, plates, and shells. In 1969, Prof. A.W. Leissa published the excellent monograph *Vibration of Plates*, in which theoretical and experimental results of approximately 500 research papers and reports were presented. And in 1973, he organized and summarized approximately 1,000 references in the field of shell vibrations and published another famous monograph entitled *Vibration of Shells*. New survey shows that the literature on the vibrations of beams, plates, and shells has expanded rapidly since then. Based on the Google Scholar search tool, the numbers of article related to the following keywords from 1973 up to 2014 are: 315,000 items for "vibration & beam," 416,000 items for "vibration & plates," and 101,000 items for "vibration & shell." This clearly reveals the importance of the vibration analysis of beams, plates, and shells.

Undeniably, significant advances in the vibration analysis of beams, plates, and shells have been achieved over the past four decades. Many accurate and efficient computational methods have also been developed, such as the Ritz method, differential quadrature method (DQM), Galerkin method, wave propagation approach, multiquadric radial basis function method (MRBFM), meshless method, finite element method (FEM), discrete singular convolution approach (DSC), etc. Furthermore, a large variety of classical and modern theories have been proposed by researchers, such as the classical structure theories (CSTs), the first-order shear deformation theories (FSDTs), and the higher order shear deformation theories (HSDTs).

However, after the review of the literature in this subject, it appears that most of the books deal with a technique that is only suitable for a particular type of classical boundary conditions (i.e., simply supported supports, clamped boundaries, free edges, shear-diaphragm restrains and their combinations), which typically requires constant modifications of the solution procedures and corresponding computation codes to adapt to different boundary cases. This will result in very tedious calculations and be easily inundated with various boundary conditions in practical applications since the boundary conditions of a beam, plate, or shell may not always be classical in nature, a variety of possible boundary restraining cases, including classical boundary conditions, elastic restraints, and their combinations may be encountered. In addition, with the development of new industries and modern processes, laminated beams, plates, and shells composed of composite laminas are extensively used in many fields of modern engineering practices such as space vehicles, civil constructions, and deep-sea engineering equipments to satisfy special functional requirements due to their outstanding bending rigidity, high strength-weight and stiffness-weight ratios, excellent vibration characteristics, and good fatigue properties. The vibration results of laminated beams, plates, and shells are far from complete. It is necessary and of great significance to develop a unified, efficient, and accurate method which is capable of universally dealing with laminated beams, plates, and shells with general boundary conditions. Furthermore, a systematic, comprehensive, and up-to-date monograph which contains vibration results of isotropic and laminated beams, plates, and shells with various lamination schemes and general boundary conditions would be highly desirable and useful for the senior undergraduate and postgraduate students, teachers, engineers, and individual researchers in this field.

In view of these apparent voids, the present monograph presents an endeavor to complement the vibration analysis of laminated beams, plates, and shells. The title,

Structural Vibration: A Uniform Accurate Solution for Laminated Beams, Plates and Shells with General Boundary Conditions, illustrates the main aim of this book, namely:

(1) To develop an accurate semi-analytical method which is capable of dealing with the vibrations of laminated beams, plates, and shells with arbitrary lamination schemes and general boundary conditions including classical boundaries, elastic supports and their combinations, aiming to provide a unified and reasonable accurate alternative to other analytical and numerical techniques.

(2) To provide a summary of known results of laminated beams, plates, and shells with various lamination schemes and general boundary conditions, which may serve as benchmark solutions for the future research in this field.

The book is organized into eight chapters. Fundamental equations of laminated shells in the framework of classical shell theory and shear deformation shell theory are derived in detail, including the kinematic relations, stress–strain relations and stress resultants, energy functions, governing equations, and boundary conditions. The corresponding fundamental equations of laminated beams and plates are specialized from the shell ones. Following the fundamental equations, a unified modified Fourier series method is developed. Then both strong and weak form solution procedures are realized and established by combining the fundamental equations and the modified Fourier series method. Finally, numerical vibration results are presented for isotropic, orthotropic, and laminated beams, plates, and shells with various geometry and material parameters, different lamination schemes and different boundary conditions including the classical boundaries, elastic ones, and their combinations. Summarizing, the work is arranged as follows:

The theories of linear vibration of laminated beams, plates, and shells are well established. In this regard, Chap. 1 introduces the fundamental equations of laminated beams, plates, and shells in the framework of classical shell theory and the first-order shear deformation shell theory without proofs.

Chapter 2 presents a modified Fourier series method which is capable of dealing with vibrations of laminated beams, plates, and shells with general boundary conditions. In the modified Fourier series method, each displacement of a laminated beam, plate, or shell, regardless of boundary conditions, is invariantly expressed as a new form of trigonometric series expansions in which several supplementary terms are introduced to ensure and accelerate the convergence of the series expansion. Then one can seek the solutions either in strong form solution procedure or the weak form one. These two solution procedures are fully illustrated in this chapter.

Chapters 3–8 deal with laminated beams, plates, and cylindrical, conical, spherical and shallow shells, respectively. In each chapter, corresponding fundamental equations in the framework of classical and shear deformation theories for the general dynamic analysis are developed first, which can be useful for potential readers. Following the fundamental equations, numerous free vibration results are presented for various configurations including different boundary conditions, laminated sequences, and geometry and material properties.

Finally, the authors would like to record their appreciation to the National Natural Science Foundation of China (Grant Nos. 51175098, 51279035, and 10802024) and the Fundamental Research Funds for the Central Universities of China (No. HEUCFQ1401) for partially funding this work.

Contents

Chapter 1
Fundamental Equations of Laminated Beams, Plates and Shells

Beams, plates and shells are named according to their size or/and shape features. Shells have all the features of plates except an additional one-curvature (Leissa 1969, 1973). Therefore, the plates, on the other hand, can be viewed as special cases of shells having no curvature. Beams are one-dimensional counterparts of plates (straight beams) or shells (curved beams) with one dimension relatively greater in comparison to the other two dimensions. This chapter introduces the fundamental equations (including kinematic relations, stress-strain relations and stress resultants, energy functions, governing equations and boundary conditions) of laminated shells in the framework of the classical shell theory (CST) and the shear deformation shell theory (SDST) without proofs due to the fact that they have been well established. The corresponding equations of laminated beams and plates are specialized from the shell ones.

1.1 Three-Dimensional Elasticity Theory in Curvilinear Coordinates

Consider a three-dimensional (3D) shell segment with total thickness h as shown in Fig. 1.1, a 3D orthogonal coordinate system (α, β and z) located on the middle surface is used to describe the geometry dimensions and deformations of the shell, in which co-ordinates along the meridional, circumferential and normal directions are represented by α, β and z, respectively. R_α and R_β are the mean radii of curvature in the α and β directions on the middle surface ($z = 0$). U, V and W separately indicate the displacement variations of the shell in the α, β and z directions. The strain-displacement relations of the three-dimensional theory of elasticity in orthogonal curvilinear coordinate system are (Leissa 1973; Soedel 2004; Carrera et al. 2011):

© Science Press, Beijing and Springer-Verlag Berlin Heidelberg 2015
G. Jin et al., *Structural Vibration*, DOI 10.1007/978-3-662-46364-2_1

Fig. 1.1 Notations in shell coordinate system (α, β and z)

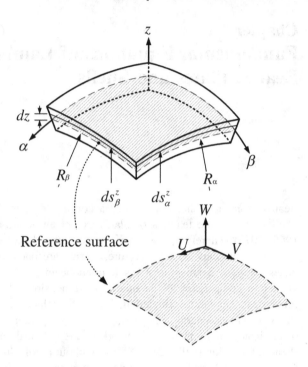

$$\varepsilon_\alpha = \frac{1}{(1+z/R_\alpha)}\left(\frac{1}{A}\frac{\partial U}{\partial \alpha} + \frac{V}{AB}\frac{\partial A}{\partial \beta} + \frac{W}{R_\alpha}\right)$$

$$\varepsilon_\beta = \frac{1}{(1+z/R_\beta)}\left(\frac{U}{AB}\frac{\partial B}{\partial \alpha} + \frac{1}{B}\frac{\partial V}{\partial \beta} + \frac{W}{R_\beta}\right)$$

$$\varepsilon_z = \frac{\partial W}{\partial z}$$

$$\gamma_{\alpha\beta} = \frac{A(1+z/R_\alpha)}{B(1+z/R_\beta)}\frac{\partial}{\partial \beta}\left[\frac{U}{A(1+z/R_\alpha)}\right]$$

$$+ \frac{B(1+z/R_\beta)}{A(1+z/R_\alpha)}\frac{\partial}{\partial \alpha}\left[\frac{V}{B(1+z/R_\beta)}\right]$$

$$\gamma_{\alpha z} = \frac{1}{A(1+z/R_\alpha)}\frac{\partial W}{\partial \alpha} + A(1+z/R_\alpha)\frac{\partial}{\partial z}\left[\frac{U}{A(1+z/R_\alpha)}\right]$$

$$\gamma_{\beta z} = \frac{1}{B(1+z/R_\beta)}\frac{\partial W}{\partial \beta} + B(1+z/R_\beta)\frac{\partial}{\partial z}\left[\frac{V}{B(1+z/R_\beta)}\right]$$

$$(1.1a\text{–}f)$$

where the quantities A and B are the Lamé parameters of the shell. They are determined by the shell characteristics and the selected orthogonal coordinate system. The detail definitions of them are given in Sect. 1.4. The lengths in the α and β directions of the shell segment at distance dz from the shell middle surface are (see Fig. 1.1):

$$ds_\alpha^z = A\left(1 + \frac{z}{R_\alpha}\right)d\alpha \quad ds_\beta^z = B\left(1 + \frac{z}{R_\beta}\right)d\beta \tag{1.2}$$

The above equations contain the fundamental strain-displacement relations of a 3D body in curvilinear coordinate system. They are specialized to those of CST and FSDT by introducing several assumptions and simplifications.

1.2   Fundamental Equations of Thin Laminated Shells

According to Eq. (1.1), it can be seen that the 3D strain-displacement equations of a shell are rather complicated when written in curvilinear coordinate system. Typically, researchers simplify the 3D shell equations into the 2D ones by making certain assumptions to eliminate the coordinate in the thickness direction. Based on different assumptions and simplifications, various sub-category classical theories of thin shells were developed, such as the Reissner-Naghdi's linear shell theory, Donner-Mushtari's theory, Flügge's theory, Sanders' theory and Goldenveizer-Novozhilov's theory, etc. In this book, we focus on shells composed of arbitrary numbers of composite layers which are bonded together rigidly. When the total thickness of a laminated shell is less than 0.05 of the wavelength of the deformation mode or radius of curvature, the classical theories of thin shells originally developed for single-layered isotropic shells can be readily extended to the laminated ones. Leissa (1973) showed that most thin shell theories yield similar results. In this section, the fundamental equations of the Reissner-Naghdi's linear shell theory are extended to thin laminated shells due to that it offers the simplest, the most accurate and consistent equations for laminated thin shells (Qatu 2004).

1.2.1   Kinematic Relations

In the classical theory of thin shells, the four assumptions made by Love (1944) are universally accepted to be valid for a *first approximation shell theory* (Rao 2007):

1. *The thickness of the shell is small compared with the other dimensions.*
2. *Strains and displacements are sufficiently small so that the quantities of second- and higher-order magnitude in the strain-displacement relations may be neglected in comparison with the first-order terms.*
3. *The transverse normal stress is small compared with the other normal stress components and may be neglected.*
4. *Normals to the undeformed middle surface remain straight and normal to the deformed middle surface and suffer no extension.*

The first assumption defines that the shell is thin enough so that the deepness terms z/R_α and z/R_β can be neglected compared to unity in the strain-displacement relations (i.e., $z/R_\alpha \ll 1$ and $z/R_\beta \ll 1$). The second assumption ensures that the differential equations will be linear. The fourth assumption is also known as Kirchhoff's hypothesis. This assumption leads to zero transverse shear strains and zero transverse normal strain ($\gamma_{\alpha z} = 0$, $\gamma_{\beta z} = 0$ and $\varepsilon_z = 0$). Taking these assumptions into consideration, the 3D strain-displacement relations of shells in orthogonal curvilinear coordinate system can be reduced to those of 2D classical thin shells as:

$$\varepsilon_\alpha = \frac{1}{A}\frac{\partial U}{\partial \alpha} + \frac{V}{AB}\frac{\partial A}{\partial \beta} + \frac{W}{R_\alpha}$$

$$\varepsilon_\beta = \frac{U}{AB}\frac{\partial B}{\partial \alpha} + \frac{1}{B}\frac{\partial V}{\partial \beta} + \frac{W}{R_\beta} \qquad (1.3\text{a--c})$$

$$\gamma_{\alpha\beta} = \frac{A}{B}\frac{\partial}{\partial \beta}\left[\frac{U}{A}\right] + \frac{B}{A}\frac{\partial}{\partial \alpha}\left[\frac{V}{B}\right]$$

According to the Kirchhoff hypothesis, the displacement variations in the α, β and z directions are restricted to the following linear relationships (Leissa 1973):

$$U(\alpha, \beta, z) = u(\alpha, \beta) + z\phi_\alpha(\alpha, \beta)$$

$$V(\alpha, \beta, z) = v(\alpha, \beta) + z\phi_\beta(\alpha, \beta) \qquad (1.4)$$

$$W(\alpha, \beta, z) = w(\alpha, \beta)$$

where u, v and w are the displacement components on the middle surface in the α, β and z directions. ϕ_α, ϕ_β represent the rotations of transverse normal respect to β- and α-axes, respectively. They are determined by substituting Eq. (1.4) into Eq. (1.1e, f) and letting $\gamma_{\alpha z} = 0$, $\gamma_{\beta z} = 0$, i.e.:

$$\phi_\alpha = \frac{u}{R_\alpha} - \frac{1}{A}\frac{\partial w}{\partial \alpha} \qquad \phi_\beta = \frac{v}{R_\beta} - \frac{1}{B}\frac{\partial w}{\partial \beta} \qquad (1.5)$$

Substituting Eqs. (1.4) and (1.5) into Eq. (1.3), the strain-displacement relations of thin shells can be rewritten as:

$$\varepsilon_\alpha = \varepsilon_\alpha^0 + z\chi_\alpha$$

$$\varepsilon_\beta = \varepsilon_\beta^0 + z\chi_\beta \qquad (1.6)$$

$$\gamma_{\alpha\beta} = \gamma_{\alpha\beta}^0 + z\chi_{\alpha\beta}$$

where ε_α^0, ε_β^0 and $\gamma_{\alpha\beta}^0$ denote the normal and shear strains in the middle surface. χ_α, χ_β and $\chi_{\alpha\beta}$ are the corresponding curvature and twist changes. They are written in terms of shell displacements u, v and w as:

$$\varepsilon_\alpha^0 = \frac{1}{A}\frac{\partial u}{\partial \alpha} + \frac{v}{AB}\frac{\partial A}{\partial \beta} + \frac{w}{R_\alpha}$$

$$\varepsilon_\beta^0 = \frac{1}{B}\frac{\partial v}{\partial \beta} + \frac{u}{AB}\frac{\partial B}{\partial \alpha} + \frac{w}{R_\beta}$$

$$\gamma_{\alpha\beta}^0 = \frac{1}{A}\frac{\partial v}{\partial \alpha} - \frac{u}{AB}\frac{\partial A}{\partial \beta} + \frac{1}{B}\frac{\partial u}{\partial \beta} - \frac{v}{AB}\frac{\partial B}{\partial \alpha}$$

$$\chi_\alpha = \frac{1}{A}\frac{\partial \phi_\alpha}{\partial \alpha} + \frac{\phi_\beta}{AB}\frac{\partial A}{\partial \beta} \tag{1.7}$$

$$\chi_\beta = \frac{1}{B}\frac{\partial \phi_\beta}{\partial \beta} + \frac{\phi_\alpha}{AB}\frac{\partial B}{\partial \alpha}$$

$$\chi_{\alpha\beta} = \frac{1}{A}\frac{\partial \phi_\beta}{\partial \alpha} - \frac{\phi_\alpha}{AB}\frac{\partial A}{\partial \beta} + \frac{1}{B}\frac{\partial \phi_\alpha}{\partial \beta} - \frac{\phi_\beta}{AB}\frac{\partial B}{\partial \alpha}$$

Equation (1.7) constitutes the strain-displacement relations of a thin shell in curvilinear coordinates.

1.2.2 Stress-Strain Relations and Stress Resultants

With the development of new industries and modern processes, composite materials are extensively used in many fields of modern engineering practices such as aircraft and spacecraft, civil constructions and deep-ocean engineering to satisfy special functional requirements due to their outstanding bending rigidity, high strength-weight and stiffness-weight ratios, excellent vibration characteristics and good fatigue properties. For instance, more than 20 % of the A380's airframe is composite materials.

Typically, composite materials are made of reinforcement material distributed in matrix material. There commonly exist three types of composite materials (Reddy 2003; Ye 2003): (1) fiber composites, in which the reinforcements are in the form of fibers. The fibers can be continuous or discontinuous, unidirectional, bidirectional, woven or randomly distributed; (2) particle composites, which are composed of macro size particles of reinforcement in a matrix of another, such as concrete; (3) laminated composites, which consist of layers of various materials, including composites of the first two types. As for many other kinds of composite structures, beams, plates and shells composed of arbitrary numbers of unidirectional fiber reinforced layers with different fiber orientations (see Fig. 1.2) are most frequently used in the engineering applications and are the mutual concern of researchers in this field as well. In such cases, by appropriately orientating the fibers in each lamina of the structure, desired strength and stiffness parameters can be achieved. As a consequence, this book is devoted to the vibration analysis of laminated beams, plates and shells made of this type of laminated composite.

Fig. 1.2 A laminated shell
made up of composite layers
with different fiber
orientations

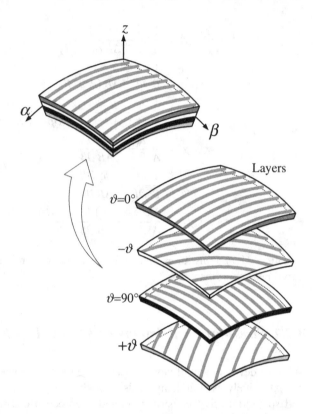

Layers

$\vartheta=0°$

$-\vartheta$

$\vartheta=90°$

$+\vartheta$

In this section we primarily study the stress-strain relations of a unidirectional
fiber reinforced layer, which is the basic building block of a composite laminated
structure. A unidirectional fiber reinforced layer can be treated as an orthotropic
material whose material symmetry planes are parallel and transverse to the fiber
direction (Reddy 2003). See Fig. 1.3, suppose the laminated shell is constructed by
N unidirectional fiber-reinforced layers which are bonded together rigidly. The
principal coordinates of the composite material in the kth layer are denoted by 1, 2
and 3, in which the coordinate axes 1 and 2 are taken to be parallel and transverse to
the fiber orientation. The 3 axis is parallel to the normal direction of the shell. The
angle between the material axis 1 (or 2) and the α axis (or β) is denoted by ϑ^k and Z_{k+1}
and Z_k are the distances from the top surface and the bottom surface of the layer to
the referenced middle surface, respectively. Thus, according to generalized Hooke's
law, the corresponding stress-strain relations in the kth layer of the laminated shell
can be written as:

$$\left\{ \begin{array}{c} \sigma_\alpha \\ \sigma_\beta \\ \tau_{\alpha\beta} \end{array} \right\}_k = \begin{bmatrix} Q_{11}^k & Q_{12}^k & Q_{16}^k \\ Q_{12}^k & Q_{22}^k & Q_{26}^k \\ Q_{16}^k & Q_{26}^k & Q_{66}^k \end{bmatrix} \left\{ \begin{array}{c} \varepsilon_\alpha \\ \varepsilon_\beta \\ \gamma_{\alpha\beta} \end{array} \right\}_k \tag{1.8}$$

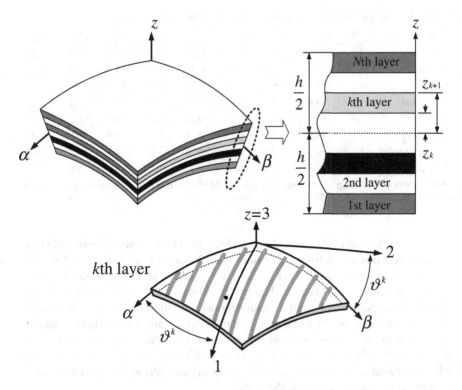

Fig. 1.3 Lamination, geometry and material coordinate systems of a laminated shell

where σ_α and σ_β are the normal stresses in the α, β directions, respectively. $\tau_{\alpha\beta}$ represents the corresponding shear stress. The constants $\overline{Q}_{ij}^k$ ($i, j = 1, 2$ and 6) are elastic stiffness coefficients of the layer which are found from following equations:

$$\begin{bmatrix} \overline{Q}_{11}^k & \overline{Q}_{12}^k & \overline{Q}_{16}^k \\ \overline{Q}_{12}^k & \overline{Q}_{22}^k & \overline{Q}_{26}^k \\ \overline{Q}_{16}^k & \overline{Q}_{26}^k & \overline{Q}_{66}^k \end{bmatrix} = \mathbf{T} \begin{bmatrix} Q_{11}^k & Q_{12}^k & 0 \\ Q_{12}^k & Q_{22}^k & 0 \\ 0 & 0 & Q_{66}^k \end{bmatrix} \mathbf{T}^T \qquad (1.9)$$

where superscript T represents the transposition operator. The material constants $Q_{ij}^k(i, j = 1, 2$ and 6) are defined in terms of the material properties of the orthotropic layer:

$$\begin{aligned} Q_{11}^k &= \frac{E_1}{1 - \mu_{12}\mu_{21}} \\ Q_{12}^k &= \frac{\mu_{12}E_2}{1 - \mu_{12}\mu_{21}} \\ Q_{22}^k &= \frac{E_2}{1 - \mu_{12}\mu_{21}} \\ Q_{66}^k &= G_{12} \end{aligned} \qquad (1.10)$$

Table 1.1 Values of the material parameters for several materials (Reddy 2003)

Material	E_1	E_2	E_3	μ_{12}	μ_{13}	μ_{23}	G_{12}	G_{13}	G_{23}
Aluminum	73.1	73.1	73.1	0.33	0.33	0.33	23.3	23.3	23.3
Copper	124.1	124.1	124.1	0.33	0.33	0.33	44.1	44.1	44.1
Steel	206.8	206.8	206.8	0.29	0.29	0.29	77.5	77.5	77.5
Gr.-Ep (As)	137.9	9.0	9.0	0.3	0.3	0.49	7.1	7.1	6.2
Gr.-Ep (T)	131.0	10.3	10.3	0.22	0.22	0.49	6.9	6.2	6.2
Gr.-Ep (1)	53.8	17.9	17.9	0.25	0.25	0.34	9.0	9.0	3.4
Gr.-Ep (2)	38.6	8.3	9.0	0.26	0.26	0.34	4.1	4.1	3.4
Br.-Ep	206.8	20.7	20.7	0.3	0.25	0.25	6.9	6.9	4.1

Moduli are in GPa; Gr.-Ep (AS) = graphite-epoxy (AS13501); Gr.-Ep (T) = graphite-epoxy (T300/934); GI.-Ep = glass-epoxy; Br.-Ep = boron-epoxy

E_1 and E_2 are moduli of elasticity of the unidirectional fiber-reinforced material in the 1 and 2 directions, respectively. μ_{12} is the major Poisson's ratio. The other Poisson's ratio μ_{21} is determined by equation $\mu_{12}E_2 = \mu_{21}E_1$ and G_{12} is shear modulus. An isotropic shell model can be obtained by letting $E_1 = E_2$, $G_{12} = E_1/(2 + 2\mu_{12})$. The material parameters of the unidirectional fiber-reinforced material can be determined experimentally using an appropriate test specimen made up the material. The values of the material parameters for several materials are given in Table 1.1 (Reddy 2003). **T** is the transformation matrix which represents the relation of the principal material coordinate system (1, 2, 3) of the kth layer with respect to the shell coordinate system (α, β, z). It is defined as:

$$\mathbf{T} = \begin{bmatrix} \cos^2 \vartheta^k & \sin^2 \vartheta^k & -2 \sin \vartheta^k \cos \vartheta^k \\ \sin^2 \vartheta^k & \cos^2 \vartheta^k & 2 \sin \vartheta^k \cos \vartheta^k \\ \sin \vartheta^k \cos \vartheta^k & -\sin \vartheta^k \cos \vartheta^k & \cos^2 \vartheta^k - \sin^2 \vartheta^k \end{bmatrix} \quad (1.11)$$

Performing the matrix multiplication in Eq. (1.9), the elastic stiffness coefficients constants $\overline{Q_{ij}^k}$ can be written as:

$$\overline{Q_{11}^k} = Q_{11}^k \cos^4 \vartheta^k + \left(2Q_{12}^k + 4Q_{66}^k\right) \cos^2 \vartheta^k \sin^2 \vartheta^k + Q_{22}^k \sin^4 \vartheta^k$$

$$\overline{Q_{12}^k} = \left(Q_{11}^k + Q_{22}^k - 4Q_{66}^k\right) \cos^2 \vartheta^k \sin^2 \vartheta^k + Q_{12}^k \left(\cos^4 \vartheta^k + \sin^4 \vartheta^k\right)$$

$$\overline{Q_{16}^k} = \left(Q_{11}^k - Q_{12}^k - 2Q_{66}^k\right) \cos^3 \vartheta^k \sin \vartheta^k + \left(Q_{12}^k - Q_{22}^k + 2Q_{66}^k\right) \cos \vartheta^k \sin^3 \vartheta^k$$

$$\overline{Q_{22}^k} = Q_{11}^k \sin^4 \vartheta^k + Q_{22}^k \cos^4 \vartheta^k + \left(2Q_{12}^k + 4Q_{66}^k\right) \cos^2 \vartheta^k \sin^2 \vartheta^k$$

$$\overline{Q_{26}^k} = \left(Q_{11}^k - Q_{12}^k - 2Q_{66}^k\right) \cos \vartheta^k \sin^3 \vartheta^k + \left(Q_{12}^k - Q_{22}^k + 2Q_{66}^k\right) \cos^3 \vartheta^k \sin \vartheta^k$$

$$\overline{Q_{66}^k} = Q_{66}^k \left(\cos^4 \vartheta^k + \sin^4 \vartheta^k\right) + \left(Q_{11}^k - 2Q_{12}^k + Q_{22}^k - 2Q_{66}^k\right) \cos^2 \vartheta^k \sin^2 \vartheta^k$$

$$(1.12)$$

The force and moment resultants of thin laminated shells are obtained by carrying the integration of stresses over the cross-section, from layer to layer:

$$\begin{bmatrix} N_\alpha \\ N_\beta \\ N_{\alpha\beta} \end{bmatrix} = \int_{-h/2}^{h/2} \begin{bmatrix} \sigma_\alpha \\ \sigma_\beta \\ \tau_{\alpha\beta} \end{bmatrix} dz \quad \begin{bmatrix} M_\alpha \\ M_\beta \\ M_{\alpha\beta} \end{bmatrix} = \int_{-h/2}^{h/2} \begin{bmatrix} \sigma_\alpha \\ \sigma_\beta \\ \tau_{\alpha\beta} \end{bmatrix} z dz \quad (1.13)$$

Performing the matrix integration in Eq. (1.13), the force and moment resultants can be defined in terms of the middle surface strains and curvature changes as:

$$\begin{bmatrix} N_\alpha \\ N_\beta \\ N_{\alpha\beta} \\ M_\alpha \\ M_\beta \\ M_{\alpha\beta} \end{bmatrix} = \begin{bmatrix} A_{11} & A_{12} & A_{16} & B_{11} & B_{12} & B_{16} \\ A_{12} & A_{22} & A_{26} & B_{12} & B_{22} & B_{26} \\ A_{16} & A_{26} & A_{66} & B_{16} & B_{26} & B_{66} \\ B_{11} & B_{12} & B_{16} & D_{11} & D_{12} & D_{16} \\ B_{12} & B_{22} & B_{26} & D_{12} & D_{22} & D_{26} \\ B_{16} & B_{26} & B_{66} & D_{16} & D_{26} & D_{66} \end{bmatrix} \begin{bmatrix} \varepsilon_\alpha^0 \\ \varepsilon_\beta^0 \\ \gamma_{\alpha\beta}^0 \\ \chi_\alpha \\ \chi_\beta \\ \chi_{\alpha\beta} \end{bmatrix} \quad (1.14)$$

where N_α, N_β and $N_{\alpha\beta}$ are the normal and shear force resultants and M_α, M_β and $M_{\alpha\beta}$ denote the bending and twisting moment resultants. A_{ij}, B_{ij}, and D_{ij} are the stiffness coefficients arising from the piecewise integration over the shell thickness:

$$A_{ij} = \sum_{k=1}^{N} \overline{Q_{ij}^k}(z_{k+1} - z_k)$$

$$B_{ij} = \frac{1}{2}\sum_{k=1}^{N} \overline{Q_{ij}^k}(z_{k+1}^2 - z_k^2) \quad (1.15)$$

$$D_{ij} = \frac{1}{3}\sum_{k=1}^{N} \overline{Q_{ij}^k}(z_{k+1}^3 - z_k^3)$$

Notably, when a thin shell is laminated symmetrically with respect to its middle surface, the constants B_{ij} equal to zero. This will sufficiently reduce the complexity of the stress-strain relations, energy functions, governing equations and boundary conditions of the shell.

1.2.3 Energy Functions

The energy functions are very convenient in deriving the governing equations and boundary conditions of a structure. In addition, they are important in searching approximate solutions of practical problems. The strain energy (U_s) of a thin laminated shell during vibration can be defined in terms of the middle surface strains and curvature changes and stress resultants as:

$$U_s = \frac{1}{2} \int_\alpha \int_\beta \left\{ \begin{array}{l} N_\alpha \varepsilon_\alpha^0 + N_\beta \varepsilon_\beta^0 + N_{\alpha\beta}\gamma_{\alpha\beta}^0 \\ + M_\alpha \chi_\alpha + M_\beta \chi_\beta + M_{\alpha\beta}\chi_{\alpha\beta} \end{array} \right\} ABd\alpha d\beta \qquad (1.16)$$

Substituting Eq. (1.7) into Eq. (1.16), the above equation can be rewritten as:

$$U_s = \frac{1}{2} \int_\alpha \int_\beta \left\{ \begin{array}{l} N_\alpha \left(B\frac{\partial u}{\partial \alpha} + v\frac{\partial A}{\partial \beta} + \frac{ABw}{R_\alpha} \right) \\ + N_\beta \left(A\frac{\partial v}{\partial \beta} + u\frac{\partial B}{\partial \alpha} + \frac{ABw}{R_\beta} \right) + N_{\alpha\beta} \\ \left(B\frac{\partial v}{\partial \alpha} - u\frac{\partial A}{\partial \beta} + A\frac{\partial u}{\partial \beta} - v\frac{\partial B}{\partial \alpha} \right) \\ + M_\alpha \left(B\frac{\partial \phi_\alpha}{\partial \alpha} + \phi_\beta \frac{\partial A}{\partial \beta} \right) \\ + M_\beta \left(A\frac{\partial \phi_\beta}{\partial \beta} + \phi_\alpha \frac{\partial B}{\partial \alpha} \right) + M_{\alpha\beta} \\ \left(B\frac{\partial \phi_\beta}{\partial \alpha} - \phi_\alpha \frac{\partial A}{\partial \beta} + A\frac{\partial \phi_\alpha}{\partial \beta} - \phi_\beta \frac{\partial B}{\partial \alpha} \right) \end{array} \right\} d\alpha d\beta \qquad (1.17)$$

The kinetic energy (T) of the shell is written as:

$$T = \frac{1}{2} \int_\alpha \int_\beta I_0 \left\{ \left(\frac{\partial u}{\partial t} \right)^2 + \left(\frac{\partial v}{\partial t} \right)^2 + \left(\frac{\partial w}{\partial t} \right)^2 \right\} ABd\alpha d\beta \qquad (1.18)$$

where the inertia terms are:

$$I_0 = \sum_{k=1}^{N} \int_{z_k}^{z_{k+1}} \rho^k dz \qquad (1.19)$$

in which ρ^k represents the mass of the kth layer per unit area. The external work (W_e) can be expressed as:

$$W_e = \int_\alpha \int_\beta \left\{ q_\alpha u + q_\beta v + q_z w \right\} ABd\alpha d\beta \qquad (1.20)$$

where q_α, q_β and q_z denote the external loads in the α, β and z directions, respectively.

Laminated shells with a variety of possible boundary conditions, such as free edges, shear-diaphragm restraints, simply-supported supports, clamped boundaries, elastic constraints and their combinations can be encountered in practice. Since the main focus of this book is to complement the vibration studies of composite laminated structures with general boundary conditions, in order to satisfy the request, at each of the four boundaries of a laminated shell, the general boundary conditions are implemented by using the artificial spring boundary technique in which three groups of linear springs (k_u, k_v, k_w) and one group of rotational springs (K_w) which are distributed uniformly along the boundary are introduced to

Fig. 1.4 Notations of artificial spring boundary technique of a thin laminated shell

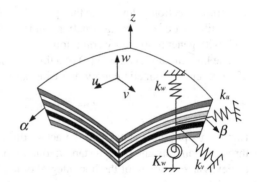

separately simulate the given or typical boundary conditions expressed in the form of axial force resultant, tangential force resultant, transverse force resultant, and the flexural moment resultant, see Fig. 1.4. The stiffness of the boundary springs can take any value from zero to infinity to better model many real-world restraint conditions. For instance, the clamped boundary conditions are essentially obtained by setting the spring stiffness substantially larger than the bending rigidity of the involved shell. Specifically, symbols k_ψ^u, k_ψ^v, k_ψ^w and K_ψ^w ($\psi = \alpha 0, \beta 0, \alpha 1$ and $\beta 1$) are used to indicate the stiffness (per unit length, unless otherwise stated, N/m and N/rad are utilized as the units of the stiffness about the linear springs and rotational springs, respectively) of the boundary springs at the boundaries $\alpha = 0$, $\beta = 0$, $\alpha = L_\alpha$ and $\beta = L_\beta$, respectively (L_α and L_β represent the lengths of the shell in the α and β directions, respectively). Therefore, the deformation strain energy about the boundary springs (U_{sp}) can be defined as:

$$U_{sp} = \frac{1}{2} \int_\beta \left\{ \begin{array}{l} \left[k_{\alpha 0}^u u^2 + k_{\alpha 0}^v v^2 + k_{\alpha 0}^w w^2 + K_{\alpha 0}^w (\partial w / A \partial \alpha)^2 \right]_{\alpha=0} \\ + \left[k_{\alpha 1}^u u^2 + k_{\alpha 1}^v v^2 + k_{\alpha 1}^w w^2 + K_{\alpha 1}^w (\partial w / A \partial \alpha)^2 \right]_{\alpha=L_\alpha} \end{array} \right\} B d\beta$$
$$+ \frac{1}{2} \int_\alpha \left\{ \begin{array}{l} \left[k_{\beta 0}^u u^2 + k_{\beta 0}^v v^2 + k_{\beta 0}^w w^2 + K_{\beta 0}^w (\partial w / B \partial \beta)^2 \right]_{\beta=0} \\ + \left[k_{\beta 1}^u u^2 + k_{\beta 1}^v v^2 + k_{\beta 1}^w w^2 + K_{\beta 1}^w (\partial w / B \partial \beta)^2 \right]_{\beta=L_\beta} \end{array} \right\} A d\alpha \quad (1.21)$$

The energy functions presented here will be applied to derive the governing equations and boundary conditions of thin laminated shells in the next section.

1.2.4 Governing Equations and Boundary Conditions

Hamilton's principle is a generalized principle of the principle of virtual displacement to dynamics of systems. It is very convenient in deriving the governing equations and boundary conditions of a structure and is covered in many textbooks (Reddy 2002).

The present section is devoted to the application of Hamilton's principle in finding the compatible set of governing equations and boundary conditions of a thin laminated shell with general boundary conditions.

The Lagrangian functional (L) of a thin laminated shell can be expressed in terms of strain energy, kinetic energy and external work as (Qatu 2004; Reddy 2002):

$$L = T - U_s - U_{sp} + W_e \tag{1.22}$$

Hamilton's principle states that the actual displacements of a structure actually go through from instant 0 to instant t; out of many possible paths, is that which achieves an extremum of the line integral of the Lagrangian functional (Qatu 2004), namely:

$$\delta \int_0^t \left(T - U_s - U_{sp} + W_e \right) dt = 0 \tag{1.23}$$

Substituting Eqs. (1.17)–(1.21) into Eq. (1.23) and applying the Hamilton's principle yields:

$$
\int_0^t \int_\alpha \int_\beta \left\{ \begin{array}{l} I_0\left(\frac{\partial u}{\partial t}\frac{\partial \delta u}{\partial t} + \frac{\partial v}{\partial t}\frac{\partial \delta v}{\partial t} + \frac{\partial w}{\partial t}\frac{\partial \delta w}{\partial t} \right) \\ + q_\alpha \delta u + q_\beta \delta v + q_z \delta w \end{array} \right\} AB d\alpha d\beta dt
$$

$$
= \int_0^t \int_\beta \left\{ \begin{array}{l} \left(k_{\alpha 0}^u u \delta u + k_{\alpha 0}^v v \delta v + k_{\alpha 0}^w w \delta w + K_{\alpha 0}^w \frac{\partial w}{A \partial \alpha}\frac{\partial \delta w}{A \partial \alpha} \right)_{\alpha=0} \\ + \left(k_{\alpha 1}^u u \delta u + k_{\alpha 1}^v v \delta v + k_{\alpha 1}^w w \delta w + K_{\alpha 1}^w \frac{\partial w}{A \partial \alpha}\frac{\partial \delta w}{A \partial \alpha} \right)_{\alpha=L_\alpha} \end{array} \right\} Bd\beta dt
$$

$$
+ \int_0^t \int_\alpha \left\{ \begin{array}{l} \left(k_{\beta 0}^u u \delta u + k_{\beta 0}^v v \delta v + k_{\beta 0}^w w \delta w + K_{\beta 0}^w \frac{\partial w}{B \partial \beta}\frac{\partial \delta w}{B \partial \beta} \right)_{\beta=0} \\ + \left(k_{\beta 1}^u u \delta u + k_{\beta 1}^v v \delta v + k_{\beta 1}^w w \delta w + K_{\beta 1}^w \frac{\partial w}{B \partial \beta}\frac{\partial \delta w}{B \partial \beta} \right)_{\beta=L_\beta} \end{array} \right\} Ad\alpha dt
$$

$$
+ \int_0^t \int_\alpha \int_\beta \left\{ \begin{array}{l} N_\alpha \left(B\frac{\partial \delta u}{\partial \alpha} + \delta v \frac{\partial A}{\partial \beta} + \frac{AB\delta w}{R_\alpha} \right) \\ + N_\beta \left(A\frac{\partial \delta v}{\partial \beta} + \delta u \frac{\partial B}{\partial \alpha} + \frac{AB\delta w}{R_\beta} \right) + N_{\alpha\beta} \\ \times \left(B\frac{\partial \delta v}{\partial \alpha} - \delta u \frac{\partial A}{\partial \beta} + A\frac{\partial \delta u}{\partial \beta} - \delta v \frac{\partial B}{\partial \alpha} \right) \\ + M_\alpha \left(B\frac{\partial \delta \phi_\alpha}{\partial \alpha} + \delta \phi_\beta \frac{\partial A}{\partial \beta} \right) \\ + M_\beta \left(A\frac{\partial \delta \phi_\beta}{\partial \beta} + \delta \phi_\alpha \frac{\partial B}{\partial \alpha} \right) + M_{\alpha\beta} \\ \times \left(B\frac{\partial \delta \phi_\beta}{\partial \alpha} - \delta \phi_\alpha \frac{\partial A}{\partial \beta} + A\frac{\partial \delta \phi_\alpha}{\partial \beta} - \delta \phi_\beta \frac{\partial B}{\partial \alpha} \right) \end{array} \right\} d\alpha d\beta dt \tag{1.24}
$$

Substituting Eq. (1.5) into Eq. (1.24) and integrating by parts to relieve the virtual displacements δu, δv and δw, such as:

$$\int_0^t \frac{\partial u}{\partial t}\frac{\partial \delta u}{\partial t}dt = \frac{\partial u}{\partial t}\delta u\Big|_0^t - \int_0^t \left(\frac{\partial^2 u}{\partial t^2}\delta u\right)dt;$$

$$\int_\alpha \int_\beta BN_\alpha \frac{\partial \delta u}{\partial \alpha}d\alpha d\beta = \int_\beta \left\{ BN_\alpha \delta u\Big|_0^{L_\alpha} - \int_\alpha \frac{\partial(BN_\alpha)}{\partial \alpha}\delta u d\alpha \right\}d\beta;$$

$$\int_\alpha \int_\beta M_\alpha \frac{\partial A}{\partial \beta}\delta\phi_\beta d\alpha d\beta = \int_\alpha \left\{ \begin{array}{l} -\frac{M_\alpha}{B}\frac{\partial A}{\partial \beta}\delta w\Big|_0^{L_\beta} + \int_\beta \left\{\frac{M_\alpha}{R_\beta}\frac{\partial A}{\partial \beta}\delta v\right\}d\beta \\[2mm] +\int_\beta \left\{\frac{\partial}{\partial \beta}\left(\frac{M_\alpha}{B}\frac{\partial A}{\partial \beta}\right)\delta w\right\}d\beta \end{array} \right\}d\alpha$$

$$\int_\alpha \int_\beta BM_\alpha \frac{\partial \delta\phi_\alpha}{\partial \alpha}d\alpha d\beta = \int_\alpha \int_\beta \left\{\frac{BM_\alpha}{R_\alpha}\frac{\partial \delta u}{\partial \alpha} - BM_\alpha \frac{\partial}{\partial \alpha}\left(\frac{\partial \delta w}{A\partial \alpha}\right)\right\}d\alpha d\beta \quad (1.25)$$

$$= \int_\beta \left\{\frac{BM_\alpha}{R_\alpha}\delta u\Big|_0^{L_\alpha} - \int_\alpha \frac{\partial}{\partial \alpha}\left(\frac{BM_\alpha}{R_\alpha}\right)\delta u d\alpha \right\}d\beta$$

$$+ \int_\beta \left\{ \begin{array}{l} -BM_\alpha\frac{\partial \delta w}{A\partial \alpha}\Big|_0^{L_\alpha} + \frac{\partial(BM_\alpha)}{A\partial \alpha}\delta w\Big|_0^{L_\alpha} \\[2mm] -\int_\alpha \frac{\partial}{\partial \alpha}\left(\frac{\partial(BM_\alpha)}{A\partial \alpha}\right)\delta w d\alpha \end{array} \right\}d\beta$$

$$\int_0^t \int_\beta \left\{M_{\alpha\beta}\frac{\partial \delta w}{\partial \beta}\Big|_0^{L_\alpha}\right\}d\beta dt = \int_0^t \left\{M_{\alpha\beta}\delta w\Big|_0^{L_\alpha}\Big|_0^{L_\beta} - \int_\beta \frac{\partial M_{\alpha\beta}}{\partial \beta}\delta w\Big|_0^{L_\alpha}d\beta \right\}dt$$

Proceeding with all terms of Eq. (1.24) in this fashion and letting

$$\begin{aligned} Q_\alpha &= \frac{\partial(BM_\alpha)}{AB\partial \alpha} + \frac{\partial(AM_{\alpha\beta})}{AB\partial \beta} + \frac{M_{\alpha\beta}}{AB}\frac{\partial A}{\partial \beta} - \frac{M_\beta}{AB}\frac{\partial B}{\partial \alpha} \\[2mm] Q_\beta &= \frac{\partial(AM_\beta)}{AB\partial \beta} + \frac{\partial(BM_{\alpha\beta})}{AB\partial \alpha} + \frac{M_{\alpha\beta}}{AB}\frac{\partial B}{\partial \alpha} - \frac{M_\alpha}{AB}\frac{\partial A}{\partial \beta} \end{aligned} \qquad (1.26)$$

we get

$$\int_\alpha \int_\beta \left\{ I_0 \frac{\partial u}{\partial t} \delta u\big|_0^t + I_0 \frac{\partial v}{\partial t} \delta v\big|_0^t + I_0 \frac{\partial w}{\partial t} \delta w\big|_0^t \right\} AB d\alpha d\beta$$

$$+ \int_0^t 2M_{\alpha\beta} \delta w\big|_0^{L_\alpha}\big|_0^{L_\beta} dt$$

$$+ \int_0^t \int_\alpha \int_\beta \left\{ \begin{array}{l} \frac{\partial(BN_\alpha)}{\partial\alpha} + \frac{\partial(AN_{\alpha\beta})}{\partial\beta} + N_{\alpha\beta}\frac{\partial A}{\partial\beta} \\ -N_\beta\frac{\partial B}{\partial\alpha} + AB\left(\frac{Q_\alpha}{R_\alpha} + q_\alpha - I_0\frac{\partial^2 u}{\partial t^2}\right) \end{array} \right\} \delta u d\alpha d\beta dt$$

$$+ \int_0^t \int_\alpha \int_\beta \left\{ \begin{array}{l} \frac{\partial(BN_{\alpha\beta})}{\partial\alpha} + \frac{\partial(AN_\beta)}{\partial\beta} + N_{\alpha\beta}\frac{\partial B}{\partial\alpha} \\ -N_\alpha\frac{\partial A}{\partial\beta} + AB\left(\frac{Q_\beta}{R_\beta} + q_\beta - I_0\frac{\partial^2 v}{\partial t^2}\right) \end{array} \right\} \delta v d\alpha d\beta dt$$

$$+ \int_0^t \int_\alpha \int_\beta \left\{ \begin{array}{l} -AB\left(\frac{N_\alpha}{R_\alpha} + \frac{N_\beta}{R_\beta}\right) + \frac{\partial(BQ_\alpha)}{\partial\alpha} \\ +\frac{\partial(AQ_\beta)}{\partial\beta} + AB\left(q_z - I_0\frac{\partial^2 w}{\partial t^2}\right) \end{array} \right\} \delta w d\alpha d\beta dt$$

$$- \int_0^t \int_\beta \left\{ \begin{array}{l} \left(N_\alpha + \frac{M_\alpha}{R_\alpha} + k_{\alpha1}^u u\right)\delta u\big|_{L_\alpha} - \left(N_\alpha + \frac{M_\alpha}{R_\alpha} - k_{\alpha0}^u u\right)\delta u\big|_0 \\ +\left(N_{\alpha\beta} + \frac{M_{\alpha\beta}}{R_\beta} + k_{\alpha1}^v v\right)\delta v\big|_{L_\alpha} - \left(N_{\alpha\beta} + \frac{M_{\alpha\beta}}{R_\beta} - k_{\alpha0}^v v\right)\delta v\big|_0 \\ +\left(Q_\alpha + \frac{\partial M_{\alpha\beta}}{B\partial\beta} + k_{\alpha1}^w w\right)\delta w\big|_{L_\alpha} - \left(Q_\alpha + \frac{\partial M_{\alpha\beta}}{B\partial\beta} - k_{\alpha0}^w w\right)\delta w\big|_0 \\ +\left(-M_\alpha + K_{\alpha1}^w \frac{\partial w}{A\partial\alpha}\right)\frac{\partial\delta w}{A\partial\alpha}\big|_{L_\alpha} - \left(-M_\alpha - K_{\alpha0}^w \frac{\partial w}{A\partial\alpha}\right)\frac{\partial\delta w}{A\partial\alpha}\big|_0 \end{array} \right\} B d\beta dt$$

$$- \int_0^t \int_\alpha \left\{ \begin{array}{l} \left(N_{\alpha\beta} + \frac{M_{\alpha\beta}}{R_\alpha} + k_{\beta1}^u u\right)\delta u\big|_{L_\beta} - \left(N_{\alpha\beta} + \frac{M_{\alpha\beta}}{R_\alpha} - k_{\beta0}^u u\right)\delta u\big|_0 \\ +\left(N_\beta + \frac{M_\beta}{R_\beta} + k_{\beta1}^v v\right)\delta v\big|_{L_\beta} - \left(N_\beta + \frac{M_\beta}{R_\beta} - k_{\beta0}^v v\right)\delta v\big|_0 \\ +\left(Q_\beta + \frac{\partial M_{\alpha\beta}}{A\partial\alpha} + k_{\beta1}^w w\right)\delta w\big|_{L_\beta} - \left(Q_\beta + \frac{\partial M_{\alpha\beta}}{A\partial\alpha} - k_{\beta0}^w w\right)\delta w\big|_0 \\ +\left(-M_\beta + K_{\beta1}^w \frac{\partial w}{B\partial\beta}\right)\frac{\partial\delta w}{B\partial\beta}\big|_{L_\beta} - \left(-M_\beta - K_{\beta0}^w \frac{\partial w}{B\partial\beta}\right)\frac{\partial\delta w}{B\partial\beta}\big|_0 \end{array} \right\} A d\alpha dt$$

$$= 0 \tag{1.27}$$

Since the virtual displacements δu, δv and δw are arbitrary, Eq. (1.27) can be satisfied only is the coefficients of the virtual displacements are zero. Thus, the governing equations of thin laminated shells are:

$$\frac{\partial(BN_\alpha)}{\partial\alpha} + \frac{\partial(AN_{\alpha\beta})}{\partial\beta} + N_{\alpha\beta}\frac{\partial A}{\partial\beta} - N_\beta\frac{\partial B}{\partial\alpha} + \frac{ABQ_\alpha}{R_\alpha} + ABq_\alpha = ABI_0\frac{\partial^2 u}{\partial t^2}$$

$$\frac{\partial(BN_{\alpha\beta})}{\partial\alpha} + \frac{\partial(AN_\beta)}{\partial\beta} + N_{\alpha\beta}\frac{\partial B}{\partial\alpha} - N_\alpha\frac{\partial A}{\partial\beta} + \frac{ABQ_\beta}{R_\beta} + ABq_\beta = ABI_0\frac{\partial^2 v}{\partial t^2} \tag{1.28a--c}$$

$$- AB\left(\frac{N_\alpha}{R_\alpha} + \frac{N_\beta}{R_\beta}\right) + \frac{\partial(BQ_\alpha)}{\partial\alpha} + \frac{\partial(AQ_\beta)}{\partial\beta} + ABq_z = ABI_0\frac{\partial^2 w}{\partial t^2}$$

and the following boundary conditions are obtained for thin shells for $\alpha = $ constant ($\alpha = 0$ and $\alpha = L_\alpha$):

$$
\alpha = 0 : \begin{cases} N_\alpha + \frac{M_\alpha}{R_\alpha} - k_{\alpha0}^u u = 0 \\ N_{\alpha\beta} + \frac{M_{\alpha\beta}}{R_\beta} - k_{\alpha0}^v v = 0 \\ Q_\alpha + \frac{\partial M_{\alpha\beta}}{B\partial\beta} - k_{\alpha0}^w w = 0 \\ -M_\alpha - K_{\alpha0}^w \frac{\partial w}{A\partial\alpha} = 0 \end{cases}
\qquad
\alpha = L_\alpha : \begin{cases} N_\alpha + \frac{M_\alpha}{R_\alpha} + k_{\alpha1}^u u = 0 \\ N_{\alpha\beta} + \frac{M_{\alpha\beta}}{R_\beta} + k_{\alpha1}^v v = 0 \\ Q_\alpha + \frac{\partial M_{\alpha\beta}}{B\partial\beta} + k_{\alpha1}^w w = 0 \\ -M_\alpha + K_{\alpha1}^w \frac{\partial w}{A\partial\alpha} = 0 \end{cases}
\tag{1.29}
$$

Similarly equations can be obtained for $\beta = $ constant ($\beta = 0$ and $\beta = L_\beta$):

$$
\beta = 0 : \begin{cases} N_{\alpha\beta} + \frac{M_{\alpha\beta}}{R_\alpha} - k_{\beta0}^u u = 0 \\ N_\beta + \frac{M_\beta}{R_\beta} - k_{\beta0}^v v = 0 \\ Q_\beta + \frac{\partial M_{\alpha\beta}}{A\partial\alpha} - k_{\beta0}^w w = 0 \\ -M_\beta - K_{\beta0}^w \frac{\partial w}{B\partial\beta} = 0 \end{cases}
\qquad
\beta = L_\beta : \begin{cases} N_{\alpha\beta} + \frac{M_{\alpha\beta}}{R_\alpha} + k_{\beta1}^u u = 0 \\ N_\beta + \frac{M_\beta}{R_\beta} + k_{\beta1}^v v = 0 \\ Q_\beta + \frac{\partial M_{\alpha\beta}}{A\partial\alpha} + k_{\beta1}^w w = 0 \\ -M_\beta + K_{\beta1}^w \frac{\partial w}{B\partial\beta} = 0 \end{cases}
\tag{1.30}
$$

Alternately, the governing equations and boundary conditions of thin laminated shells can be derived by physical arguments, namely, by taking a differential element of the shell and requiring the sum of the external and internal forces and moments in the α, β and z directions to be zero.

For general thin shells and unsymmetrically laminated thin plates, each boundary can exist four possible combinations for each classical boundary condition (free, simply-supported, and clamped) (Qatu 2004). For example, at each boundary of $\alpha = $ constant, the possible combinations for each classical boundary condition are given in Table 1.2, similar classifications can be obtained for boundaries $\beta = $ constant.

It is obvious that there exists a huge number of possible combinations of boundary conditions for a thin laminated shell, particularly when the shells are open. In the open literature, it appears that most of the studies have mostly dealt with a technique that is only suitable for a particular type of boundary conditions (i.e., F, SD, S, C and their combinations) which typically require constant modifications of the solution procedures to adapt to different boundary cases. Therefore, the use of the existing solution procedures will result in very tedious calculations and be easily inundated with various boundary conditions in practical applications. By using the artificial spring boundary technique, the stiffness of the boundary springs can take any value from zero to infinity to better model many real-world restraint conditions. Taking edge $\alpha = 0$ for example, the F (completely free), S (simply-supported), SD (shear-diaphragm) and C (completely clamped) boundary conditions which are widely encountered in the engineering applications and are of particular interest can be readily realized by assigning the stiffness of the boundary springs at proper values as follows:

Table 1.2 Possible classical boundary conditions for thin shell at each boundary of α = constant

Boundary type	Conditions
Free boundary conditions	
F	$N_\alpha + \frac{M_\alpha}{R_\alpha} = N_{\alpha\beta} + \frac{M_{\alpha\beta}}{R_\beta} = Q_\alpha + \frac{\partial M_{\alpha\beta}}{B\partial\beta} = M_\alpha = 0$
F2	$u = N_{\alpha\beta} + \frac{M_{\alpha\beta}}{R_\beta} = Q_\alpha + \frac{\partial M_{\alpha\beta}}{B\partial\beta} = M_\alpha = 0$
F3	$N_\alpha + \frac{M_\alpha}{R_\alpha} = v = Q_\alpha + \frac{\partial M_{\alpha\beta}}{B\partial\beta} = M_\alpha = 0$
F4	$u = v = Q_\alpha + \frac{\partial M_{\alpha\beta}}{B\partial\beta} = M_\alpha = 0$
Simply supported boundary conditions	
S	$u = v = w = M_\alpha = 0$
SD	$N_\alpha + \frac{M_\alpha}{R_\alpha} = v = w = M_\alpha = 0$
S3	$u = N_{\alpha\beta} + \frac{M_{\alpha\beta}}{R_\beta} = w = M_\alpha = 0$
S4	$N_\alpha + \frac{M_\alpha}{R_\alpha} = N_{\alpha\beta} + \frac{M_{\alpha\beta}}{R_\beta} = w = M_\alpha = 0$
Clamped boundary conditions	
C	$u = v = w = \frac{\partial w}{A\partial\alpha} = 0$
C2	$N_\alpha + \frac{M_\alpha}{R_\alpha} = v = w = \frac{\partial w}{A\partial\alpha} = 0$
C3	$u = N_{\alpha\beta} + \frac{M_{\alpha\beta}}{R_\beta} = w = \frac{\partial w}{A\partial\alpha} = 0$
C4	$N_\alpha + \frac{M_\alpha}{R_\alpha} = N_{\alpha\beta} + \frac{M_{\alpha\beta}}{R_\beta} = w = \frac{\partial w}{A\partial\alpha} = 0$

$$
\begin{aligned}
F&: k_{\alpha 0}^u = k_{\alpha 0}^v = k_{\alpha 0}^w = K_{\alpha 0}^w = 0 \\
SD&: k_{\alpha 0}^v = k_{\alpha 0}^w = 10^7 D, \ k_{\alpha 0}^u = K_{\alpha 0}^w = 0 \\
S&: k_{\alpha 0}^u = k_{\alpha 0}^v = k_{\alpha 0}^w = 10^7 D, \ K_{\alpha 0}^w = 0 \\
C&: k_{\alpha 0}^u = k_{\alpha 0}^v = k_{\alpha 0}^w = K_{\alpha 0}^w = 10^7 D
\end{aligned}
\tag{1.31}
$$

where $D = E_1 h^3/12(1 - \mu_{12}\mu_{21})$ is the flexural stiffness of the shell. The appropriateness of defining these types of boundary conditions in terms of boundary spring components had been proved by several researches (Jin et al. 2013a; Ye et al. 2013, 2014c).

Equations (1.3a–c)–(1.31) describe the fundamental equations of thin laminated shells in the framework of classical thin shell theory. It is applicable for thin isotropic and laminated composite shells with general boundary conditions. These equations will be specialized later for laminated beams.

1.3 Fundamental Equations of Thick Laminated Shells

In the previous section, the fundamental equations of thin laminated shells have been derived according to Love's assumptions. When the shell thickness is less than 1/50 of the wave length of the deformation mode and/or radius of curvature, the thin shell theory is generally acceptable (Qatu 2004). Although sufficiently accurate vibration results for thin shells can be achieved using the thin shell theory with appropriate solution procedures, however, it is inadequate for the vibration analysis of laminated shells which are rather thick or when they are made from materials with a high degree of anisotropic. Love's first assumption assumed that the thickness of the shell is small compared with the other dimensions, the deepness terms z/R_α and z/R_β are less important and they can be neglected in the thin shell theory. In such case, symmetric stress resultants are obtained, namely $N_{\alpha\beta} = N_{\beta\alpha}$, $M_{\alpha\beta} = M_{\beta\alpha}$. This is not true for the shells that are not spherical. Love's fourth assumption was that normals to the undeformed middle surface remain straight and normal to the deformed middle surface and suffer no extension. The shear deformation is taken to be zero. This assumption allowed us to obtain ϕ_α and ϕ_β as functions of the displacement components u, v and w. For thick shells, shear deformation can no longer be neglected and normal to the undeformed middle surface is no longer normal to the deformed middle surface, the shear deformation and rotary inertia effects become significant in thick and deep shells. In such case, this assumption should be relaxed by including transverse shear strains in the theory for shells with higher thickness ratios or when they are made from materials with a high degree of anisotropic.

In the shear deformation shell theories, the middle surface displacements of a shell can be expanded in terms of shell thickness of a first or a higher order. In the case of first order expansion, the theories are referred to as first-order shear deformation theories (Qatu 2004). Under the remaining assumptions and restrictions as in the thin laminated shell theory, fundamental equations of laminated shells in the framework of the first-order shear deformation shell theory in which both the shear deformation and the rotary inertia effects as well as the deepness terms z/R_α and z/R_β are considered are derived in this section.

1.3.1 Kinematic Relations

Assuming that normals to the undeformed middle surface remain straight but not normal to the deformed middle surface, the displacement field in the shell space can be expressed in terms of the middle surface displacements and rotation components as:

$$
\begin{aligned}
U(\alpha, \beta, z) &= u(\alpha, \beta) + z\phi_\alpha(\alpha, \beta) \\
V(\alpha, \beta, z) &= v(\alpha, \beta) + z\phi_\beta(\alpha, \beta) \\
W(\alpha, \beta, z) &= w(\alpha, \beta)
\end{aligned}
\tag{1.32}
$$

where u, v and w are the displacement components at the middle surface in the α, β and z directions. ϕ_α and ϕ_β represent the rotations of transverse normal respect to β- and α-axes, respectively, see Fig. 1.1. It should be pointed out that, unlike those of the classical thin shell theories, ϕ_α and ϕ_β are not functions of u, v and w, they are treated as independent variables. The third of Eq. (1.32) yields $\varepsilon_z = \partial w/\partial z = 0$. Substituting Eq. (1.32) into Eq. (1.1), the strains at any point in the shells can be defined in terms of middle surface strains and curvature changes as:

$$
\begin{aligned}
\varepsilon_\alpha &= \frac{1}{(1+z/R_\alpha)}\left(\varepsilon_\alpha^0 + z\chi_\alpha\right) \\
\varepsilon_\beta &= \frac{1}{(1+z/R_\beta)}\left(\varepsilon_\beta^0 + z\chi_\beta\right) \\
\gamma_{\alpha\beta} &= \frac{1}{(1+z/R_\alpha)}\left(\gamma_{\alpha\beta}^0 + z\chi_{\alpha\beta}\right) + \frac{1}{(1+z/R_\beta)}\left(\gamma_{\beta\alpha}^0 + z\chi_{\beta\alpha}\right) \\
\gamma_{\alpha z} &= \frac{\gamma_{\alpha z}^0}{(1+z/R_\alpha)} \\
\gamma_{\beta z} &= \frac{\gamma_{\beta z}^0}{(1+z/R_\beta)}
\end{aligned}
\tag{1.33}
$$

where ε_α^0, ε_β^0, $\gamma_{\alpha\beta}^0$ and $\gamma_{\beta\alpha}^0$ denote the normal and shear strains in the reference surface. χ_α, χ_β, $\chi_{\alpha\beta}$ and $\chi_{\beta\alpha}$ are the curvature and twist changes; $\gamma_{\alpha z}^0$ and $\gamma_{\beta z}^0$ represent the transverse shear strains. The middle surface strains and curvature changes are:

$$
\begin{aligned}
\varepsilon_\alpha^0 &= \frac{1}{A}\frac{\partial u}{\partial \alpha} + \frac{v}{AB}\frac{\partial A}{\partial \beta} + \frac{w}{R_\alpha}, & \chi_\alpha &= \frac{1}{A}\frac{\partial \phi_\alpha}{\partial \alpha} + \frac{\phi_\beta}{AB}\frac{\partial A}{\partial \beta} \\
\varepsilon_\beta^0 &= \frac{1}{B}\frac{\partial v}{\partial \beta} + \frac{u}{AB}\frac{\partial B}{\partial \alpha} + \frac{w}{R_\beta}, & \chi_\beta &= \frac{1}{B}\frac{\partial \phi_\beta}{\partial \beta} + \frac{\phi_\alpha}{AB}\frac{\partial B}{\partial \alpha} \\
\gamma_{\alpha\beta}^0 &= \frac{1}{A}\frac{\partial v}{\partial \alpha} - \frac{u}{AB}\frac{\partial A}{\partial \beta}, & \chi_{\alpha\beta} &= \frac{1}{A}\frac{\partial \phi_\beta}{\partial \alpha} - \frac{\phi_\alpha}{AB}\frac{\partial A}{\partial \beta} \\
\gamma_{\beta\alpha}^0 &= \frac{1}{B}\frac{\partial u}{\partial \beta} - \frac{v}{AB}\frac{\partial B}{\partial \alpha}, & \chi_{\beta\alpha} &= \frac{1}{B}\frac{\partial \phi_\alpha}{\partial \beta} - \frac{\phi_\beta}{AB}\frac{\partial A}{\partial \alpha} \\
\gamma_{\alpha z}^0 &= \frac{1}{A}\frac{\partial w}{\partial \alpha} - \frac{u}{R_\alpha} + \phi_\alpha \\
\gamma_{\beta z}^0 &= \frac{1}{B}\frac{\partial w}{\partial \beta} - \frac{v}{R_\beta} + \phi_\beta
\end{aligned}
\tag{1.34}
$$

The above equations constitute the fundamental strain-displacement relations of a thick 2D shell in curvilinear coordinates.

1.3.2 Stress-Strain Relations and Stress Resultants

According to the generalized Hooke's law, the corresponding stress-strain relations in the kth layer of thick laminated shell can be written as (see, Fig. 1.3):

$$
\left\{
\begin{array}{c}
\sigma_\alpha \\
\sigma_\beta \\
\tau_{\beta z} \\
\tau_{\alpha z} \\
\tau_{\alpha\beta}
\end{array}
\right\}_k
=
\begin{bmatrix}
\overline{Q^k_{11}} & \overline{Q^k_{12}} & 0 & 0 & \overline{Q^k_{16}} \\
\overline{Q^k_{12}} & \overline{Q^k_{22}} & 0 & 0 & \overline{Q^k_{26}} \\
0 & 0 & \overline{Q^k_{44}} & \overline{Q^k_{45}} & 0 \\
0 & 0 & \overline{Q^k_{45}} & \overline{Q^k_{55}} & 0 \\
\overline{Q^k_{16}} & \overline{Q^k_{26}} & 0 & 0 & \overline{Q^k_{66}}
\end{bmatrix}
\left\{
\begin{array}{c}
\varepsilon_\alpha \\
\varepsilon_\beta \\
\gamma_{\beta z} \\
\gamma_{\alpha z} \\
\gamma_{\alpha\beta}
\end{array}
\right\}_k
\tag{1.35}
$$

where σ_α and σ_β represent the normal stresses in the α, β directions, $\tau_{\alpha\beta}$, $\tau_{\alpha z}$ and $\tau_{\beta z}$ are the corresponding shear stress components. The elastic stiffness coefficients $\overline{Q^k_{ij}}$ ($i, j = 1, 2, 4, 5$ and 6) are defined by following equation:

$$
\begin{bmatrix}
\overline{Q^k_{11}} & \overline{Q^k_{12}} & 0 & 0 & \overline{Q^k_{16}} \\
\overline{Q^k_{12}} & \overline{Q^k_{22}} & 0 & 0 & \overline{Q^k_{26}} \\
0 & 0 & \overline{Q^k_{44}} & \overline{Q^k_{45}} & 0 \\
0 & 0 & \overline{Q^k_{45}} & \overline{Q^k_{55}} & 0 \\
\overline{Q^k_{16}} & \overline{Q^k_{26}} & 0 & 0 & \overline{Q^k_{66}}
\end{bmatrix}
= \mathbf{T}
\begin{bmatrix}
Q^k_{11} & Q^k_{12} & 0 & 0 & 0 \\
Q^k_{12} & Q^k_{22} & 0 & 0 & 0 \\
0 & 0 & Q^k_{44} & 0 & 0 \\
0 & 0 & 0 & Q^k_{55} & 0 \\
0 & 0 & 0 & 0 & Q^k_{66}
\end{bmatrix}
\mathbf{T}^\mathrm{T}
\tag{1.36}
$$

where superscript T represents the transposition operator. The material elastic stiffness coefficients Q^k_{ij} of the kth layer are written in terms of the material properties of the layer:

$$
Q^k_{11} = \frac{E_1}{1 - \mu_{12}\mu_{21}}, \qquad Q^k_{44} = G_{23}
$$

$$
Q^k_{12} = \frac{\mu_{12}E_2}{1 - \mu_{12}\mu_{21}}, \qquad Q^k_{55} = G_{13}
\tag{1.37}
$$

$$
Q^k_{22} = \frac{E_2}{1 - \mu_{12}\mu_{21}}, \qquad Q^k_{66} = G_{12}
$$

where E_1 and E_2 are moduli of elasticity of the unidirectional fiber-reinforced material in the 1 and 2 directions, respectively. μ_{12} is the major Poisson's ratio. The Poisson's ratio μ_{21} are determined by equation $\mu_{12}E_2 = \mu_{21}E_1$. G_{12}, G_{13} and G_{23} are shear moduli. It should be noted that by letting $E_1 = E_2$, $G_{12} = G_{13} = G_{23} = E_1/(2 + 2\mu_{12})$, the present analysis can be readily used to analyze isotropic thick shells. $\mathbf{T}$ is the transformation matrix:

$$
\mathbf{T} =
\begin{bmatrix}
\cos^2 \vartheta^k & \sin^2 \vartheta^k & 0 & 0 & -2 \sin \vartheta^k \cos \vartheta^k \\
\sin^2 \vartheta^k & \cos^2 \vartheta^k & 0 & 0 & 2 \sin \vartheta^k \cos \vartheta^k \\
0 & 0 & \cos \vartheta^k & \sin \vartheta^k & 0 \\
0 & 0 & -\sin \vartheta^k & \cos \vartheta^k & 0 \\
\sin \vartheta^k \cos \vartheta^k & -\sin \vartheta^k \cos \vartheta^k & 0 & 0 & \cos^2 \vartheta^k - \sin^2 \vartheta^k
\end{bmatrix}
\tag{1.38}
$$

where ϑ^k is the included angle between the principal direction of the layer and the α-axis. Performing the matrix multiplication in Eq. (1.36), the elastic stiffness coefficients constants $\overline{Q^k_{ij}}$ ($i, j = 1, 2, 4, 5$ and 6) can be written as:

$$\overline{Q^k_{11}} = Q^k_{11}\cos^4\vartheta^k + (2Q^k_{12} + 4Q^k_{66})\cos^2\vartheta^k\sin^2\vartheta^k + Q^k_{22}\sin^4\vartheta^k$$

$$\overline{Q^k_{12}} = (Q^k_{11} + Q^k_{22} - 4Q^k_{66})\cos^2\vartheta^k\sin^2\vartheta^k + Q^k_{12}(\cos^4\vartheta^k + \sin^4\vartheta^k)$$

$$\overline{Q^k_{16}} = (Q^k_{11} - Q^k_{12} - 2Q^k_{66})\cos^3\vartheta^k\sin\vartheta^k + (Q^k_{12} - Q^k_{22} + 2Q^k_{66})\cos\vartheta^k\sin^3\vartheta^k$$

$$\overline{Q^k_{22}} = Q^k_{11}\sin^4\vartheta^k + Q^k_{22}\cos^4\vartheta^k + (2Q^k_{12} + 4Q^k_{66})\cos^2\vartheta^k\sin^2\vartheta^k$$

$$\overline{Q^k_{26}} = (Q^k_{11} - Q^k_{12} - 2Q^k_{66})\cos\vartheta^k\sin^3\vartheta^k + (Q^k_{12} - Q^k_{22} + 2Q^k_{66})\cos^3\vartheta^k\sin\vartheta^k \quad (1.39)$$

$$\overline{Q^k_{66}} = Q^k_{66}(\cos^4\vartheta^k + \sin^4\vartheta^k) + (Q^k_{11} - 2Q^k_{12} + Q^k_{22} - 2Q^k_{66})\cos^2\vartheta^k\sin^2\vartheta^k$$

$$\overline{Q^k_{44}} = Q^k_{44}\cos^2\vartheta^k + Q^k_{55}\sin^2\vartheta^k$$

$$\overline{Q^k_{45}} = (Q^k_{55} - Q^k_{44})\cos\vartheta^k\sin\vartheta^k$$

$$\overline{Q^k_{55}} = Q^k_{55}\cos^2\vartheta^k + Q^k_{44}\sin^2\vartheta^k$$

The force and moment resultants of laminated thick shells can be obtained by carrying the integration of stresses over the cross-section, from layer to layer (see Eq. (1.2), Figs. 1.5 and 1.6):

$$
\begin{bmatrix} N_\alpha \\ N_{\alpha\beta} \\ Q_\alpha \end{bmatrix} = \int\limits_{-h/2}^{h/2} \begin{bmatrix} \sigma_\alpha \\ \tau_{\alpha\beta} \\ \tau_{\alpha z} \end{bmatrix}\left(1+\tfrac{z}{R_\beta}\right)dz, \quad
\begin{bmatrix} M_\alpha \\ M_{\alpha\beta} \end{bmatrix} = \int\limits_{-h/2}^{h/2} \begin{bmatrix} \sigma_\alpha \\ \tau_{\alpha\beta} \end{bmatrix}\left(1+\tfrac{z}{R_\beta}\right)z\,dz
$$

$$
\begin{bmatrix} N_\beta \\ N_{\beta\alpha} \\ Q_\beta \end{bmatrix} = \int\limits_{-h/2}^{h/2} \begin{bmatrix} \sigma_\beta \\ \tau_{\beta\alpha} \\ \tau_{\beta z} \end{bmatrix}\left(1+\tfrac{z}{R_\alpha}\right)dz, \quad
\begin{bmatrix} M_\beta \\ M_{\beta\alpha} \end{bmatrix} = \int\limits_{-h/2}^{h/2} \begin{bmatrix} \sigma_\beta \\ \tau_{\beta\alpha} \end{bmatrix}\left(1+\tfrac{z}{R_\alpha}\right)z\,dz
\tag{1.40}
$$

where N_α, N_β, $N_{\alpha\beta}$ and $N_{\beta\alpha}$ are the normal and shear force resultants and M_α, M_β, $M_{\alpha\beta}$ and $M_{\beta\alpha}$ denote the bending and twisting moment resultants. Q_α and Q_β are the transverse shear force resultants. It should be stressed that although $\tau_{\alpha\beta}$ equal to $\tau_{\beta\alpha}$ from the symmetry of the stress tensor, it is obvious form Eq. (1.40) that when the

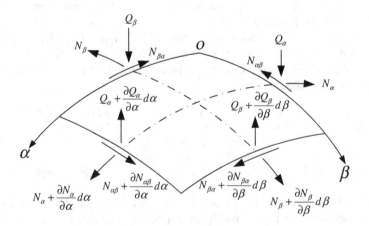

Fig. 1.5 Notations of force resultants in shell coordinate system (Qatu 2004)

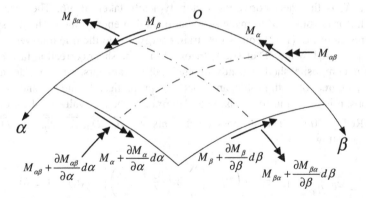

Fig. 1.6 Notations of moment resultants in shell coordinate system

shells are not spherical, the shear force resultants $N_{\alpha\beta}$ and $N_{\beta\alpha}$ are not equal. Similarly, the twisting moment resultants $M_{\alpha\beta}$ and $M_{\beta\alpha}$ are not equal too. In addition, for rectangular plates, $z/R_\alpha = z/R_\beta = 0$ automatically satisfied, these force and moment resultants will be equal. Performing the integration operation in Eq. (1.40), the force and moment resultants can be written in terms of the middle surface strains and curvature changes as:

$$
\begin{bmatrix} N_\alpha \\ N_\beta \\ N_{\alpha\beta} \\ N_{\beta\alpha} \\ M_\alpha \\ M_\beta \\ M_{\alpha\beta} \\ M_{\beta\alpha} \end{bmatrix}
=
\begin{bmatrix}
\overline{A}_{11} & A_{12} & \overline{A}_{16} & A_{16} & \overline{B}_{11} & B_{12} & \overline{B}_{16} & B_{16} \\
A_{12} & \overline{A}_{22} & A_{26} & \overline{A}_{26} & B_{12} & \overline{B}_{22} & B_{26} & \overline{B}_{26} \\
\overline{A}_{16} & A_{26} & \overline{A}_{66} & A_{66} & \overline{B}_{16} & B_{26} & \overline{B}_{66} & B_{66} \\
A_{16} & \overline{A}_{26} & A_{66} & \overline{A}_{66} & B_{16} & \overline{B}_{26} & B_{66} & \overline{B}_{66} \\
\overline{B}_{11} & B_{12} & \overline{B}_{16} & B_{16} & \overline{D}_{11} & D_{12} & \overline{D}_{16} & D_{16} \\
B_{12} & \overline{B}_{22} & B_{26} & \overline{B}_{26} & D_{12} & \overline{D}_{22} & D_{26} & \overline{D}_{26} \\
\overline{B}_{16} & B_{26} & \overline{B}_{66} & B_{66} & \overline{D}_{16} & D_{26} & \overline{D}_{66} & D_{66} \\
B_{16} & \overline{B}_{26} & B_{66} & \overline{B}_{66} & D_{16} & \overline{D}_{26} & D_{66} & \overline{D}_{66}
\end{bmatrix}
\begin{bmatrix} \varepsilon_\alpha^0 \\ \varepsilon_\beta^0 \\ \gamma_{\alpha\beta}^0 \\ \gamma_{\beta\alpha}^0 \\ \chi_\alpha \\ \chi_\beta \\ \chi_{\alpha\beta} \\ \chi_{\beta\alpha} \end{bmatrix}
\tag{1.41}
$$

$$
\begin{bmatrix} Q_\beta \\ Q_\alpha \end{bmatrix}
=
\begin{bmatrix} \overline{\overline{A}}_{44} & A_{45} \\ A_{45} & \overline{\overline{A}}_{55} \end{bmatrix}
\begin{bmatrix} \gamma_{\beta z}^0 \\ \gamma_{\alpha z}^0 \end{bmatrix}
\tag{1.42}
$$

The stiffness coefficients A_{ij}, B_{ij} and D_{ij} are defined as follows:

$$
A_{ij} = \sum_{k=1}^{N} \overline{Q}_{ij}^k (z_{k+1} - z_k), \qquad (i,j = 1,2,6)
$$

$$
A_{ij} = K_s \sum_{k=1}^{N} \overline{Q}_{ij}^k (z_{k+1} - z_k), \quad (i,j = 4,5)
$$

$$
B_{ij} = \tfrac{1}{2} \sum_{k=1}^{N} \overline{Q}_{ij}^k (z_{k+1}^2 - z_k^2), \quad (i,j = 1,2,4,5,6)
$$

$$
D_{ij} = \tfrac{1}{3} \sum_{k=1}^{N} \overline{Q}_{ij}^k (z_{k+1}^3 - z_k^3), \quad (i,j = 1,2,4,5,6)
$$

$$\tag{1.43}$$

in which K_s is the shear correction factor, typically taken at 5/6. The shear correction factor is computed to make sure that the strain energy due to the transverse shear stresses in Eq. (1.42) equals the strain energy due to the true transverse shear stresses predicted by the 3D elasticity theory. Since the shear correction factor for a laminated composite shell depends on the shell parameters, such as lamination schemes, degree of orthotropic and material properties, boundary and loading conditions, it is still an unsolved issue to determinate the true value of shear correct factor (Reddy 2003). The stiffness coefficients $\overline{A}_{ij}$, $\overline{B}_{ij}$, $\overline{D}_{ij}$, $\overline{\overline{A}}_{ij}$, $\overline{\overline{B}}_{ij}$ and $\overline{\overline{D}}_{ij}$ are defined as follows:

$$\overline{A}_{ij} = \sum_{k=1}^{N} \overline{Q}_{ij}^k \left[\frac{R_\alpha}{R_\beta} (z_{k+1} - z_k) + \left(R_\alpha - \frac{R_\alpha^2}{R_\beta} \right) \ln\left(\frac{R_\alpha + z_{k+1}}{R_\alpha + z_k} \right) \right], \quad (i,j = 1,2,6)$$

$$\overline{A}_{ij} = K_s \sum_{k=1}^{N} \overline{Q}_{ij}^k \left[\frac{R_\alpha}{R_\beta} (z_{k+1} - z_k) + \left(R_\alpha - \frac{R_\alpha^2}{R_\beta} \right) \ln\left(\frac{R_\alpha + z_{k+1}}{R_\alpha + z_k} \right) \right], \quad (i,j = 4,5)$$

$$\overline{B}_{ij} = \sum_{k=1}^{N} \overline{Q}_{ij}^k \left[\begin{array}{l} \left(R_\alpha - \frac{R_\alpha^2}{R_\beta} \right)(z_{k+1} - z_k) + \frac{R_\alpha}{2R_\beta}\left(z_{k+1}^2 - z_k^2 \right) \\ + \left(\frac{R_\alpha^3}{R_\beta} - R_\alpha^2 \right) \ln\left(\frac{R_\alpha+z_{k+1}}{R_\alpha+z_k} \right) \end{array} \right] \qquad (1.44)$$

$$\overline{D}_{ij} = \sum_{k=1}^{N} \overline{Q}_{ij}^k \left[\begin{array}{l} \left(\frac{R_\alpha^3}{R_\beta} - R_\alpha^2 \right)(z_{k+1} - z_k) + \left(\frac{R_\alpha}{2} - \frac{R_\alpha^2}{2R_\beta} \right)\left(z_{k+1}^2 - z_k^2 \right) \\ + \frac{R_\alpha}{3R_\beta}\left(z_{k+1}^3 - z_k^3 \right) + \left(R_\alpha^3 - \frac{R_\alpha^4}{R_\beta} \right) \ln\left(\frac{R_\alpha+z_{k+1}}{R_\alpha+z_k} \right) \end{array} \right]$$

$$\overline{\overline{A}}_{ij} = \sum_{k=1}^{N} \overline{Q}_{ij}^k \left[\frac{R_\beta}{R_\alpha} (z_{k+1} - z_k) + \left(R_\beta - \frac{R_\beta^2}{R_\alpha} \right) \ln\left(\frac{R_\beta + z_{k+1}}{R_\beta + z_k} \right) \right], \quad (i,j = 1,2,6)$$

$$\overline{\overline{A}}_{ij} = K_s \sum_{k=1}^{N} \overline{Q}_{ij}^k \left[\frac{R_\beta}{R_\alpha} (z_{k+1} - z_k) + \left(R_\beta - \frac{R_\beta^2}{R_\alpha} \right) \ln\left(\frac{R_\beta + z_{k+1}}{R_\beta + z_k} \right) \right], \quad (i,j = 4,5)$$

$$\overline{\overline{B}}_{ij} = \sum_{k=1}^{N} \overline{Q}_{ij}^k \left[\begin{array}{l} \left(R_\beta - \frac{R_\beta^2}{R_\alpha} \right)(z_{k+1} - z_k) + \frac{R_\beta}{2R_\alpha}\left(z_{k+1}^2 - z_k^2 \right) \\ + \left(\frac{R_\beta^3}{R_\alpha} - R_\beta^2 \right) \ln\left(\frac{R_\beta+z_{k+1}}{R_\beta+z_k} \right) \end{array} \right] \qquad (1.45)$$

$$\overline{\overline{D}}_{ij} = \sum_{k=1}^{N} \overline{Q}_{ij}^k \left[\begin{array}{l} \left(\frac{R_\beta^3}{R_\alpha} - R_\beta^2 \right)(z_{k+1} - z_k) + \left(\frac{R_\beta}{2} - \frac{R_\beta^2}{2R_\alpha} \right)\left(z_{k+1}^2 - z_k^2 \right) \\ + \frac{R_\beta}{3R_\alpha}\left(z_{k+1}^3 - z_k^3 \right) + \left(R_\beta^3 - \frac{R_\beta^4}{R_\alpha} \right) \ln\left(\frac{R_\beta+z_{k+1}}{R_\beta+z_k} \right) \end{array} \right]$$

The above equations include the effects of deepness terms z/R_α and z/R_β, thus creates much complexities in carrying out the integration and the corresponding programming. When the SDST is applied to thin and slightly thick shells, the z/R_α and z/R_β terms can be neglected (Qu et al. 2013a; Jin et al. 2013b, 2014a; Ye et al. 2014a, b). In such case, Eqs. (1.41) and (1.42) is degenerated as:

$$
\begin{bmatrix}
N_\alpha \\
N_\beta \\
N_{\alpha\beta} \\
N_{\beta\alpha} \\
M_\alpha \\
M_\beta \\
M_{\alpha\beta} \\
M_{\beta\alpha}
\end{bmatrix}
=
\begin{bmatrix}
A_{11} & A_{12} & A_{16} & A_{16} & B_{11} & B_{12} & B_{16} & B_{16} \\
A_{12} & A_{22} & A_{26} & A_{26} & B_{12} & B_{22} & B_{26} & B_{26} \\
A_{16} & A_{26} & A_{66} & A_{66} & B_{16} & B_{26} & B_{66} & B_{66} \\
A_{16} & A_{26} & A_{66} & A_{66} & B_{16} & B_{26} & B_{66} & B_{66} \\
B_{11} & B_{12} & B_{16} & B_{16} & D_{11} & D_{12} & D_{16} & D_{16} \\
B_{12} & B_{22} & B_{26} & B_{26} & D_{12} & D_{22} & D_{26} & D_{26} \\
B_{16} & B_{26} & B_{66} & B_{66} & D_{16} & D_{26} & D_{66} & D_{66} \\
B_{16} & B_{26} & B_{66} & B_{66} & D_{16} & D_{26} & D_{66} & D_{66}
\end{bmatrix}
\begin{bmatrix}
\varepsilon_\alpha^0 \\
\varepsilon_\beta^0 \\
\gamma_{\alpha\beta}^0 \\
\gamma_{\beta\alpha}^0 \\
\chi_\alpha \\
\chi_\beta \\
\chi_{\alpha\beta} \\
\chi_{\beta\alpha}
\end{bmatrix}
\tag{1.46}
$$

$$
\begin{bmatrix} Q_\beta \\ Q_\alpha \end{bmatrix}
=
\begin{bmatrix} A_{44} & A_{45} \\ A_{45} & A_{55} \end{bmatrix}
\begin{bmatrix} \gamma_{\beta z}^0 \\ \gamma_{\alpha z}^0 \end{bmatrix}
\tag{1.47}
$$

In addition, for thick spherical shell ($R_\alpha = R_\beta$) and flat plates ($R_\alpha = R_\beta = \infty$), their stress resultants to middle surface strain-curvature change relations are in the same form as Eqs. (1.46) and (1.47).

1.3.3 Energy Functions

The strain energy (U_s) of thick laminated shells during vibration can be defined in terms of the shell strains and corresponding stresses as:

$$
U_s = \frac{1}{2} \int_\alpha \int_\beta \int_{-h/2}^{h/2}
\left\{
\begin{array}{l}
\sigma_\alpha \varepsilon_\alpha + \sigma_\beta \varepsilon_\beta \\
+\tau_{\alpha\beta}\gamma_{\alpha\beta} + \tau_{\alpha z}\gamma_{\alpha z} \\
+\tau_{\beta z}\gamma_{\beta z}
\end{array}
\right\}
\left(1 + \frac{z}{R_\alpha}\right)\left(1 + \frac{z}{R_\beta}\right) ABdzd\beta d\alpha
\tag{1.48}
$$

Substituting Eq. (1.40) into Eq. (1.48), the strain energy (U_s) function of the thick shells can be written in terms of the middle surface strains, curvature changes and the stress resultants as:

$$
U_s = \frac{1}{2} \int_\alpha \int_\beta
\left\{
\begin{array}{l}
N_\alpha \varepsilon_\alpha^0 + N_\beta \varepsilon_\beta^0 + N_{\alpha\beta}\gamma_{\alpha\beta}^0 + N_{\beta\alpha}\gamma_{\beta\alpha}^0 \\
+M_\alpha \chi_\alpha + M_\beta \chi_\beta + M_{\alpha\beta}\chi_{\alpha\beta} \\
+M_{\beta\alpha}\chi_{\beta\alpha} + Q_\alpha \gamma_{\alpha z}^0 + Q_\beta \gamma_{\beta z}^0
\end{array}
\right\} ABd\beta d\alpha
\tag{1.49}
$$

Substituting Eq. (1.34) into Eq. (1.49), the strain energy of thick laminated shells can be rewritten as:

$$
U_s = \frac{1}{2} \int_\alpha \int_\beta
\left\{
\begin{array}{l}
N_\alpha \left(B \frac{\partial u}{\partial \alpha} + v \frac{\partial A}{\partial \beta} + \frac{ABw}{R_\alpha} \right) \\[4pt]
+ N_\beta \left(A \frac{\partial v}{\partial \beta} + u \frac{\partial B}{\partial \alpha} + \frac{ABw}{R_\beta} \right) \\[4pt]
+ N_{\alpha\beta} \left(B \frac{\partial v}{\partial \alpha} - u \frac{\partial A}{\partial \beta} \right) + N_{\beta\alpha} \left(A \frac{\partial u}{\partial \beta} - v \frac{\partial B}{\partial \alpha} \right) \\[4pt]
+ M_\alpha \left(B \frac{\partial \phi_\alpha}{\partial \alpha} + \phi_\beta \frac{\partial A}{\partial \beta} \right) + M_\beta \left(A \frac{\partial \phi_\beta}{\partial \beta} + \phi_\alpha \frac{\partial B}{\partial \alpha} \right) \\[4pt]
+ M_{\alpha\beta} \left(B \frac{\partial \phi_\beta}{\partial \alpha} - \phi_\alpha \frac{\partial A}{\partial \beta} \right) + M_{\beta\alpha} \left(A \frac{\partial \phi_\alpha}{\partial \beta} - \phi_\beta \frac{\partial B}{\partial \alpha} \right) \\[4pt]
+ Q_\alpha \left(B \frac{\partial w}{\partial \alpha} - \frac{ABu}{R_\alpha} + AB\phi_\alpha \right) \\[4pt]
+ Q_\beta \left(A \frac{\partial w}{\partial \beta} - \frac{ABv}{R_\beta} + AB\phi_\beta \right)
\end{array}
\right\} d\beta d\alpha \qquad (1.50)
$$

The kinetic energy (T) of the shell is written as:

$$
T = \int_\alpha \int_\beta \int_{-h/2}^{h/2} \frac{\rho_z}{2}
\left\{
\begin{array}{l}
\left(\frac{\partial u}{\partial t} + z \frac{\partial \phi_\alpha}{\partial t} \right)^2 \\[4pt]
+ \left(\frac{\partial v}{\partial t} + z \frac{\partial \phi_\beta}{\partial t} \right)^2 \\[4pt]
+ \left(\frac{\partial w}{\partial t} \right)^2
\end{array}
\right\}
\left(1 + \frac{z}{R_\alpha} \right) \left(1 + \frac{z}{R_\beta} \right) AB\, dz\, d\beta\, d\alpha
$$

$$
= \frac{1}{2} \int_\alpha \int_\beta
\left\{
\begin{array}{l}
\overline{I_0} \left(\frac{\partial u}{\partial t} \right)^2 + \overline{I_0} \left(\frac{\partial v}{\partial t} \right)^2 + \overline{I_0} \left(\frac{\partial w}{\partial t} \right)^2 \\[4pt]
+ 2\overline{I_1} \frac{\partial u}{\partial t} \frac{\partial \phi_\alpha}{\partial t} + 2\overline{I_1} \frac{\partial v}{\partial t} \frac{\partial \phi_\beta}{\partial t} \\[4pt]
+ \overline{I_2} \left(\frac{\partial \phi_\alpha}{\partial t} \right)^2 + \overline{I_2} \left(\frac{\partial \phi_\beta}{\partial t} \right)^2
\end{array}
\right\} AB\, d\beta\, d\alpha \qquad (1.51)
$$

where the inertia terms are:

$$
\overline{I_0} = I_0 + \frac{I_1}{R_\alpha} + \frac{I_1}{R_\beta} + \frac{I_2}{R_\alpha R_\beta}
$$

$$
\overline{I_1} = I_1 + \frac{I_2}{R_\alpha} + \frac{I_2}{R_\beta} + \frac{I_3}{R_\alpha R_\beta}
$$

$$
\overline{I_2} = I_2 + \frac{I_3}{R_\alpha} + \frac{I_3}{R_\beta} + \frac{I_4}{R_\alpha R_\beta} \qquad (1.52)
$$

$$
[I_0, I_1, I_2, I_3, I_4] = \sum_{k=1}^{N} \int_{z_k}^{z_{k+1}} \rho^k [1, z, z^2, z^3, z^4]\, dz
$$

in which ρ^k is the mass of the kth layer per unit surface area. The external work is (W_e) expressed as:

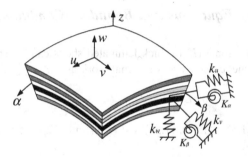

Fig. 1.7 Notations of artificial spring boundary technique for thick laminated shells

$$W_e = \int\limits_{\alpha} \int\limits_{\beta} \{q_\alpha u + q_\beta v + q_z w + m_\alpha \phi_\alpha + m_\beta \phi_\beta\} ABd\alpha d\beta \qquad (1.53)$$

where q_α, q_β and q_z denote the external loads. m_α and m_β are the external couples in the middle surface of the shell.

Using the artificial spring boundary technique similar to that described earlier, the general boundary conditions of a laminated thick shell are implemented by introducing three groups of translational springs (k_u, k_v and k_w) and two groups of rotational springs (K_α and K_β) which are distributed uniformly along the boundary at each shell boundary (see Fig. 1.7). Therefore, arbitrary boundary conditions of the shell can be generated by assigning the translational springs and rotational springs at proper rigidities. Specifically, symbols k_ψ^u, k_ψ^v, k_ψ^w, K_ψ^α and K_ψ^β ($\psi = \alpha 0, \beta 0,$ $\alpha 1$ and $\beta 1$) are used to indicate the rigidities (per unit length) of the boundary springs at the boundaries of $\alpha = 0$, $\beta = 0$, $\alpha = L_\alpha$ and $\beta = L_\beta$, respectively. Therefore, the deformation strain energy about the boundary springs (U_{sp}) is defined as:

$$U_{sp} = \frac{1}{2} \int\limits_{\beta} \left\{ \begin{array}{l} \left[k_{\alpha 0}^u u^2 + k_{\alpha 0}^v v^2 + k_{\alpha 0}^w w^2 + K_{\alpha 0}^\alpha \phi_\alpha^2 + K_{\alpha 0}^\beta \phi_\beta^2\right]_{\alpha=0} \\ + \left[k_{\alpha 1}^u u^2 + k_{\alpha 1}^v v^2 + k_{\alpha 1}^w w^2 + K_{\alpha 1}^\alpha \phi_\alpha^2 + K_{\alpha 1}^\beta \phi_\beta^2\right]_{\alpha=L_\alpha} \end{array} \right\} Bd\beta$$

$$+ \frac{1}{2} \int\limits_{\alpha} \left\{ \begin{array}{l} \left[k_{\beta 0}^u u^2 + k_{\beta 0}^v v^2 + k_{\beta 0}^w w^2 + K_{\beta 0}^\alpha \phi_\alpha^2 + K_{\beta 0}^\beta \phi_\beta^2\right]_{\beta=0} \\ + \left[k_{\beta 1}^u u^2 + k_{\beta 1}^v v^2 + k_{\beta 1}^w w^2 + K_{\beta 1}^\alpha \phi_\alpha^2 + K_{\beta 1}^\beta \phi_\beta^2\right]_{\beta=L_\beta} \end{array} \right\} Ad\alpha \qquad (1.54)$$

The energy expressions presented here will be used to derive the governing equations and boundary conditions of thick laminated shells in the next section by applying Hamilton's principle as described earlier.

1.3.4 Governing Equations and Boundary Conditions

The Lagrangian functional (L) of thick laminated shells can be expressed in terms of strain energy, kinetic energy and external work as:

$$L = T - U_s - U_{sp} + W_e \tag{1.55}$$

Substituting Eqs. (1.50), (1.51), (1.53) and (1.54) into Eq. (1.55) and applying the Hamilton's principle

$$\delta \int_0^t \left(T - U_s - U_{sp} + W_e \right) dt = 0 \tag{1.56}$$

yields

$$\int_0^t \int_\alpha \int_\beta \left\{ \begin{array}{l} \overline{I}_0\left(\frac{\partial u}{\partial t}\frac{\partial \delta u}{\partial t} + \frac{\partial v}{\partial t}\frac{\partial \delta v}{\partial t} + \frac{\partial w}{\partial t}\frac{\partial \delta w}{\partial t}\right) \\[4pt] +\overline{I}_2\left(\frac{\partial \phi_\alpha}{\partial t}\frac{\partial \delta\phi_\alpha}{\partial t} + \frac{\partial \phi_\beta}{\partial t}\frac{\partial \delta\phi_\beta}{\partial t}\right) \\[4pt] +\overline{I}_1\left(\frac{\partial u}{\partial t}\frac{\partial \delta\phi_\alpha}{\partial t} + \frac{\partial \delta u}{\partial t}\frac{\partial \phi_\alpha}{\partial t} + \frac{\partial v}{\partial t}\frac{\partial \delta\phi_\beta}{\partial t} + \frac{\partial \delta v}{\partial t}\frac{\partial \phi_\beta}{\partial t}\right) \\[4pt] +q_\alpha\delta u + q_\beta\delta v + q_z\delta w + m_\alpha\delta\phi_\alpha + m_\beta\delta\phi_\beta \end{array} \right\} AB\,d\alpha\,d\beta\,dt$$

$$= \int_0^t \int_\alpha \int_\beta \left\{ \begin{array}{l} N_\alpha\left(B\frac{\partial \delta u}{\partial \alpha} + \delta v\frac{\partial A}{\partial \beta} + \frac{AB\delta w}{R_\alpha}\right) \\[4pt] +N_\beta\left(A\frac{\partial \delta v}{\partial \beta} + \delta u\frac{\partial B}{\partial \alpha} + \frac{AB\delta w}{R_\beta}\right) \\[4pt] +N_{\alpha\beta}\left(B\frac{\partial \delta v}{\partial \alpha} - \delta u\frac{\partial A}{\partial \beta}\right) + N_{\beta\alpha}\left(A\frac{\partial \delta u}{\partial \beta} - \delta v\frac{\partial B}{\partial \alpha}\right) \\[4pt] +M_\alpha\left(B\frac{\partial \delta\phi_\alpha}{\partial \alpha} + \delta\phi_\beta\frac{\partial A}{\partial \beta}\right) + M_\beta\left(A\frac{\partial \delta\phi_\beta}{\partial \beta} + \delta\phi_\alpha\frac{\partial B}{\partial \alpha}\right) \\[4pt] +M_{\alpha\beta}\left(B\frac{\partial \delta\phi_\beta}{\partial \alpha} - \delta\phi_\alpha\frac{\partial A}{\partial \beta}\right) + M_{\beta\alpha}\left(A\frac{\partial \delta\phi_\alpha}{\partial \beta} - \delta\phi_\beta\frac{\partial B}{\partial \alpha}\right) \\[4pt] +Q_\alpha\left(B\frac{\partial \delta w}{\partial \alpha} - \frac{AB}{R_\alpha}\delta u + AB\delta\phi_\alpha\right) \\[4pt] +Q_\beta\left(A\frac{\partial \delta w}{\partial \beta} - \frac{AB}{R_\beta}\delta v + AB\delta\phi_\beta\right) \end{array} \right\} d\alpha\,d\beta\,dt$$

$$+ \int_0^t \int_\beta \left\{ \begin{array}{l} \left(k_{\alpha 0}^u u\delta u + k_{\alpha 0}^v v\delta v + k_{\alpha 0}^w w\delta w + K_{\alpha 0}^\alpha \phi_\alpha\delta\phi_\alpha + K_{\alpha 0}^\beta \phi_\beta\delta\phi_\beta\right)_{\alpha=0} \\[4pt] +\left(k_{\alpha 1}^u u\delta u + k_{\alpha 1}^v v\delta v + k_{\alpha 1}^w w\delta w + K_{\alpha 1}^\alpha \phi_\alpha\delta\phi_\alpha + K_{\alpha 1}^\beta \phi_\beta\delta\phi_\beta\right)_{\alpha=L_\alpha} \end{array} \right\} Bd\beta\,dt$$

$$+ \int_0^t \int_\alpha \left\{ \begin{array}{l} \left(k_{\beta 0}^u u\delta u + k_{\beta 0}^v v\delta v + k_{\beta 0}^w w\delta w + K_{\beta 0}^\alpha \phi_\alpha\delta\phi_\alpha + K_{\beta 0}^\beta \phi_\beta\delta\phi_\beta\right)_{\beta=0} \\[4pt] +\left(k_{\beta 1}^u u\delta u + k_{\beta 1}^v v\delta v + k_{\beta 1}^w w\delta w + K_{\beta 1}^\alpha \phi_\alpha\delta\phi_\alpha + K_{\beta 1}^\beta \phi_\beta\delta\phi_\beta\right)_{\beta=L_\beta} \end{array} \right\} Ad\alpha\,dt$$

$$\tag{1.57}$$

Integrating by parts to relieve the virtual displacements δu, δv, δw, $\delta \phi$ and $\delta \phi_\beta$ such as:

$$\int_0^t \frac{\partial u}{\partial t} \frac{\partial \delta u}{\partial t} dt = \frac{\partial u}{\partial t} \delta u \Big|_0^t - \int_0^t \left(\frac{\partial^2 u}{\partial t^2} \delta u \right) dt$$

$$\int_\alpha \int_\beta BN_\alpha \frac{\partial \delta u}{\partial \alpha} d\alpha d\beta = \int_\beta \left\{ BN_\alpha \delta u \Big|_0^{L_\alpha} - \int_\alpha \frac{\partial (BN_\alpha)}{\partial \alpha} \delta u d\alpha \right\} d\beta$$

$$\int_\alpha \int_\beta BM_\alpha \frac{\partial \delta \phi_\alpha}{\partial \alpha} d\alpha d\beta = \int_\beta \left\{ BM_\alpha \delta \phi_\alpha \Big|_0^{L_\alpha} - \int_\alpha \frac{\partial (BM_\alpha)}{\partial \alpha} \delta \phi_\alpha d\alpha \right\} d\beta \qquad (1.58)$$

Proceeding with all terms of Eq. (1.57) in this fashion, we get the following governing equations:

$$\frac{\partial (BN_\alpha)}{\partial \alpha} + N_{\alpha\beta} \frac{\partial A}{\partial \beta} - N_\beta \frac{\partial B}{\partial \alpha} + \frac{\partial (AN_{\beta\alpha})}{\partial \beta} + \frac{ABQ_\alpha}{R_\alpha} + ABq_\alpha$$

$$= AB \left(\bar{I}_0 \frac{\partial^2 u}{\partial t^2} + \bar{I}_1 \frac{\partial^2 \phi_\alpha}{\partial t^2} \right)$$

$$\frac{\partial (BN_{\alpha\beta})}{\partial \alpha} - N_\alpha \frac{\partial A}{\partial \beta} + N_{\beta\alpha} \frac{\partial B}{\partial \alpha} + \frac{\partial (AN_\beta)}{\partial \beta} + \frac{ABQ_\beta}{R_\beta} + ABq_\beta$$

$$= AB \left(\bar{I}_0 \frac{\partial^2 v}{\partial t^2} + \bar{I}_1 \frac{\partial^2 \phi_\beta}{2} \right)$$

$$- AB \left(\frac{N_\alpha}{R_\alpha} + \frac{N_\beta}{R_\beta} \right) + \frac{\partial (BQ_\alpha)}{\partial \alpha} + \frac{\partial (AQ_\beta)}{\partial \beta} + ABq_z = AB \left(\bar{I}_0 \frac{\partial^2 w}{\partial t^2} \right) \qquad (1.59)$$

$$\frac{\partial (BM_\alpha)}{\partial \alpha} + M_{\alpha\beta} \frac{\partial A}{\partial \beta} - M_\beta \frac{\partial B}{\partial \alpha} + \frac{\partial (AM_{\beta\alpha})}{\partial \beta} - ABQ_\alpha + ABm_\alpha$$

$$= AB \left(\bar{I}_1 \frac{\partial^2 u}{\partial t^2} + \bar{I}_2 \frac{\partial^2 \phi_\alpha}{\partial t^2} \right)$$

$$\frac{\partial (BM_{\alpha\beta})}{\partial \alpha} - M_\alpha \frac{\partial A}{\partial \beta} + M_{\beta\alpha} \frac{\partial B}{\partial \alpha} + \frac{\partial (AM_\beta)}{\partial \beta} - ABQ_\beta + ABm_\beta$$

$$= AB \left(\bar{I}_1 \frac{\partial^2 v}{\partial t^2} + \bar{I}_2 \frac{\partial^2 \phi_\beta}{\partial t^2} \right)$$

On each boundary of α = constant, five boundary equations are obtained:

$$\alpha = 0: \begin{cases} N_\alpha - k_{\alpha 0}^u u = 0 \\ N_{\alpha\beta} - k_{\alpha 0}^v v = 0 \\ Q_\alpha - k_{\alpha 0}^w w = 0 \\ M_\alpha - K_{\alpha 0}^\alpha \phi_\alpha = 0 \\ M_{\alpha\beta} - K_{\alpha 0}^\beta \phi_\beta = 0 \end{cases} \qquad \alpha = L_\alpha: \begin{cases} N_\alpha + k_{\alpha 1}^u u = 0 \\ N_{\alpha\beta} + k_{\alpha 1}^v v = 0 \\ Q_\alpha + k_{\alpha 1}^w w = 0 \\ M_\alpha + K_{\alpha 1}^\alpha \phi_\alpha = 0 \\ M_{\alpha\beta} + K_{\alpha 1}^\beta \phi_\beta = 0 \end{cases} \qquad (1.60)$$

Similarly, for boundaries β = 0 and $\beta = L_\beta$, the corresponding boundary conditions are:

$$\beta = 0: \begin{cases} N_{\beta\alpha} - k_{\beta 0}^u u = 0 \\ N_\beta - k_{\beta 0}^v v = 0 \\ Q_\beta - k_{\beta 0}^w w = 0 \\ M_{\beta\alpha} - K_{\beta 0}^\alpha \phi_\alpha = 0 \\ M_\beta - K_{\beta 0}^\beta \phi_\beta = 0 \end{cases} \qquad \beta = L_\beta: \begin{cases} N_{\beta\alpha} + k_{\beta 1}^u u = 0 \\ N_\beta + k_{\beta 1}^v v = 0 \\ Q_\beta + k_{\beta 1}^w w = 0 \\ M_{\beta\alpha} + K_{\beta 1}^\alpha \phi_\alpha = 0 \\ M_\beta + K_{\beta 1}^\beta \phi_\beta = 0 \end{cases} \qquad (1.61)$$

For general thick shells and unsymmetrically laminated thick plates, each boundary can exist eight possible combinations for each of the classical boundary conditions (free, simple supported, clamped (Qatu 2004)). For example, at each boundary of α = constant, the 24 possible classical boundary conditions are given in Table 1.3, similar classifications can be obtained for boundaries β = constant.

In the engineering applications, the F (completely free), S (simply supported), SD (shear diaphragm supported) and C (completely clamped) boundary conditions are widely encountered and are of particular interest. Using the artificial spring boundary technique as described earlier, the stiffness of the boundary springs can take any value from zero to infinity to better model many real-world restraint conditions. Taking edge α = 0 for example, the corresponding spring stiffness for the mentioned classical boundary conditions can be defined in terms of boundary spring rigidities as:

$$\begin{aligned} &F: k_{\alpha 0}^u = k_{\alpha 0}^v = k_{\alpha 0}^w = K_{\alpha 0}^\alpha = K_{\alpha 0}^\beta = 0 \\ &SD: k_{\alpha 0}^v = k_{\alpha 0}^w = K_{\alpha 0}^\beta = 10^7 D, \ k_{\alpha 0}^u = K_{\alpha 0}^\alpha = 0 \\ &S: k_{\alpha 0}^u = k_{\alpha 0}^v = k_{\alpha 0}^w = K_{\alpha 0}^\beta = 10^7 D, \ K_{\alpha 0}^\alpha = 0 \\ &C: k_{\alpha 0}^u = k_{\alpha 0}^v = k_{\alpha 0}^w = K_{\alpha 0}^\alpha = K_{\alpha 0}^\beta = 10^7 D \end{aligned} \qquad (1.62a)$$

The appropriateness of defining these types of boundary conditions in terms of boundary spring components had been proved by several researches (Jin et al. 2013b, 2014a; Su et al. 2014a, b, c; Ye et al. 2014a, b).

Equations (1.32)–(1.62a) describe the first-order shear deformation theory for 2D thick laminated shells. It is applicable for isotropic and laminated composite thick shells (the shell thickness is less than 1/10 of the wave length of the

Table 1.3 Possible classical boundary conditions for thick shell at each boundary of $\alpha = $ constant

Boundary type	Conditions
Free boundary conditions	
F	$N_\alpha = N_{\alpha\beta} = Q_\alpha = M_\alpha = M_{\alpha\beta} = 0$
F2	$u = N_{\alpha\beta} = Q_\alpha = M_\alpha = M_{\alpha\beta} = 0$
F3	$N_\alpha = v = Q_\alpha = M_\alpha = M_{\alpha\beta} = 0$
F4	$u = v = Q_\alpha = M_\alpha = M_{\alpha\beta} = 0$
F5	$N_\alpha = N_{\alpha\beta} = Q_\alpha = M_\alpha = \phi_\beta = 0$
F6	$u = N_{\alpha\beta} = Q_\alpha = M_\alpha = \phi_\beta = 0$
F7	$N_\alpha = v = Q_\alpha = M_\alpha = \phi_\beta = 0$
F8	$u = v = Q_\alpha = M_\alpha = \phi_\beta = 0$
Simply supported boundary conditions	
S	$u = v = w = M_\alpha = \phi_\beta = 0$
SD	$N_\alpha = v = w = M_\alpha = \phi_\beta = 0$
S3	$u = N_{\alpha\beta} = w = M_\alpha = \phi_\beta = 0$
S4	$N_\alpha = N_{\alpha\beta} = w = M_\alpha = \phi_\beta = 0$
S5	$u = v = w = M_\alpha = M_{\alpha\beta} = 0$
S6	$N_\alpha = v = w = M_\alpha = M_{\alpha\beta} = 0$
S7	$u = N_{\alpha\beta} = w = M_\alpha = M_{\alpha\beta} = 0$
S8	$N_\alpha = N_{\alpha\beta} = w = M_\alpha = M_{\alpha\beta} = 0$
Clamped boundary conditions	
C	$u = v = w = \phi_\alpha = \phi_\beta = 0$
C2	$N_\alpha = v = w = \phi_\alpha = \phi_\beta = 0$
C3	$u = N_{\alpha\beta} = w = \phi_\alpha = \phi_\beta = 0$
C4	$N_\alpha = N_{\alpha\beta} = w = \phi_\alpha = \phi_\beta = 0$
C5	$u = v = w = \phi_\alpha = M_{\alpha\beta} = 0$
C6	$N_\alpha = v = w = \phi_\alpha = M_{\alpha\beta} = 0$
C7	$u = N_{\alpha\beta} = w = \phi_\alpha = M_{\alpha\beta} = 0$
C8	$N_\alpha = N_{\alpha\beta} = w = \phi_\alpha = M_{\alpha\beta} = 0$

deformation mode and/or radii of curvature) with general boundary conditions. This theory will be specialized later for laminated beams, plates and various closed and open shells (Chaps. 3–8).

1.4 Lamé Parameters for Plates and Shells

The fundamental equations of laminated shells presented in previous sections can be readily specialized to some of the frequently encountered structural elements such as flat plates, cylindrical shells, conical shells, spherical shells and shallow shells by substituting their Lamé parameters into the equations. Lamé parameters for plates and shells are considered in this section.

In the global Cartesian coordinate system (X, Y, Z), the point vector of a point $P\ (\alpha, \beta, 0)$ in the middle surface of a shell, $\mathbf{r}$, can be written in terms of orthogonal shell coordinates α and β as (see Fig. 1.8):

$$\mathbf{r} = \mathbf{r}(\alpha, \beta) \tag{1.63}$$

and in the global Cartesian coordinate system (X, Y, Z), the coordinates can be written as functions of α and β as:

$$X = X(\alpha, \beta) \quad Y = Y(\alpha, \beta) \quad Z = Z(\alpha, \beta) \tag{1.64}$$

Therefore, the point vector $\mathbf{r}$ can be rewritten as:

$$\mathbf{r} = X(\alpha, \beta)\mathbf{i} + Y(\alpha, \beta)\mathbf{j} + Z(\alpha, \beta)\mathbf{k} \tag{1.65}$$

where $\mathbf{i}$, $\mathbf{j}$ and $\mathbf{k}$ represent the unit vectors along the X, Y and Z coordinates, respectively. The incensement of the vector $\mathbf{r}$ in moving from point $P\ (\alpha, \beta, 0)$ to point $Q\ (\alpha + d\alpha, \beta + d\beta, 0)$ is defined as:

$$d\mathbf{r} = \frac{\partial \mathbf{r}}{\partial \alpha} d\alpha + \frac{\partial \mathbf{r}}{\partial \beta} d\beta \tag{1.66}$$

Thus, the square of the distance ds between point $(\alpha, \beta, 0)$ and point $(\alpha + d\alpha, \beta + d\beta, 0)$ on the middle surface is:

$$(ds)^2 = d\mathbf{r} \cdot d\mathbf{r} = \frac{\partial \mathbf{r}}{\partial \alpha} \cdot \frac{\partial \mathbf{r}}{\partial \alpha} d\alpha^2 + 2\frac{\partial \mathbf{r}}{\partial \alpha} \cdot \frac{\partial \mathbf{r}}{\partial \beta} d\alpha d\beta + \frac{\partial \mathbf{r}}{\partial \beta} \cdot \frac{\partial \mathbf{r}}{\partial \beta} d\beta^2 \tag{1.67}$$

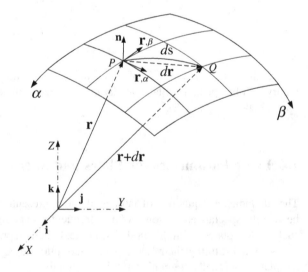

Fig. 1.8 Coordinate systems of the shell's middle surface

Letting

$$A^2 = \frac{\partial \mathbf{r}}{\partial \alpha} \cdot \frac{\partial \mathbf{r}}{\partial \alpha} = \left|\frac{\partial \mathbf{r}}{\partial \alpha}\right|^2, \qquad B^2 = \frac{\partial \mathbf{r}}{\partial \beta} \cdot \frac{\partial \mathbf{r}}{\partial \beta} = \left|\frac{\partial \mathbf{r}}{\partial \beta}\right|^2 \qquad (1.68)$$

and considering the coordinates α and β are orthogonal, so that

$$\frac{\partial \mathbf{r}}{\partial \alpha} \cdot \frac{\partial \mathbf{r}}{\partial \beta} = 0, \qquad (ds)^2 = A^2 d\alpha^2 + B^2 d\beta^2 \qquad (1.69)$$

where A and B are the so called Lamé parameters. Equation (1.67) can be rewritten as:

$$(ds)^2 = ds_1^2 + ds_2^2, \qquad ds_1 = A d\alpha, \qquad ds_2 = B d\beta \qquad (1.70)$$

It can be seen that the Lamé parameters relate a change in the arc length in the middle surface to the changes in the curvilinear coordinates of the shell.

In engineering practices, the rectangular plates, sectorial plates, cylindrical shells, conical shells and spherical shells are the most commonly used structural elements. The Lamé parameters of these structural elements are given below.

(a) *Rectangular plates*

Let $\alpha = x$ and $\beta = y$ as shown in Fig. 1.9. $\mathbf{i}$, $\mathbf{j}$ and $\mathbf{k}$ denote the unit vectors along the x, y and z coordinates. Any point P in the middle surface of the rectangular plate is defined by the vector

$$\mathbf{r} = x\mathbf{i} + y\mathbf{j} \qquad (1.71)$$

Thus, the Lamé parameters of rectangular plates can be obtained as

$$\frac{\partial \mathbf{r}}{\partial \alpha} = \frac{\partial \mathbf{r}}{\partial x} = \mathbf{i} \quad A = \left|\frac{\partial \mathbf{r}}{\partial \alpha}\right| = |\mathbf{i}| = 1$$
$$\frac{\partial \mathbf{r}}{\partial \beta} = \frac{\partial \mathbf{r}}{\partial y} = \mathbf{j} \quad B = \left|\frac{\partial \mathbf{r}}{\partial \beta}\right| = |\mathbf{j}| = 1 \qquad (1.72)$$

Fig. 1.9 Rectangular plates

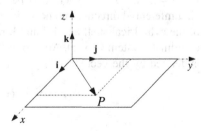

Fig. 1.10 Sectorial plates

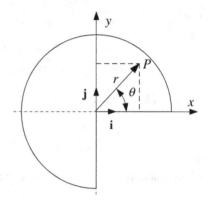

(b) *Sectorial (annular or circular) plates*

Let $\alpha = r$ and $\beta = \theta$ as shown in Fig. 1.10. $\mathbf{i}$ and $\mathbf{j}$ denote the unit vectors along the x and y coordinates. Any point P in the middle surface of the sectorial plate is defined by the vector

$$\mathbf{r} = r\cos\theta \cdot \mathbf{i} + r\sin\theta \cdot \mathbf{j} \tag{1.73}$$

Thus, the Lamé parameters of sectorial plates can be obtained as

$$\frac{\partial \mathbf{r}}{\partial \alpha} = \frac{\partial \mathbf{r}}{\partial r} = \cos\theta \cdot \mathbf{i} + \sin\theta \cdot \mathbf{j}$$
$$A = \left|\frac{\partial \mathbf{r}}{\partial r}\right| = |\cos\theta \cdot \mathbf{i} + \sin\theta \cdot \mathbf{j}| = 1 \tag{1.74}$$

$$\frac{\partial \mathbf{r}}{\partial \beta} = \frac{\partial \mathbf{r}}{\partial \theta} = -r\sin\theta \cdot \mathbf{i} + r\cos\theta \cdot \mathbf{j}$$
$$B = \left|\frac{\partial \mathbf{r}}{\partial \theta}\right| = |-r\sin\theta \cdot \mathbf{i} + r\cos\theta \cdot \mathbf{j}| = r \tag{1.75}$$

(c) *Cylindrical shells*

Let $\alpha = x$ and $\beta = \theta$, where α is parallel to the axis of revolution and β is in the circumferential direction as shown in Fig. 1.11. R is the radius of the middle surface of the cylindrical shell and $\mathbf{i}$, $\mathbf{j}$ and $\mathbf{k}$ represent the unit vectors along the Cartesian coordinate system (x, y, z). Any point P in the middle surface of the cylindrical shell is defined by the vector

$$\mathbf{r} = x\mathbf{i} + R\cos\theta \cdot \mathbf{j} + R\sin\theta \cdot \mathbf{k} \tag{1.76}$$

Fig. 1.11 Cylindrical shells

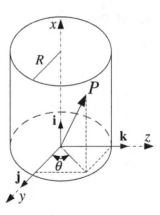

Thus, the Lamé parameters of cylindrical shells can be obtained as

$$\frac{\partial \mathbf{r}}{\partial \alpha} = \frac{\partial \mathbf{r}}{\partial x} = \mathbf{i} \quad A = \left|\frac{\partial \mathbf{r}}{\partial \alpha}\right| = |\mathbf{i}| = 1$$

$$\frac{\partial \mathbf{r}}{\partial \beta} = \frac{\partial \mathbf{r}}{\partial \theta} = -R \sin \theta \cdot \mathbf{j} + R \cos \theta \cdot \mathbf{k} \tag{1.77}$$

$$B = \left|\frac{\partial \mathbf{r}}{\partial \beta}\right| = |-R \sin \theta \cdot \mathbf{j} + R \cos \theta \cdot \mathbf{k}| = R$$

(d) *Conical shells*

Let $\alpha = x$ and $\beta = \theta$. Figure 1.12 shows the geometry and the coordinate system (x, θ, z) for the middle surface of conical shells, in which x is measured along the generator of the cone starting at the vertex, θ is in the circumferential direction and z is perpendicular to the middle surface. φ is the semi-vertex angle of the conical shell and R_0 denotes the middle surface radius of the conical shell at its small edge. $\mathbf{i}, \mathbf{j}$ and $\mathbf{k}$ represent the unit vectors along the Cartesian coordinate system (X, Y, Z). Any point P in the middle surface of the conical shell can be defined by the vector

$$\mathbf{r} = x \cos \varphi \mathbf{i} + x \sin \varphi \cos \theta \cdot \mathbf{j} + x \sin \varphi \sin \theta \cdot \mathbf{k} \tag{1.78}$$

Thus, the Lamé parameters of the conical shell can be obtained as

$$\frac{\partial \mathbf{r}}{\partial \alpha} = \frac{\partial \mathbf{r}}{\partial x} = \cos \varphi \mathbf{i} + \sin \varphi \cos \theta \cdot \mathbf{j} + \sin \varphi \sin \theta \cdot \mathbf{k}$$

$$A = \left|\frac{\partial \mathbf{r}}{\partial \alpha}\right| = |\cos \varphi \mathbf{i} + \sin \varphi \cos \theta \cdot \mathbf{j} + \sin \varphi \sin \theta \cdot \mathbf{k}| = 1 \tag{1.79}$$

Fig. 1.12 Conical shells

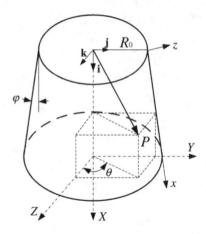

$$\frac{\partial \mathbf{r}}{\partial \beta} = \frac{\partial \mathbf{r}}{\partial \theta} = -x \sin \varphi \sin \theta \cdot \mathbf{j} + x \sin \varphi \cos \theta \cdot \mathbf{k}$$

$$B = \left| \frac{\partial \mathbf{r}}{\partial \beta} \right| = \left| -x \sin \varphi \sin \theta \cdot \mathbf{j} + x \sin \varphi \cos \theta \cdot \mathbf{k} \right| = x \sin \varphi$$

$$(1.80)$$

(e) Spherical shells

Let $\alpha = \varphi$ and $\beta = \theta$, where φ and θ are taken in the meridional and circumferential directions of the spherical shell, respectively, see Fig. 1.13. $\mathbf{i}$, $\mathbf{j}$ and $\mathbf{k}$ represent the unit vectors along the Cartesian coordinate system (x, y, z). R denotes the radius of the middle surface of the spherical shell. Any point P in the middle surface of the spherical shell is defined by the vector

Fig. 1.13 Spherical shells

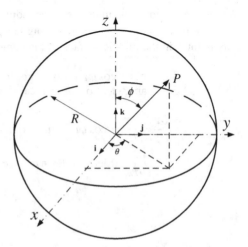

$$\mathbf{r} = R\cos\phi\mathbf{i} + R\sin\phi\cos\theta\cdot\mathbf{j} + R\sin\phi\sin\theta\cdot\mathbf{k} \tag{1.81}$$

Thus, the Lamé parameters of spherical shells can be obtained as

$$\frac{\partial\mathbf{r}}{\partial\alpha} = \frac{\partial\mathbf{r}}{\partial\phi} = -R\sin\phi\mathbf{i} + R\cos\phi\cos\theta\cdot\mathbf{j} + R\cos\phi\sin\theta\cdot\mathbf{k}$$

$$A = \left|\frac{\partial\mathbf{r}}{\partial\alpha}\right| = |-R\sin\phi\mathbf{i} + R\cos\phi\cos\theta\cdot\mathbf{j} + R\cos\phi\sin\theta\cdot\mathbf{k}| = R \tag{1.82}$$

$$\frac{\partial\mathbf{r}}{\partial\beta} = \frac{\partial\mathbf{r}}{\partial\theta} = -R\sin\phi\sin\theta\cdot\mathbf{j} + R\sin\phi\cos\theta\cdot\mathbf{k}$$

$$B = \left|\frac{\partial\mathbf{r}}{\partial\beta}\right| = |-R\sin\phi\sin\theta\cdot\mathbf{j} + R\sin\phi\cos\theta\cdot\mathbf{k}| = R\sin\phi \tag{1.83}$$

Chapter 2
Modified Fourier Series and Rayleigh-Ritz Method

Although the governing equations and associated boundary equations for laminated beams, plates and shells presented in Chap. 1 show the possibility of seeking their exact solutions of vibration, however, it is commonly believed that very few exact solutions are possible for plate and shell vibration problems. For instance, an exact solution is available only for rectangular plates which are simply supported along, at least, one pair of opposite edges, and one has to resort to an approximate solution for other boundary conditions (Zhang and Li 2009). It is important for engineering applications to have available approaches that give accurate solutions for cases that cannot be solved accurately.

In recent decades, many accurate and efficient experimental and computational methods have been developed for the vibration analysis of laminated beams, plates and shells, such as the scaled down models and similitude theory, Ritz method, differential quadrature method (DQM), Galerkin method, wave propagation approach, multiquadric radial basis function method, meshless method, finite element method (FEM), discrete singular convolution approach (DSC), etc. It should be stressed that most of these methods were applied firstly to isotropic structures, and were subsequently extended to study the dynamic behaviors of the anisotropic and laminated composite ones. However, it appears that most of the existing methods are only suitable for a particular type of boundary conditions which typically require constant modifications of the solution procedures to adapt to different boundary cases. Therefore, the use of the existing solution procedures will result in very tedious calculations and be easily inundated with various boundary conditions in practical applications due to the fact that the boundary conditions of a beam, plate or shell may not always be classical in nature, a variety of possible boundary restraining cases, including classical boundary conditions, elastic restraints and their combinations can be encountered in practice. For example, even just considering the four simplest classical boundary conditions (i.e., F, SD, S and C), one should realize that there can constitute 256 combinations of different boundary conditions for a thin shell (four edges) or unsymmetrically laminated thin plate. Furthermore, the possible combinations of classical boundary conditions of a general thick open shell or unsymmetrically laminated thick plate can be as many as 331,776 types. The finite element method (FEM) has dominated engineering

© Science Press, Beijing and Springer-Verlag Berlin Heidelberg 2015
G. Jin et al., *Structural Vibration*, DOI 10.1007/978-3-662-46364-2_2

computations since its invention and its application has expanded to a variety of engineering fields. The FEM overcomes the difficulties in dealing with various boundary conditions, however, there still exist some drawbacks due to its mesh-based interpolation. For instance, it suffers heavily from mesh distortion in large deformation and intensive remeshing requirements in dealing with the structures with complex geometries and discontinuities. In addition, the computational demands increase with structural and material complexity and with analysis frequency range (Price et al. 1998; Liew et al. 2011). It is necessary and of great significance to develop a unified, efficient and accurate method which is capable of universally dealing with laminated beams, plates and shells with general boundary conditions.

The present chapter deals with a unified modified Fourier series method which is capable of universally dealing with laminated beams, plates and shells with general boundary conditions. The accurate modified Fourier series solutions of isotropic, anisotropic and laminated beams, plates and shells can be obtained by using both strong and weak form solution procedures as described in the following sections.

2.1 Modified Fourier Series

For vibration problems of beams, plates and shells, the admissible functions are often expressed in the form of Fourier series expansions because of their orthogonality and completeness, as well as their excellent stability in numerical calculations. Furthermore, vibrations are naturally expressible as waves, which are normally described by Fourier series (Li 2000). However, the conventional Fourier series expression will generally has a convergence problem along the boundary edges except for a few simple boundary conditions, thus limiting the applications of Fourier series to only a few ideal boundary conditions. Mathematically, when the displacements of a shell (2D) are periodically extended as standard Fourier series onto the entire α–β surface, discontinuities potentially exist in original displacements and their derivatives at the edges. In such case, the Fourier series expansions cannot be differentiated term-by-term, and thus the solution may not converge or converge slowly. Recognizing the fact that the convergence rate for the Fourier series expansion of a periodic function is directly related to its smoothness, Li (2000, 2002) proposed a modified Fourier series method for the vibration analysis of isotropic Euler Bernoulli beams with general elastic boundary conditions.

In this book, this method is further developed and extended to the vibration analysis of laminated composite beams, plates and shells with general boundary conditions and arbitrary lamination schemes, aiming to provide a unified and reasonable accurate alternative to other analytical and numerical techniques. The method will be briefly explained in this section for the completeness of the book.

2.1.1 Traditional Fourier Series Solutions

To fully illustrate the basic idea of the modified Fourier series method, we consider the longitude and transverse vibrations of a classical straight beam with length L, uniform thickness h and width b as shown in Fig. 2.1. The two-dimensional rect-angular coordinate system (x, z) is used to describe the geometry dimensions and deformations of the beam, in which co-ordinates along the axial and thickness directions are represented by x and z, respectively.

Letting $\alpha = x$, $A = 1$, according to Eqs. (1.7), (1.14) and (1.28), the governing equations for free vibration of a generally laminated composite beam are obtained as:

$$A_{11} \frac{\partial^2 u}{\partial x^2} - B_{11} \frac{\partial^3 w}{\partial x^3} = -\omega^2 I_0 u$$
$$B_{11} \frac{\partial^3 u}{\partial x^3} - D_{11} \frac{\partial^4 w}{\partial x^4} = -\omega^2 I_0 w$$

(2.1a, b)

where ω represent the natural frequencies of the beam. Suppose the classical beam considered here is made from isotropic materials, therefore, the B_{11} terms become zero. In such case, the longitude and transverse vibrations of the beam are decoupled. Subsequently, Eq. (2.1) is rewritten as:

$$A_{11} \frac{\partial^2 u}{\partial x^2} = -\omega^2 I_0 u$$
$$D_{11} \frac{\partial^4 w}{\partial x^4} = \omega^2 I_0 w$$

(2.2a, b)

The solution of Eq. (2.2) is often desired to be expanded in the form of either Fourier sine series or Fourier cosine series. Take the transverse vibration problem for example (Eq. 2.2b), mathematically, the displacement $w(x)$ can be expanded as Fourier series only contains the cosine terms by making the even extension of $w(x)$ from the interval $[0, L]$ onto the interval $[-L, 0]$, as shown in Fig. 2.2 (Xu 2011):

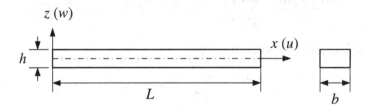

Fig. 2.1 Notations of a classical straight beam

Fig. 2.2 An illustration of the possible discontinuities of the displacement at the ends

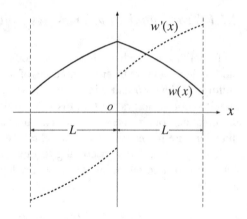

$$w(x) = \sum_{m=0}^{\infty} A_m \cos \lambda_m x \quad 0 \le x \le L \tag{2.3}$$

where A_m are the expansion coefficients, $\lambda_m = m\pi/L$. According to Eq. (2.2b), it is obvious that the transverse displacement $w(x)$ is required to have up to the fourth-derivative ($w''''(x)$). The Fourier cosine series is able to correctly converge to $w(x)$ at any point over $[0, L]$. However, its first-derivative $w'(x)$ and third- derivative $w'''(x)$ are odd functions over $[-L, L]$ leading to a jump at end locations (see Fig. 2.2). Thus, their Fourier series expansions (sine series) will accordingly have a convergence problem due to the discontinuity at end points. Moreover, the displacement function $w(x)$ of the beam given in Eq. (2.3) may not be differentiated term-by-term. The reasons are given below (Tolstov 1976):

Theorem 1 *Let $f(x)$ be a continuous function defined on $[0, L]$ with an absolutely integrable derivative, and let $f(x)$ be expanded in Fourier sine series*

$$f(x) = \sum_{m=1}^{\infty} a_m \sin \lambda_m x, \quad 0 < x < L \quad (\lambda_m = m\pi/L) \tag{2.4}$$

then

$$f'(x) = \frac{f(L) - f(0)}{L} + \sum_{m=1}^{\infty} \left(\frac{2}{L}[(-1)^m f(L) - f(0)] + a_m \lambda_m \right) \cos \lambda_m x \tag{2.5}$$

Apparently, when $f(L) = f(0) = 0$,

$$f'(x) = \sum_{m=1}^{\infty} a_m \lambda_m \cos \lambda_m x \tag{2.6}$$

The theorem reveals that a sine series can be differentiated term-by-term only if $f(L) = f(0) = 0$.

Theorem 2 *Let f(x) be a continuous function defined on* [0, L] *with an absolutely integrable derivative, and let f(x) be expanded in Fourier cosine series*

$$f(x) = \sum_{m=0}^{\infty} b_m \cos \lambda_m x, \quad 0 < x < L \quad (\lambda_m = m\pi/L) \tag{2.7}$$

then

$$f'(x) = -\sum_{m=1}^{\infty} b_m \lambda_m \sin \lambda_m x \tag{2.8}$$

The theorem reveals that a cosine series can always be differentiated term-by-term.

Theorem 3 *Let f(x) be a continuous function of period 2L, which has n derivatives, where n−1 derivatives are continuous and the mth derivative is absolutely integrable (the mth derivative may not exist at certain points). Then, the Fourier series of all m derivatives can be obtained by term-by-term differentiation of the Fourier series of f(x), where all the series, except possibly the last, converge to the corresponding derivatives. Moreover, the Fourier coefficients of the function f(x) satisfy the relations*

$$\lim_{n \to \infty} a_m \lambda_m^n = \lim_{n \to \infty} b_m \lambda_m^n = 0 \tag{2.9}$$

With these in mind, for the cases when the beam is elastically supported, we have

$$w'(x) = -\sum_{m=1}^{\infty} \lambda_m A_m \sin \lambda_m x, \quad 0 < x < L \tag{2.10}$$

$$w''(x) = \frac{w'(L) - w'(0)}{L}$$
$$+ \sum_{m=1}^{\infty} \left(\frac{2}{L} [(-1)^m w'(L) - w'(0)] - A_m \lambda_m^2 \right) \cos \lambda_m x, \quad 0 \le x \le L \tag{2.11}$$

$$w'''(x) = -\sum_{m=1}^{\infty} \left(\frac{2}{L} [(-1)^m w'(L) - w'(0)] \lambda_m - A_m \lambda_m^3 \right) \sin \lambda_m x, \quad 0 < x < L \tag{2.12}$$

$$w''''(x) = \frac{w'''(L) - w'''(0)}{L}$$
$$+ \sum_{m=1}^{\infty} \left(\begin{array}{c} \frac{2}{L} [(-1)^m w'''(L) - w'''(0)] \\ -\frac{2}{L} [(-1)^m w'(L) - w'(0)] \lambda_m^2 + A_m \lambda_m^4 \end{array} \right) \cos \lambda_m x, \quad 0 \le x \le L \tag{2.13}$$

and the Fourier coefficient A_m satisfies

$$\lim_{m \to \infty} A_m \lambda_m = 0 \tag{2.14}$$

Combining Eqs. (2.2b) and (2.13) results in

$$
\begin{aligned}
D_{11} \frac{w'''(L) - w'''(0)}{L} &+ \sum_{m=1}^{\infty} \left(\frac{2D_{11}}{L} [(-1)^m w'''(L) - w'''(0)] \right) \cos \lambda_m x \\
&+ \sum_{m=1}^{\infty} \left(\frac{2D_{11}}{L} [(-1)^m w'(L) - w'(0)] \lambda_m^2 \right) \\
&+ \left(D_{11} \lambda_m^4 - \omega^2 I_0 \right) A_m \Bigg) \cos \lambda_m x = \omega^2 I_0
\end{aligned}
\tag{2.15}
$$

Obviously, it is a big challenge to obtain the natural frequencies and determine the expansion coefficients from Eq. (2.15).

Alternatively, one may prefer to expand the beam displacement $w(x)$ in the form of Fourier sine series. In such case

$$w(x) = \sum_{m=1}^{\infty} A_m \sin \lambda_m x, \quad 0 < x < L \tag{2.16}$$

Then

$$
\begin{aligned}
w'(x) = \frac{w(L) - w(0)}{L} \\
+ \sum_{m=1}^{\infty} \left(\frac{2}{L} [(-1)^m w(L) - w(0)] + A_m \lambda_m \right) \cos \lambda_m x, \quad 0 \le x \le L
\end{aligned}
\tag{2.17}
$$

$$w''(x) = -\sum_{m=1}^{\infty} \left(\frac{2}{L} [(-1)^m w(L) - w(0)] \lambda_m + A_m \lambda_m^2 \right) \sin \lambda_m x, \quad 0 < x < L \tag{2.18}$$

$$
\begin{aligned}
w'''(x) = \frac{w''(L) - w''(0)}{L} \\
+ \sum_{m=1}^{\infty} \left(\begin{array}{l} \frac{2}{L} [(-1)^m w''(L) - w''(0)] \\ -\frac{2}{L} [(-1)^m w(L) - w(0)] \lambda_m^2 - A_m \lambda_m^3 \end{array} \right) \cos \lambda_m x, \quad 0 \le x \le L
\end{aligned}
\tag{2.19}
$$

$$w''''(x) = -\sum_{m=1}^{\infty} \left(\begin{array}{l} \frac{2}{L} [(-1)^m w''(L) - w''(0)] \lambda_m \\ -\frac{2}{L} [(-1)^m w(L) - w(0)] \lambda_m^3 - A_m \lambda_m^4 \end{array} \right) \sin \lambda_m x, \quad 0 < x < L \tag{2.20}$$

and the Fourier coefficient A_m satisfies

$$\lim_{m \to \infty} A_m = 0 \tag{2.21}$$

Combining Eqs. (2.2b) and (2.20) results in

$$[(-1)^m w''(L) - w''(0)]\lambda_m - [(-1)^m w(L)$$
$$- w(0)]\lambda_m^3 = \frac{A_m L}{2D_{11}} \left(\omega^2 I_0 - D_{11}\lambda_m^4\right) \tag{2.22}$$

and

$$A_m = \frac{2D_{11}\lambda_m}{L\left(D_{11}\lambda_m^4 - \omega^2 I\right)_0} \left(\begin{matrix} [(-1)^m w''(L) - w''(0)] \\ -[(-1)^m w(L) - w(0)]\lambda_{mm}^2 \end{matrix} \right) \tag{2.23}$$

Mathematically, the natural frequencies are simply obtained by requiring the determinant of the coefficient matrix to vanish (Wang and Lin 1996). Such a procedure involves solving a non-linear equation, which may not always be an easy job numerically (Li 2000).

In conclusion, a beam with simply supported boundary conditions, the Fourier sine series can be used to determine the vibrations of the beam readily due to the fact that all the required derivatives of the displacement function can be directly obtained from the Fourier sine series through term-by-term differentiation. For other boundary conditions, however, a Fourier series tends to become slow converged, if it converges at all, and its derivatives may not be so easily obtained (Li 2000). In order to overcome these difficulties and satisfy the general boundary conditions, a modified Fourier series method was proposed by Li (2000), in which several supplementary terms are introduced into the Fourier series expansion to remove any potential discontinuities of the original displacements and their derivatives throughout the entire solution domain including the boundaries and then to effectively enhance the convergence of the results. This modified Fourier series method is briefly illustrated in following section.

2.1.2 One-Dimensional Modified Fourier Series Solutions

Unlike in the traditional Fourier methods, the transverse displacement $w(x)$ of the beam is expanded into a standard Fourier cosine series plus an sufficiently smooth auxiliary polynomial function defined over $[0, L]$ as:

$$w(x) = W(x) + P(x), \quad \text{and} \quad W(x) = \sum_{m=0}^{\infty} A_m \cos \lambda_m x \tag{2.24}$$

where A_m are the expansion coefficients, $\lambda_m = m\pi/L$. The sufficiently smooth auxiliary polynomial function $P(x)$ is selected to remove all the discontinuities potentially associated with the first-order and third-order derivatives at the boundaries. By setting

$$
\begin{aligned}
P'(0) &= w'(0) = \varsigma_{10} \quad P'(L) = w'(L) = \varsigma_{11} \\
P'''(0) &= w'''(0) = \varsigma_{30} \quad P'''(L) = w'''(L) = \varsigma_{31}
\end{aligned}
\tag{2.25}
$$

Such requirements can be readily satisfied by choosing simple polynomials as follows (Zhang and Li 2009; Du 2009):

$$
P(x) = \begin{bmatrix} P_1(x) \\ P_2(x) \\ P_3(x) \\ P_4(x) \end{bmatrix}^T \begin{bmatrix} \varsigma_{10} \\ \varsigma_{11} \\ \varsigma_{30} \\ \varsigma_{31} \end{bmatrix}
\tag{2.26a}
$$

and

$$
\begin{bmatrix} P_1(x) \\ P_2(x) \\ P_3(x) \\ P_4(x) \end{bmatrix} = \begin{bmatrix} \frac{9L}{4\pi}\sin\left(\frac{\pi x}{2L}\right) - \frac{L}{12\pi}\sin\left(\frac{3\pi x}{2L}\right) \\ -\frac{9L}{4\pi}\cos\left(\frac{\pi x}{2L}\right) - \frac{L}{12\pi}\cos\left(\frac{3\pi x}{2L}\right) \\ \frac{L^3}{\pi^3}\sin\left(\frac{\pi x}{2L}\right) - \frac{L^3}{3\pi^3}\sin\left(\frac{3\pi x}{2L}\right) \\ -\frac{L^3}{\pi^3}\cos\left(\frac{\pi x}{2L}\right) - \frac{L^3}{3\pi^3}\cos\left(\frac{3\pi x}{2L}\right) \end{bmatrix}
\tag{2.26b}
$$

It should be pointed out that in actual calculation, the boundary values $\varsigma_{10}, \varsigma_{11}, \varsigma_{30}$ and ς_{31} can be treated as undetermined coefficient associated with the auxiliary polynomial function and solved in a strong form solution procedures or a weak form one such as Ritz method. It is easy to verify that

$$
\begin{aligned}
P'(0) &= \begin{bmatrix} 1 \\ 0 \\ 0 \\ 0 \end{bmatrix}^T \begin{bmatrix} \varsigma_{10} \\ \varsigma_{11} \\ \varsigma_{30} \\ \varsigma_{31} \end{bmatrix} \quad P'(L) = \begin{bmatrix} 0 \\ 1 \\ 0 \\ 0 \end{bmatrix}^T \begin{bmatrix} \varsigma_{10} \\ \varsigma_{11} \\ \varsigma_{30} \\ \varsigma_{31} \end{bmatrix} \\[2ex]
P'''(0) &= \begin{bmatrix} 0 \\ 0 \\ 1 \\ 0 \end{bmatrix}^T \begin{bmatrix} \varsigma_{10} \\ \varsigma_{11} \\ \varsigma_{30} \\ \varsigma_{31} \end{bmatrix} \quad P'''(L) = \begin{bmatrix} 0 \\ 0 \\ 0 \\ 1 \end{bmatrix}^T \begin{bmatrix} \varsigma_{10} \\ \varsigma_{11} \\ \varsigma_{30} \\ \varsigma_{31} \end{bmatrix}
\end{aligned}
\tag{2.27}
$$

so that

$$
W'(0) = W'(L) = 0 \quad W'''(0) = W'''(L) = 0
\tag{2.28}
$$

Essentially, $W(x)$ represents a residual beam displacement which is continuous over $[0, L]$ and has zero-slopes at the both ends as shown in Fig. 2.3. Apparently,

Fig. 2.3 An illustration of the modified Fourier method (Xu 2010)

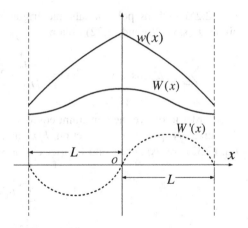

the cosine series representation of $W(x)$ is able to converge correctly to the function itself and its first derivative at every point (including the boundaries) on the beam. Analogously, discontinuities potentially associated with the third-order derivative can be removed as well. In addition, the residual beam displacement $W(x)$ has at least three continuous derivatives, then all the required differentiations can be simply carried out term-by-term basically. In such case, we have

$$w'(x) = -\sum_{m=1}^{\infty} \lambda_m A_m \sin \lambda_m x + P'(x) \tag{2.29}$$

$$w''(x) = -\sum_{m=1}^{\infty} A_m \lambda_m^2 \cos \lambda_m x + P''(x) \tag{2.30}$$

$$w'''(x) = \sum_{m=1}^{\infty} A_m \lambda_m^3 \sin \lambda_m x + P'''(x) \tag{2.31}$$

$$w''''(x) = \sum_{m=1}^{\infty} A_m \lambda_m^4 \cos \lambda_m x + P''''(x) \tag{2.32}$$

and the Fourier coefficient A_m satisfies

$$\lim_{m \to \infty} A_m \lambda_m^4 = 0 \tag{2.33}$$

Comparing Eq. (2.33) with (2.21), it can be found that the modified Fourier series solution converges at a much faster speed. It should be stressed that the form of auxiliary polynomial function given in Eq. (2.26a, b) should be understood as a continuous function that satisfies Eq. (2.25), its form is not a concern with respect to the convergence of the series solution (Li 2004). Actually, any function satisfies

Eq. (2.25) such as polynomials and trigonometric functions can be used. Combining Eqs. (2.2b) and (2.32) obtains

$$\sum_{m=1}^{\infty} A_m \lambda_m^4 \cos \lambda_m x + P''''(x) = \frac{\omega^2 I_0}{D_{11}} \left(\sum_{m=0}^{\infty} A_m \cos \lambda_m x + P(x) \right) \tag{2.34}$$

In order to derive the constraint equations for the unknown Fourier coefficients, the auxiliary polynomial function $P(x)$ and its four-order derivative $P''''(x)$ in Eq. (2.32) are expanded into Fourier cosine series, namely

$$P(x) = \sum_{m=0}^{\infty} B_m \cos \lambda_m x$$
$$P''''(x) = \sum_{m=0}^{\infty} C_m \cos \lambda_m x \tag{2.35}$$

where

$$B_m = \frac{\int_0^L P(x) \cos \lambda_m x dx}{\int_0^L (\cos \lambda_m x)^2 dx}$$
$$C_m = \frac{\int_0^L P''''(x) \cos \lambda_m x dx}{\int_0^L (\cos \lambda_m x)^2 dx} \tag{2.36}$$

Substituting Eq. (2.35) into Eq. (2.34), we have

$$C_0 + \sum_{m=1}^{\infty} \left(A_m \lambda_m^4 + C_m \right) \cos \lambda_m x = \sum_{m=0}^{\infty} \rho_D \omega^2 (A_m + B_m) \cos \lambda_m x \tag{2.37}$$

where $\rho_D = I_0/D_{11}$, Therefore

$$C_0 - \rho_D \omega^2 (A_0 + B_0) = 0$$
$$A_m \lambda_m^4 + C_m - \rho_D \omega^2 (A_m + B_m) = 0 \quad m = 1, 2, \ldots \tag{2.38}$$

According to Eq. (1.29), the general boundary conditions for the beam can be written as

$$k_{x0}^w w(0) = D_{11} w'''(0) \quad k_{x1}^w w(L) = -D_{11} w'''(L)$$
$$K_{x0}^w w'(0) = D_{11} w''(0) \quad K_{x1}^w w'(L) = -D_{11} w''(L) \tag{2.39}$$

Substituting Eq. (2.24) into Eq. (2.39) the boundary conditions of the beam can be rewritten as

$$k_{x0}^w \left(\sum_{m=0}^{\infty} A_m + P(0) \right) = D_{11} P'''(0)$$

$$k_{x1}^w \left(\sum_{m=0}^{\infty} (-1)^m A_m + P(L) \right) = -D_{11} P'''(L)$$

$$K_{x0}^w P'(0) = D_{11} \left(\sum_{m=0}^{\infty} -\lambda_m^2 A_m + P''(0) \right)$$ (2.40)

$$K_{x1}^w P'(L) = -D_{11} \left(\sum_{m=0}^{\infty} (-1)^{m+1} \lambda_m^2 A_m + P''(L) \right)$$

The natural frequencies and mode shapes of the beam can now be easily determined by solving Eq. (2.38) with boundary condition equations Eq. (2.40), the more detail solution procedure will be given in Sect. 2.2.

Alternatively, the transverse displacement of the beam can also be expanded into a modified Fourier sine series. In that case, the auxiliary polynomial function $P(x)$ is selected to remove all the discontinuities potentially associated with the original displacement and its second-order derivative at the boundaries. Namely, the transverse displacement $w(x)$ of the beam should be expanded into a standard Fourier sine series plus a sufficiently smooth auxiliary polynomial function defined over $[0, L]$ as:

$$w(x) = W(x) + P(x), \quad \text{where} \quad W(x) = \sum_{m=0}^{\infty} A_m \sin \lambda_m x$$ (2.41)

and

$$\begin{array}{ll} P(0) = w(0) = \varsigma_{00} & P(L) = w(L) = \varsigma_{01} \\ P''(0) = w''(0) = \varsigma_{20} & P''(L) = w''(L) = \varsigma_{21} \end{array}$$ (2.42)

Similarly, Eqs. (2.17)–(2.20) can be rewritten as

$$w'(x) = \sum_{m=1}^{\infty} A_m \lambda_m \cos \lambda_m x + P'(x)$$ (2.43)

$$w''(x) = -\sum_{m=1}^{\infty} A_m \lambda_m^2 \sin \lambda_m x + P''(x)$$ (2.44)

$$w'''(x) = -\sum_{m=1}^{\infty} A_m \lambda_m^3 \cos \lambda_m x + P'''(x) \qquad (2.45)$$

$$w''''(x) = \sum_{m=1}^{\infty} A_m \lambda_m^4 \sin \lambda_m x + P''''(x) \qquad (2.46)$$

The solution procedure is the same as those of the modified Fourier cosine series.

2.1.3 Two-Dimensional Modified Fourier Series Solutions

Using the modified Fourier series technique, in the manner similar to that described earlier, two-dimensional modified Fourier series solutions for laminated plates and shells are presented in this section. For the sake of completeness, we consider the free vibration analysis of a moderately thick. Generally laminated rectangular plate (where a and b denote its length and width) with general boundary conditions (see Fig. 2.4), the solution procedure is given step-by-step as follows.

Substituting $\alpha = x$, $\beta = y$, $A = B = 1$ and $R_\alpha = R_\beta = \infty$ into Eq. (1.59), the governing equations of the plate are written as:

$$\frac{\partial N_x}{\partial x} + \frac{\partial N_{yx}}{\partial y} = -\omega^2 (I_0 u + I_1 \phi_x)$$

$$\frac{\partial N_{xy}}{\partial x} + \frac{\partial N_y}{\partial y} = -\omega^2 (I_0 v + I_1 \phi_y)$$

$$\frac{\partial Q_x}{\partial x} + \frac{\partial Q_y}{\partial y} = -\omega^2 I_0 w \qquad (2.47a\text{--}e)$$

$$\frac{\partial M_x}{\partial x} + \frac{\partial M_{yx}}{\partial y} - Q_x = -\omega^2 (I_1 u + I_2 \phi_x)$$

$$\frac{\partial M_{xy}}{\partial x} + \frac{\partial M_y}{\partial y} - Q_y = -\omega^2 (I_1 v + I_2 \phi_\beta)$$

Fig. 2.4 A generally laminated moderately thick rectangular plate

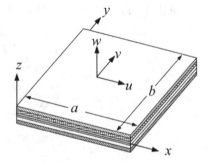

Similarly, substituting $\alpha = x$, $\beta = y$, $A = B = 1$ and $R_\alpha = R_\beta = \infty$ into Eqs. (1.34), (1.46) and (1.47) and then substituting these three equations into Eq. (2.47) yields

$$\left(\begin{bmatrix} L_{11} & L_{12} & L_{13} & L_{14} & L_{15} \\ L_{21} & L_{22} & L_{23} & L_{24} & L_{25} \\ L_{31} & L_{32} & L_{33} & L_{34} & L_{35} \\ L_{41} & L_{42} & L_{43} & L_{44} & L_{45} \\ L_{51} & L_{52} & L_{53} & L_{54} & L_{55} \end{bmatrix} - \omega^2 \begin{bmatrix} M_{11} & 0 & 0 & M_{14} & 0 \\ 0 & M_{22} & 0 & 0 & M_{25} \\ 0 & 0 & M_{33} & 0 & 0 \\ M_{41} & 0 & 0 & M_{44} & 0 \\ 0 & M_{52} & 0 & 0 & M_{55} \end{bmatrix} \right) \begin{bmatrix} u \\ v \\ w \\ \phi_x \\ \phi_y \end{bmatrix} = \begin{bmatrix} 0 \\ 0 \\ 0 \\ 0 \\ 0 \end{bmatrix} \quad (2.48)$$

where the coefficients of the linear operator $L(L_{ij} = L_{ji}, M_{ij} = M_{ji})$ are given below:

$$L_{11} = A_{11} \frac{\partial^2}{\partial x^2} + 2A_{16} \frac{\partial^2}{\partial x \partial y} + A_{66} \frac{\partial^2}{\partial y^2}$$

$$L_{12} = A_{16} \frac{\partial^2}{\partial x^2} + (A_{12} + A_{66}) \frac{\partial^2}{\partial x \partial y} + A_{26} \frac{\partial^2}{\partial y^2}$$

$$L_{13} = 0 \quad L_{23} = 0$$

$$L_{14} = B_{11} \frac{\partial^2}{\partial x^2} + 2B_{16} \frac{\partial^2}{\partial x \partial y} + B_{66} \frac{\partial^2}{\partial y^2}$$

$$L_{15} = B_{16} \frac{\partial^2}{\partial x^2} + (B_{12} + B_{66}) \frac{\partial^2}{\partial x \partial y} + B_{26} \frac{\partial^2}{\partial y^2}$$

$$L_{22} = A_{66} \frac{\partial^2}{\partial x^2} + 2A_{26} \frac{\partial^2}{\partial x \partial y} + A_{22} \frac{\partial^2}{\partial y^2}$$

$$L_{24} = B_{16} \frac{\partial^2}{\partial x^2} + (B_{12} + B_{66}) \frac{\partial^2}{\partial x \partial y} + B_{26} \frac{\partial^2}{\partial y^2}$$

$$L_{25} = B_{66} \frac{\partial^2}{\partial x^2} + 2B_{26} \frac{\partial^2}{\partial x \partial y} + B_{22} \frac{\partial^2}{\partial y^2}$$

$$L_{33} = -A_{55} \frac{\partial^2}{\partial x^2} - 2A_{45} \frac{\partial^2}{\partial x \partial y} - A_{44} \frac{\partial^2}{\partial y^2}$$

$$L_{34} = -A_{55} \frac{\partial}{\partial x} - A_{45} \frac{\partial}{\partial y} \qquad (2.49)$$

$$L_{35} = -A_{45} \frac{\partial}{\partial x} - A_{44} \frac{\partial}{\partial y}$$

$$L_{44} = D_{11} \frac{\partial^2}{\partial x^2} + 2D_{16} \frac{\partial^2}{\partial x \partial y} + D_{66} \frac{\partial^2}{\partial y^2} - A_{55}$$

$$L_{45} = D_{16} \frac{\partial^2}{\partial x^2} + (D_{12} + D_{66}) \frac{\partial^2}{\partial x \partial y} + D_{26} \frac{\partial^2}{\partial y^2} - A_{45}$$

$$L_{55} = D_{66} \frac{\partial^2}{\partial x^2} + 2D_{26} \frac{\partial^2}{\partial x \partial y} + D_{22} \frac{\partial^2}{\partial y^2} - A_{44}$$

$$M_{11} = M_{22} = M_{33} = -I_0$$

$$M_{14} = M_{25} = -I_1$$

$$M_{44} = M_{55} = -I_2$$

and the general boundary conditions of the plate are:

$$
x = 0 : \begin{cases} N_x = k_{x0}^u u \\ N_{xy} = k_{x0}^v v \\ Q_x = k_{x0}^w w \\ M_x = K_{x0}^x \phi_x \\ M_{xy} = K_{x0}^y \phi_y \end{cases} \quad x = a : \begin{cases} N_x = -k_{x1}^u u \\ N_{xy} = -k_{x1}^v v \\ Q_x = -k_{x1}^w w \\ M_x = -K_{x1}^x \phi_x \\ M_{xy} = -K_{x1}^y \phi_y \end{cases}
$$
$$
y = 0 : \begin{cases} N_{yx} = k_{y0}^u u \\ N_y = k_{y0}^v v \\ Q_y = k_{y0}^w w \\ M_{yx} = K_{y0}^x \phi_x \\ M_y = K_{y0}^y \phi_y \end{cases} \quad y = b : \begin{cases} N_{yx} = -k_{y1}^u u \\ N_y = -k_{y1}^v v \\ Q_y = -k_{y1}^w w \\ M_{yx} = -K_{y1}^x \phi_x \\ M_y = -K_{y1}^y \phi_y \end{cases} \tag{2.50}
$$

Taking the plate displacement component $u(x, y)$ for example, it can be expanded into a standard double Fourier cosine series plus two sufficiently smooth auxiliary polynomial functions defined over $[0, a] \times [0, b]$ as

$$
u(x, y) = U(x, y) + P_x(x, y) + P_y(x, y) \tag{2.51a}
$$

and

$$
U(x, y) = \sum_{m=0}^{\infty} \sum_{n=0}^{\infty} A_{mn} \cos \lambda_m x \cos \lambda_n y \tag{2.51b}
$$

where A_{mn} are the expansion coefficients. $\lambda_m = m\pi/a$ and $\lambda_n = n\pi/b$. $P_x(x, y)$ and $P_y(x, y)$ denote the auxiliary polynomial functions introduced to ensure and accelerate the convergence of the series expansion of the displacement $u(x, y)$. According to Eq. (2.49), it is obvious that each of the displacements and rotation components of the plate is required to have up to the second derivatives. Therefore, the auxiliary polynomial functions $P_x(x, y)$ and $P_y(x, y)$ are selected to remove all the discontinuities potentially associated with the first-order derivatives at the boundaries. By setting

$$
\begin{aligned}
\frac{\partial P_x(0, y)}{\partial x} &= \frac{\partial u(0, y)}{\partial x} = \xi_{x0}(y) \\
\frac{\partial P_x(a, y)}{\partial x} &= \frac{\partial u(a, y)}{\partial x} = \xi_{x1}(y) \\
\frac{\partial P_y(x, 0)}{\partial y} &= \frac{\partial u(x, 0)}{\partial y} = \xi_{y0}(x) \\
\frac{\partial P_y(x, b)}{\partial y} &= \frac{\partial u(x, b)}{\partial y} = \xi_{y1}(x)
\end{aligned} \tag{2.52}
$$

where $\xi_{x0}(y)$, $\xi_{x1}(y)$, $\xi_{y0}(x)$ and $\xi_{y1}(x)$ are the unknown boundary derivatives at boundaries $x = 0$, $x = a$, $y = 0$ and $y = b$, respectively. They can be expanded in the form of Fourier cosine series as

$$\xi_{x0}(y) = \sum_{n=0}^{\infty} a_{1n} \cos \lambda_n y$$

$$\xi_{x1}(y) = \sum_{n=0}^{\infty} a_{2n} \cos \lambda_n y$$

$$\xi_{y0}(x) = \sum_{m=0}^{\infty} b_{1m} \cos \lambda_m x \qquad (2.53)$$

$$\xi_{y1}(x) = \sum_{m=0}^{\infty} b_{2m} \cos \lambda_m x$$

where a_{1n}, a_{2n}, b_{1m} and b_{2m} are the expansion coefficients. The requirements of Eq. (2.52) can be readily satisfied by choosing the auxiliary polynomial functions $P_x(x, y)$ and $P_y(x, y)$ as follows (Du 2009):

$$P_x(x,y) = \begin{bmatrix} P_1(x) \\ P_2(x) \end{bmatrix}^T \begin{bmatrix} \xi_{x0}(y) \\ \xi_{x1}(y) \end{bmatrix}, \quad \begin{bmatrix} P_1(x) \\ P_2(x) \end{bmatrix} = \begin{bmatrix} x(x/a - 1)^2 \\ x^2(x/a - 1)/a \end{bmatrix} \qquad (2.54a)$$

and

$$P_y(x,y) = \begin{bmatrix} P_1(y) \\ P_2(y) \end{bmatrix}^T \begin{bmatrix} \xi_{y0}(x) \\ \xi_{y1}(x) \end{bmatrix}, \quad \begin{bmatrix} P_1(y) \\ P_2(y) \end{bmatrix} = \begin{bmatrix} y(y/b - 1)^2 \\ y^2(y/b - 1)/b \end{bmatrix} \qquad (2.54b)$$

It is easy to verify that

$$\frac{\partial P_x(0,y)}{\partial x} = \begin{bmatrix} 1 \\ 0 \end{bmatrix}^T \begin{bmatrix} \xi_{x0}(y) \\ \xi_{x1}(y) \end{bmatrix}$$

$$\frac{\partial P_x(a,y)}{\partial x} = \begin{bmatrix} 0 \\ 1 \end{bmatrix}^T \begin{bmatrix} \xi_{x0}(y) \\ \xi_{x1}(y) \end{bmatrix}$$

$$\frac{\partial P_y(x,0)}{\partial y} = \begin{bmatrix} 1 \\ 0 \end{bmatrix}^T \begin{bmatrix} \xi_{y0}(x) \\ \xi_{y1}(x) \end{bmatrix} \qquad (2.55)$$

$$\frac{\partial P_y(x,b)}{\partial y} = \begin{bmatrix} 0 \\ 1 \end{bmatrix}^T \begin{bmatrix} \xi_{y0}(x) \\ \xi_{y1}(x) \end{bmatrix}$$

so that

$$\frac{\partial U(0,y)}{\partial x} = \frac{\partial U(a,y)}{\partial x} = 0$$

$$\frac{\partial U(x,0)}{\partial y} = \frac{\partial U(x,b)}{\partial y} = 0 \qquad (2.56)$$

Essentially, $U(x, y)$ represents a residual plate displacement which is continuous over $[0, a] \times [0, b]$ and has zero-slopes at the four boundaries. Apparently, the cosine series representation of $U(x, y)$ is able to converge correctly to the function itself and its first derivative at every point on the plate. Moreover, all the required differentiations of the residual plate displacement $U(x, y)$ can be simply carried out term-by-term. Thus, the plate displacement function of component $u(x, y)$ can be rewritten as

$$
\begin{aligned}
u(x, y) = & \sum_{m=0}^{\infty} \sum_{n=0}^{\infty} A_{mn} \cos \lambda_m x \cos \lambda_n y + \sum_{l=1}^{2} \sum_{n=0}^{\infty} a_{ln} P_l(x) \cos \lambda_n y \\
& + \sum_{l=1}^{2} \sum_{m=0}^{\infty} b_{lm} P_l(y) \cos \lambda_m x
\end{aligned}
\tag{2.57}
$$

Similarly, the other displacements and rotation components of the plate can be expanded as the two-dimensional modified Fourier series as

$$
\begin{aligned}
v(x, y) = & \sum_{m=0}^{\infty} \sum_{n=0}^{\infty} B_{mn} \cos \lambda_m x \cos \lambda_n y + \sum_{l=0}^{2} \sum_{n=0}^{\infty} c_{ln} P_l(x) \cos \lambda_n y \\
& + \sum_{l=0}^{2} \sum_{m=0}^{\infty} d_{lm} P_l(y) \cos \lambda_m x \\
w(x, y) = & \sum_{m=0}^{\infty} \sum_{n=0}^{\infty} C_{mn} \cos \lambda_m x \cos \lambda_n y + \sum_{l=0}^{2} \sum_{n=0}^{\infty} e_{ln} P_l(x) \cos \lambda_n y \\
& + \sum_{l=0}^{2} \sum_{m=0}^{\infty} f_{lm} P_l(y) \cos \lambda_m x \\
\phi_x(x, y) = & \sum_{m=0}^{\infty} \sum_{n=0}^{\infty} D_{mn} \cos \lambda_m x \cos \lambda_n y + \sum_{l=0}^{2} \sum_{n=0}^{\infty} g_{ln} P_l(x) \cos \lambda_n y \\
& + \sum_{l=0}^{2} \sum_{m=0}^{\infty} h_{lm} P_l(y) \cos \lambda_m x \\
\phi_y(x, y) = & \sum_{m=0}^{\infty} \sum_{n=0}^{\infty} E_{mn} \cos \lambda_m x \cos \lambda_n y + \sum_{l=0}^{2} \sum_{n=0}^{\infty} i_{ln} P_l(x) \cos \lambda_n y \\
& + \sum_{l=0}^{2} \sum_{m=0}^{\infty} j_{lm} P_l(y) \cos \lambda_m x
\end{aligned}
\tag{2.58}
$$

where B_{mn}, C_{mn}, D_{mn} and E_{mn} are the standard Fourier series expansion coefficients. c_{ln}, d_{lm}, e_{ln}, f_{lm} g_{ln}, h_{lm}, i_{ln} and j_{lm} are the corresponding supplement coefficients.

2.2 Strong Form Solution Procedure

With the modified Fourier series, vibration of isotropic, anisotropic and laminated beams, plates and shells can be obtained by using the strong form solution procedures as described below. Taking the previously studied laminated rectangular plate for example, the solution procedure is given step-by-step as follows.

In the actual calculations, all the five infinite modified Fourier series expressions given in Eqs. (2.57) and (2.58) need to be truncated as finite series to obtain the results with acceptable accuracy due to the limited speed, the capacity and the numerical accuracy of computers. Unless otherwise stressed, the involving terms in all the plate displacements and rotation components are uniformly taken as $m \in [0, M]$ and $n \in [0, N]$. Thus, the modified Fourier series expressions presented in Eqs. (2.57) and (2.58) can be rewritten in the matrix form as:

$$
\begin{aligned}
u(x, y) &= \mathbf{H}_{xy}\mathbf{A} + \mathbf{H}_x\mathbf{a} + \mathbf{H}_y\mathbf{b} \\
v(x, y) &= \mathbf{H}_{xy}\mathbf{B} + \mathbf{H}_x\mathbf{c} + \mathbf{H}_y\mathbf{d} \\
w(x, y) &= \mathbf{H}_{xy}\mathbf{C} + \mathbf{H}_x\mathbf{e} + \mathbf{H}_y\mathbf{f} \\
\phi_x(x, y) &= \mathbf{H}_{xy}\mathbf{D} + \mathbf{H}_x\mathbf{g} + \mathbf{H}_y\mathbf{h} \\
\phi_y(x, y) &= \mathbf{H}_{xy}\mathbf{E} + \mathbf{H}_x\mathbf{i} + \mathbf{H}_y\mathbf{j}
\end{aligned}
\tag{2.59}
$$

where

$$
\begin{aligned}
\mathbf{H}_{xy} &= [\cos \lambda_0 x \cos \lambda_0 y, \ldots, \cos \lambda_m x \cos \lambda_n y, \ldots, \cos \lambda_M x \cos \lambda_N y] \\
\mathbf{H}_x &= [P_1(x) \cos \lambda_0 y, \ldots, P_l(x) \cos \lambda_n y, \ldots, P_2(x) \cos \lambda_N y] \\
\mathbf{H}_y &= [P_1(y) \cos \lambda_0 x, \ldots, P_l(y) \cos \lambda_m x, \ldots, P_2(y) \cos \lambda_M x]
\end{aligned}
\tag{2.60}
$$

and

$$
\begin{aligned}
\mathbf{A} &= [A_{00}, \ldots, A_{mn}, \ldots, A_{MN}]^T \quad & \mathbf{a} &= [a_{10}, \ldots, a_{ln}, \ldots, a_{2N}]^T \\
\mathbf{b} &= [b_{10}, \ldots, b_{lm}, \ldots, b_{2M}]^T \\
\mathbf{B} &= [B_{00}, \ldots, B_{mn}, \ldots, B_{MN}]^T \quad & \mathbf{c} &= [c_{10}, \ldots, c_{ln}, \ldots, c_{2N}]^T \\
\mathbf{d} &= [d_{10}, \ldots, d_{lm}, \ldots, d_{2M}]^T \\
\mathbf{C} &= [C_{00}, \ldots, C_{mn}, \ldots, C_{MN}]^T \quad & \mathbf{e} &= [e_{10}, \ldots, e_{ln}, \ldots, e_{2N}]^T \\
\mathbf{f} &= [f_{10}, \ldots, f_{lm}, \ldots, f_{2M}]^T \\
\mathbf{D} &= [D_{00}, \ldots, D_{mn}, \ldots, D_{MN}]^T \quad & \mathbf{g} &= [g_{10}, \ldots, g_{ln}, \ldots, g_{2N}]^T \\
\mathbf{h} &= [h_{10}, \ldots, h_{lm}, \ldots, h_{2M}]^T \\
\mathbf{E} &= [E_{00}, \ldots, E_{mn}, \ldots, E_{MN}]^T \quad & \mathbf{i} &= [i_{10}, \ldots, i_{ln}, \ldots, i_{2N}]^T \\
\mathbf{j} &= [j_{10}, \ldots, j_{lm}, \ldots, j_{2M}]^T
\end{aligned}
\tag{2.61}
$$

Superscript T represents the transposition operator. Substituting Eq. (2.59) into Eq. (2.48) results in

$$\mathbf{L}_{xy}\mathbf{\Gamma}_{xy} + \mathbf{L}_x\mathbf{\Gamma}_x + \mathbf{L}_y\mathbf{\Gamma}_y - \omega^2\left(\mathbf{M}_{xy}\mathbf{\Gamma}_{xy} + \mathbf{M}_x\mathbf{\Gamma}_x + \mathbf{M}_y\mathbf{\Gamma}_y\right) = 0 \qquad (2.62)$$

where

$$\mathbf{L}_i = \begin{bmatrix} L_{11}\mathbf{H}_i & L_{12}\mathbf{H}_i & L_{13}\mathbf{H}_i & L_{14}\mathbf{H}_i & L_{15}\mathbf{H}_i \\ L_{21}\mathbf{H}_i & L_{22}\mathbf{H}_i & L_{23}\mathbf{H}_i & L_{24}\mathbf{H}_i & L_{25}\mathbf{H}_i \\ L_{31}\mathbf{H}_i & L_{32}\mathbf{H}_i & L_{33}\mathbf{H}_i & L_{34}\mathbf{H}_i & L_{35}\mathbf{H}_i \\ L_{41}\mathbf{H}_i & L_{42}\mathbf{H}_i & L_{43}\mathbf{H}_i & L_{44}\mathbf{H}_i & L_{45}\mathbf{H}_i \\ L_{51}\mathbf{H}_i & L_{52}\mathbf{H}_i & L_{53}\mathbf{H}_i & L_{54}\mathbf{H}_i & L_{55}\mathbf{H}_i \end{bmatrix}, \quad (i = xy, x, y) \qquad (2.63)$$

$$\mathbf{M}_i = \begin{bmatrix} M_{11}\mathbf{H}_i & 0 & 0 & M_{14}\mathbf{H}_i & 0 \\ 0 & M_{22}\mathbf{H}_i & 0 & 0 & M_{25}\mathbf{H}_i \\ 0 & 0 & M_{33}\mathbf{H}_i & 0 & 0 \\ M_{41}\mathbf{H}_i & 0 & 0 & M_{44}\mathbf{H}_i & 0 \\ 0 & M_{52}\mathbf{H}_i & 0 & 0 & M_{55}\mathbf{H}_i \end{bmatrix}, \quad (i = xy, x, y) \qquad (2.64)$$

$$\mathbf{\Gamma}_{xy} = \begin{bmatrix} \mathbf{A} \\ \mathbf{B} \\ \mathbf{C} \\ \mathbf{D} \\ \mathbf{E} \end{bmatrix}, \quad \mathbf{\Gamma}_x = \begin{bmatrix} \mathbf{a} \\ \mathbf{c} \\ \mathbf{e} \\ \mathbf{g} \\ \mathbf{i} \end{bmatrix}, \quad \mathbf{\Gamma}_y = \begin{bmatrix} \mathbf{b} \\ \mathbf{d} \\ \mathbf{f} \\ \mathbf{h} \\ \mathbf{j} \end{bmatrix} \qquad (2.65)$$

In the same way, substituting Eq. (2.59) into Eq. (2.50), the general boundary conditions of the plate can be rewritten as

$$x = 0 : \mathbf{L}_{xy}^{x0}\mathbf{\Gamma}_{xy} + \mathbf{L}_x^{x0}\mathbf{\Gamma}_x + \mathbf{L}_y^{x0}\mathbf{\Gamma}_y = 0$$
$$x = a : \mathbf{L}_{xy}^{x1}\mathbf{\Gamma}_{xy} + \mathbf{L}_x^{x1}\mathbf{\Gamma}_x + \mathbf{L}_y^{x1}\mathbf{\Gamma}_y = 0 \qquad (2.66a)$$

$$y = 0 : \mathbf{L}_{xy}^{y0}\mathbf{\Gamma}_{xy} + \mathbf{L}_x^{y0}\mathbf{\Gamma}_x + \mathbf{L}_y^{y0}\mathbf{\Gamma}_y = 0$$
$$y = b : \mathbf{L}_{xy}^{y1}\mathbf{\Gamma}_{xy} + \mathbf{L}_x^{y1}\mathbf{\Gamma}_x + \mathbf{L}_y^{y1}\mathbf{\Gamma}_y = 0 \qquad (2.66b)$$

in which

$$\mathbf{L}_i^j = \begin{bmatrix} L_{11}^j\mathbf{H}_i & L_{12}^j\mathbf{H}_i & L_{13}^j\mathbf{H}_i & L_{14}^j\mathbf{H}_i & L_{15}^j\mathbf{H}_i \\ L_{21}^j\mathbf{H}_i & L_{22}^j\mathbf{H}_i & L_{23}^j\mathbf{H}_i & L_{24}^j\mathbf{H}_i & L_{25}^j\mathbf{H}_i \\ L_{31}^j\mathbf{H}_i & L_{32}^j\mathbf{H}_i & L_{33}^j\mathbf{H}_i & L_{34}^j\mathbf{H}_i & L_{35}^j\mathbf{H}_i \\ L_{41}^j\mathbf{H}_i & L_{42}^j\mathbf{H}_i & L_{43}^j\mathbf{H}_i & L_{44}^j\mathbf{H}_i & L_{45}^j\mathbf{H}_i \\ L_{25}^j\mathbf{H}_i & L_{52}^j\mathbf{H}_i & L_{53}^j\mathbf{H}_i & L_{54}^j\mathbf{H}_i & L_{55}^j\mathbf{H}_i \end{bmatrix}, \quad \begin{pmatrix} i = xy, x, y \\ j = x0, x1, \\ y0, y1 \end{pmatrix} \qquad (2.67)$$

the coefficients of the linear operator in Eq. (2.67) are given below:

$$
\begin{aligned}
&L_{11}^{x0} = A_{11}\frac{\partial}{\partial x} + A_{16}\frac{\partial}{\partial y} - k_{x0}^u, \quad && L_{12}^{x0} = A_{12}\frac{\partial}{\partial y} + A_{16}\frac{\partial}{\partial x} \\
&L_{13}^{x0} = L_{31}^{x0} = L_{43}^{x0} = 0, \quad && L_{14}^{x0} = B_{11}\frac{\partial}{\partial x} + B_{16}\frac{\partial}{\partial y} = L_{41}^{x0} \\
&L_{15}^{x0} = B_{12}\frac{\partial}{\partial y} + B_{16}\frac{\partial}{\partial x}, \quad && L_{21}^{x0} = A_{16}\frac{\partial}{\partial x} + A_{66}\frac{\partial}{\partial y} \\
&L_{22}^{x0} = A_{26}\frac{\partial}{\partial y} + A_{66}\frac{\partial}{\partial x} - k_{x0}^v, \quad && L_{23}^{x0} = L_{32}^{x0} = L_{53}^{x0} = 0 \\
&L_{24}^{x0} = B_{16}\frac{\partial}{\partial x} + B_{66}\frac{\partial}{\partial y}, \quad && L_{25}^{x0} = B_{26}\frac{\partial}{\partial y} + B_{66}\frac{\partial}{\partial x} = L_{52}^{x0} \\
&L_{33}^{x0} = A_{45}\frac{\partial}{\partial y} + A_{55}\frac{\partial}{\partial x} - k_{x0}^w, \quad && L_{34}^{x0} = A_{55}, L_{35}^{x0} = A_{45} \\
&L_{42}^{x0} = B_{12}\frac{\partial}{\partial y} + B_{16}\frac{\partial}{\partial x}, \quad && L_{44}^{x0} = D_{11}\frac{\partial}{\partial x} + D_{16}\frac{\partial}{\partial y} - K_{x0}^x \\
&L_{45}^{x0} = D_{12}\frac{\partial}{\partial y} + D_{16}\frac{\partial}{\partial x}, \quad && L_{51}^{x0} = B_{16}\frac{\partial}{\partial x} + B_{66}\frac{\partial}{\partial y} \\
&L_{54}^{x0} = D_{16}\frac{\partial}{\partial x} + D_{66}\frac{\partial}{\partial y}, \quad && L_{55}^{x0} = D_{26}\frac{\partial}{\partial y} + D_{66}\frac{\partial}{\partial x} - K_{x0}^y
\end{aligned} \tag{2.68}
$$

and

$$
\begin{aligned}
&L_{11}^{y0} = A_{16}\frac{\partial}{\partial x} + A_{66}\frac{\partial}{\partial y} - k_{y0}^u, \quad && L_{12}^{y0} = A_{26}\frac{\partial}{\partial y} + A_{66}\frac{\partial}{\partial x} \\
&L_{13}^{y0} = L_{31}^{y0} = L_{43}^{y0} = 0, \quad && L_{14}^{y0} = B_{16}\frac{\partial}{\partial x} + B_{66}\frac{\partial}{\partial y} = L_{41}^{y0} \\
&L_{15}^{y0} = B_{26}\frac{\partial}{\partial y} + B_{66}\frac{\partial}{\partial x}, \quad && L_{21}^{y0} = A_{12}\frac{\partial}{\partial x} + A_{26}\frac{\partial}{\partial y} \\
&L_{22}^{y0} = A_{22}\frac{\partial}{\partial y} + A_{26}\frac{\partial}{\partial x} - k_{y0}^v, \quad && L_{23}^{y0} = L_{32}^{y0} = L_{53}^{y0} = 0 \\
&L_{24}^{y0} = B_{12}\frac{\partial}{\partial x} + B_{26}\frac{\partial}{\partial y}, \quad && L_{25}^{y0} = B_{22}\frac{\partial}{\partial y} + B_{26}\frac{\partial}{\partial x} = L_{52}^{y0} \\
&L_{33}^{y0} = A_{44}\frac{\partial}{\partial y} + A_{45}\frac{\partial}{\partial x} - k_{y0}^w, \quad && L_{34}^{y0} = A_{45}, \quad L_{35}^{y0} = A_{44} \\
&L_{42}^{y0} = B_{26}\frac{\partial}{\partial y} + B_{66}\frac{\partial}{\partial x}, \quad && L_{44}^{y0} = D_{16}\frac{\partial}{\partial x} + D_{66}\frac{\partial}{\partial y} - K_{y0}^x \\
&L_{45}^{y0} = D_{26}\frac{\partial}{\partial y} + D_{66}\frac{\partial}{\partial x}, \quad && L_{51}^{y0} = B_{12}\frac{\partial}{\partial x} + B_{26}\frac{\partial}{\partial y} \\
&L_{54}^{y0} = D_{12}\frac{\partial}{\partial x} + D_{26}\frac{\partial}{\partial y}, \quad && L_{55}^{y0} = D_{22}\frac{\partial}{\partial y} + D_{26}\frac{\partial}{\partial x} - K_{y0}^y
\end{aligned} \tag{2.69}
$$

And for $j = x1$ and $j = y1$, we have:

$$
\begin{aligned}
&L_{ij}^{x1} = L_{ij}^{x0}, (i \neq j); \quad && L_{ij}^{y1} = L_{ij}^{y0}, (i \neq j) \\
&L_{11}^{x1} = A_{11}\frac{\partial}{\partial x} + A_{16}\frac{\partial}{\partial y} + k_{x1}^u, \quad && L_{11}^{y1} = A_{16}\frac{\partial}{\partial x} + A_{66}\frac{\partial}{\partial y} + k_{y1}^u \\
&L_{22}^{x1} = A_{26}\frac{\partial}{\partial y} + A_{66}\frac{\partial}{\partial x} + k_{x1}^v, \quad && L_{22}^{y1} = A_{22}\frac{\partial}{\partial y} + A_{26}\frac{\partial}{\partial x} + k_{y1}^v \\
&L_{33}^{x1} = A_{45}\frac{\partial}{\partial y} + A_{55}\frac{\partial}{\partial x} + k_{x1}^w, \quad && L_{33}^{y1} = A_{44}\frac{\partial}{\partial y} + A_{45}\frac{\partial}{\partial x} + k_{y1}^w \\
&L_{44}^{x1} = D_{11}\frac{\partial}{\partial x} + D_{16}\frac{\partial}{\partial y} + K_{x1}^x, \quad && L_{44}^{y1} = D_{16}\frac{\partial}{\partial x} + D_{66}\frac{\partial}{\partial y} + K_{y1}^x \\
&L_{55}^{x1} = D_{26}\frac{\partial}{\partial y} + D_{66}\frac{\partial}{\partial x} + K_{x1}^y, \quad && L_{55}^{y1} = D_{22}\frac{\partial}{\partial y} + D_{26}\frac{\partial}{\partial x} + K_{y1}^y
\end{aligned} \tag{2.70}
$$

In order to derive the constraint equations for the unknown Fourier coefficients, all the sine terms, the auxiliary polynomial functions and their derivatives in Eqs. (2.62) and (2.66a, b) will be expanded into Fourier cosine series. Letting

$$
\begin{aligned}
\mathbf{C}_{xy} &= [\cos\lambda_0 x \cos\lambda_0 y, \ldots, \cos\lambda_m x \cos\lambda_n y, \ldots, \cos\lambda_M x \cos\lambda_N y]^T \\
\mathbf{C}_x &= [\cos\lambda_0 x, \ldots, \cos\lambda_m x, \ldots, \cos\lambda_M x]^T \\
\mathbf{C}_y &= [\cos\lambda_0 y, \ldots, \cos\lambda_n x, \ldots, \cos\lambda_N y]^T
\end{aligned} \tag{2.71}
$$

Multiplying Eq. (2.62) with $\mathbf{C}_{xy}$ in the left side and integrating it from 0 to a and 0 to b separately with respect to x and y obtains

$$\overline{\mathbf{L}}_{xy}\boldsymbol{\Gamma}_{xy} + \begin{bmatrix} \overline{\mathbf{L}}_x & \overline{\mathbf{L}}_y \end{bmatrix}\begin{bmatrix} \boldsymbol{\Gamma}_x \\ \boldsymbol{\Gamma}_y \end{bmatrix} - \omega^2\left(\overline{\mathbf{M}}_{xy}\boldsymbol{\Gamma}_{xy} + \begin{bmatrix} \overline{\mathbf{M}}_x & \overline{\mathbf{M}}_y \end{bmatrix}\begin{bmatrix} \boldsymbol{\Gamma}_x \\ \boldsymbol{\Gamma}_y \end{bmatrix}\right) = \mathbf{0} \tag{2.72}$$

where

$$\begin{aligned}
\overline{\mathbf{L}}_{xy} &= \int_0^a \int_0^b \mathbf{C}_{xy}\mathbf{L}_{xy}dydx, & \overline{\mathbf{M}}_{xy} &= \int_0^a \int_0^b \mathbf{C}_{xy}\mathbf{M}_{xy}dydx \\
\overline{\mathbf{L}}_x &= \int_0^a \int_0^b \mathbf{C}_{xy}\mathbf{L}_x dydx, & \overline{\mathbf{M}}_x &= \int_0^a \int_0^b \mathbf{C}_{xy}\mathbf{M}_x dydx \\
\overline{\mathbf{L}}_y &= \int_0^a \int_0^b \mathbf{C}_{xy}\mathbf{L}_y dydx, & \overline{\mathbf{M}}_y &= \int_0^a \int_0^b \mathbf{C}_{xy}\mathbf{M}_y dydx
\end{aligned} \tag{2.73}$$

Similarly, multiplying Eq. (2.66a) with $\mathbf{C}_y$ in the left side then integrating it from 0 to b with respect to y, and multiplying Eq. (2.66b) with $\mathbf{C}_x$ in the left side then integrating it from 0 to a with respect to x, we have

$$\begin{aligned}
x = 0 &: \overline{\mathbf{L}}_{xy}^{x0}\boldsymbol{\Gamma}_{xy} + \overline{\mathbf{L}}_x^{x0}\boldsymbol{\Gamma}_x + \overline{\mathbf{L}}_y^{x0}\boldsymbol{\Gamma}_y = 0 \\
x = a &: \overline{\mathbf{L}}_{xy}^{x1}\boldsymbol{\Gamma}_{xy} + \overline{\mathbf{L}}_x^{x1}\boldsymbol{\Gamma}_x + \overline{\mathbf{L}}_y^{x1}\boldsymbol{\Gamma}_y = 0
\end{aligned} \tag{2.74a}$$

$$\begin{aligned}
y = 0 &: \overline{\mathbf{L}}_{xy}^{y0}\boldsymbol{\Gamma}_{xy} + \overline{\mathbf{L}}_x^{y0}\boldsymbol{\Gamma}_x + \overline{\mathbf{L}}_y^{y0}\boldsymbol{\Gamma}_y = 0 \\
y = b &: \overline{\mathbf{L}}_{xy}^{y1}\boldsymbol{\Gamma}_{xy} + \overline{\mathbf{L}}_x^{y1}\boldsymbol{\Gamma}_x + \overline{\mathbf{L}}_y^{y1}\boldsymbol{\Gamma}_y = 0
\end{aligned} \tag{2.74b}$$

where

$$\begin{aligned}
\overline{\mathbf{L}}_{xy}^{x0} &= \int_0^b \mathbf{C}_y\mathbf{L}_{xy}^{x0}dy, & \overline{\mathbf{L}}_{xy}^{x1} &= \int_0^b \mathbf{C}_y\mathbf{L}_{xy}^{x1}dy \\
\overline{\mathbf{L}}_x^{x0} &= \int_0^b \mathbf{C}_y\mathbf{L}_x^{x0}dy, & \overline{\mathbf{L}}_x^{x1} &= \int_0^b \mathbf{C}_y\mathbf{L}_x^{x1}dy \\
\overline{\mathbf{L}}_y^{x0} &= \int_0^b \mathbf{C}_y\mathbf{L}_y^{x0}dy, & \overline{\mathbf{L}}_y^{x1} &= \int_0^b \mathbf{C}_y\mathbf{L}_y^{x1}dy
\end{aligned} \tag{2.75a}$$

$$\begin{aligned}
\overline{\mathbf{L}}_{xy}^{y0} &= \int_0^a \mathbf{C}_x\mathbf{L}_{xy}^{y0}dx, & \overline{\mathbf{L}}_{xy}^{y1} &= \int_0^a \mathbf{C}_x\mathbf{L}_{xy}^{y1}dx \\
\overline{\mathbf{L}}_x^{y0} &= \int_0^a \mathbf{C}_x\mathbf{L}_x^{y0}dx, & \overline{\mathbf{L}}_x^{y1} &= \int_0^a \mathbf{C}_x\mathbf{L}_x^{y1}dx \\
\overline{\mathbf{L}}_y^{y0} &= \int_0^a \mathbf{C}_x\mathbf{L}_y^{y0}dx, & \overline{\mathbf{L}}_y^{y1} &= \int_0^a \mathbf{C}_x\mathbf{L}_y^{y1}dx
\end{aligned} \tag{2.75a}$$

Thus, Eq. (2.74a, b) can be rewritten as

$$\begin{bmatrix} \boldsymbol{\Gamma}_x \\ \boldsymbol{\Gamma}_x \end{bmatrix} = -\begin{bmatrix} \overline{\mathbf{L}}_x^{x0} & \overline{\mathbf{L}}_y^{x0} \\ \overline{\mathbf{L}}_x^{x1} & \overline{\mathbf{L}}_y^{x1} \\ \overline{\mathbf{L}}_x^{y0} & \overline{\mathbf{L}}_y^{y0} \\ \overline{\mathbf{L}}_x^{y1} & \overline{\mathbf{L}}_y^{y1} \end{bmatrix}^{-1} \begin{bmatrix} \overline{\mathbf{L}}_{xy}^{x0} \\ \overline{\mathbf{L}}_{xy}^{x1} \\ \overline{\mathbf{L}}_{xy}^{y0} \\ \overline{\mathbf{L}}_{xy}^{y1} \end{bmatrix} \boldsymbol{\Gamma}_{xy} \tag{2.76}$$

Finally, combine Eqs. (2.72) and (2.76) results in

$$\left(\mathbf{K} - \omega^2 \mathbf{M}\right)\mathbf{\Gamma}_{xy} = \mathbf{0} \tag{2.77}$$

where $\mathbf{K}$ is the stiffness matrix for the plate, and $\mathbf{M}$ is the mass matrix. They are defined as

$$\mathbf{K} = \overline{\mathbf{L}}_{xy} - \begin{bmatrix} \overline{\mathbf{L}}_x & \overline{\mathbf{L}}_y \end{bmatrix} \begin{bmatrix} \overline{\mathbf{L}}_x^{x0} & \overline{\mathbf{L}}_y^{x0} \\ \overline{\mathbf{L}}_x^{x1} & \overline{\mathbf{L}}_y^{x1} \\ \overline{\mathbf{L}}_x^{y0} & \overline{\mathbf{L}}_y^{y0} \\ \overline{\mathbf{L}}_x^{y1} & \overline{\mathbf{L}}_y^{y1} \end{bmatrix}^{-1} \begin{bmatrix} \overline{\mathbf{L}}_{xy}^{x0} \\ \overline{\mathbf{L}}_{xy}^{x1} \\ \overline{\mathbf{L}}_{xy}^{y0} \\ \overline{\mathbf{L}}_{xy}^{y1} \end{bmatrix} \tag{2.78a}$$

$$\mathbf{M} = \overline{\mathbf{M}}_{xy} - \begin{bmatrix} \overline{\mathbf{M}}_x & \overline{\mathbf{M}}_y \end{bmatrix} \begin{bmatrix} \overline{\mathbf{L}}_x^{x0} & \overline{\mathbf{L}}_y^{x0} \\ \overline{\mathbf{L}}_x^{x1} & \overline{\mathbf{L}}_y^{x1} \\ \overline{\mathbf{L}}_x^{y0} & \overline{\mathbf{L}}_y^{y0} \\ \overline{\mathbf{L}}_x^{y1} & \overline{\mathbf{L}}_y^{y1} \end{bmatrix}^{-1} \begin{bmatrix} \overline{\mathbf{L}}_{xy}^{x0} \\ \overline{\mathbf{L}}_{xy}^{x1} \\ \overline{\mathbf{L}}_{xy}^{y0} \\ \overline{\mathbf{L}}_{xy}^{y1} \end{bmatrix} \tag{2.78b}$$

Mathematically, Eq. (2.77) represents a generalized eigenvalue problem from which all the natural frequencies and modes of the plate can be determined easily by solving the standard characteristic equation. Once the coefficient eigenvector $\mathbf{\Gamma}_{xy}$ is determined for a given frequency, the corresponding supplement coefficient eigenvectors $\mathbf{\Gamma}_x$ and $\mathbf{\Gamma}_y$ can be obtained. Then the displacements and rotation components of the plate can be determined by substituting these coefficients into Eqs. (2.57) and (2.58). Thus, the corresponding mode shape of the plate can be directly constructed from the determined displacement functions. Although Eq. (2.77) represents the free vibration of laminated rectangular plates, by summing the loading vector $\mathbf{F}$ on the right side of Eq. (2.77), thus, the characteristic equation for the forced vibration is readily obtained. Similarly, the present formulation can be readily applied to static analysis of laminated plates with general boundary conditions by letting $\omega = 0$ and summing the loading vector $\mathbf{F}$ on the right side of Eq. (2.77).

Although the modified Fourier series solution procedure derived herein is focused on rectangular plates, it can readily be used for other laminated structures, such as beams, cylindrical shells, conical shells, spherical shells and shallow shell, etc. The method described in this section is believed to include two main advantages: first, it is a general method which can be used to determinate the static, bending, free and forced vibration behaviors of laminated pates with arbitrary boundary conditions accurately; second, the proposed method offers an easy analysis operation for the entire restraining conditions and the change of boundary conditions from one case to another is as easy as changing structure parameters without the need of making any change to the solution procedure or modifying the basic functions as often required in other methods.

Instead of seeking a solution in strong form solution procedure as described in the previous paragraphs, all the expansion coefficients can be treated equally and independently as the generalized coordinates and solved directly from the weak form solution procedure such as Rayleigh–Ritz technique, which is the focus of the next section.

2.3 Rayleigh-Ritz Method (Weak Form Solution Procedure)

In a variety of vibration problems, exact solutions always unable to be obtained, in such cases, one has to employ approximate method. In this regard, many methods exist. Among them, the minimization of energy approaches such as the Rayleigh-Ritz method, the variational integral method and the Galerkin method are widely used in the vibration analysis of continuous systems due to the reliability of their results and efficiency in modeling and solution procedure. In this section, we focus on the Rayleigh-Ritz method. The modified Fourier series version of Rayleigh-Ritz method is presented as follows.

In the Rayleigh-Ritz method, a displacement field associated with undetermined coefficients is assumed firstly. The displacement field is then substituted into the Lagrangian energy functional (i.e., $\Pi = T - U + W$). Then the undetermined coefficients in the displacement field are determined by finding the stationary value of the energy functional, namely, minimizing the total expression of the Lagrangian energy function by taking its derivatives with respect to the undetermined coefficients and making them equal to zero. Finally, a series of equations related to corresponding coefficients can be achieved and summed up in matrix form as a standard characteristic equation. And the desired frequencies and modes of the structure can be determined easily by solving the standard characteristic equation (Qatu 2004; Reddy 2002).

The constructing of appropriate admissible displacement field is of crucial importance in the Rayleigh–Ritz procedure because the accuracy of the solution will usually depend upon how well the actual displacement can be faithfully represented by it. For vibration analysis of laminated beams, plates and shells, the admissible displacement field is often expressed in terms of beam functions under the same boundary conditions. Thus, a specially customized set of beam functions is required for each type of boundary conditions. As a result, the use of the existing solution procedures will result in very tedious calculations and be easily inundated with various boundary conditions because even only considering the classical (homogeneous) cases, one will have a total of hundreds of different combinations. Instead of the beam functions, one may also use other forms of admissible functions such as orthogonal polynomials. However, the higher order polynomials tend to become numerically unstable due to the computer round-off errors. This numerical difficulty can be avoided by expressing the displacement functions in the form of a Fourier series expansion because Fourier functions constitute a complete set and exhibit an excellent numerical stability. However, the conventional Fourier series expression will generally have a convergence problem along the boundary edges and cannot be differentiated term-by-term except for a few simple boundary conditions (see Sect. 2.1.1). These difficulties can be overcame by using the modified Fourier series. A weak form solution procedure which combining the modified Fourier series and the Rayleigh-Ritz method is given below step-by-step.

Taking the previous studied moderately thick laminated rectangular plate (Fig. 2.4) for example, letting

$$\mathbf{G}^u = [\mathbf{A}, \quad \mathbf{a}, \quad \mathbf{b}]^T$$
$$\mathbf{G}^v = [\mathbf{B}, \quad \mathbf{c}, \quad \mathbf{d}]^T$$
$$\mathbf{G}^w = [\mathbf{C}, \quad \mathbf{e}, \quad \mathbf{f}]^T$$
$$\mathbf{G}^x = [\mathbf{D}, \quad \mathbf{g}, \quad \mathbf{h}]^T$$

$$\mathbf{G}^y = [\mathbf{E}, \quad \mathbf{i}, \quad \mathbf{j}]^T$$
$$\mathbf{H} = [\mathbf{H}_{xy}, \mathbf{H}_x, \mathbf{H}_y] \tag{2.79}$$

where $\mathbf{H}_{xy}$, $\mathbf{H}_x$, $\mathbf{H}_y$, $\mathbf{A}$ to $\mathbf{E}$ and $\mathbf{a}$ to $\mathbf{j}$ are presented in Eqs. (2.60) and (2.61). Therefore, the displacement expressions of the plate can be rewritten in the vector form as:

$$u(x,y) = \mathbf{H}\mathbf{G}^u$$
$$v(x,y) = \mathbf{H}\mathbf{G}^v$$
$$w(x,y) = \mathbf{H}\mathbf{G}^w \tag{2.80}$$
$$\phi_x(x,y) = \mathbf{H}\mathbf{G}^x$$
$$\phi_y(x,y) = \mathbf{H}\mathbf{G}^y$$

For free vibration analysis, the Lagrangian energy functional (L) of the plate can be simplified and written in terms of the strain energy and kinetic energy functions as:

$$L = T - U_s - U_{sp} \tag{2.81}$$

According to Eqs. (1.50), (1.51) and (1.54). The kinetic energy and strain energy functions of the laminated plate are:

$$T = \frac{\omega^2}{2} \int_x \int_y \left\{ \begin{array}{l} I_0 u^2 + 2I_1 u\phi_x + I_2 \phi_x^2 \\ +I_0 v^2 + 2I_1 v\phi_y + I_2 \phi_y^2 + I_0 w^2 \end{array} \right\} dx dy \tag{2.82}$$

$$U_{sp} = \frac{1}{2} \int_0^b \left\{ \begin{array}{l} [k_{x0}^u u^2 + k_{x0}^v v^2 + k_{x0}^w w^2 + K_{x0}^x \phi_x^2 + K_{x0}^y \phi_y^2]_{x=0} \\ +[k_{x1}^u u^2 + k_{x1}^v v^2 + k_{x1}^w w^2 + K_{x1}^x \phi_x^2 + K_{x1}^y \phi_y^2]_{x=a} \end{array} \right\} dy$$
$$+ \frac{1}{2} \int_0^a \left\{ \begin{array}{l} [k_{y0}^u u^2 + k_{y0}^v v^2 + k_{y0}^w w^2 + K_{y0}^x \phi_x^2 + K_{y0}^y \phi_y^2]_{y=0} \\ +[k_{y1}^u u^2 + k_{y1}^v v^2 + k_{y1}^w w^2 + K_{y1}^x \phi_x^2 + K_{y1}^y \phi_y^2]_{y=b} \end{array} \right\} dx \tag{2.83}$$

$$U_s = \frac{1}{2} \int_0^a \int_0^b \left\{ \begin{array}{l} A_{11}\left(\frac{\partial u}{\partial x}\right)^2 + A_{22}\left(\frac{\partial v}{\partial y}\right)^2 + A_{66}\left(\frac{\partial v}{\partial x} + \frac{\partial u}{\partial y}\right)^2 \\[2mm] +2A_{12}\left(\frac{\partial v}{\partial y}\right)\left(\frac{\partial u}{\partial x}\right) + 2A_{16}\left(\frac{\partial v}{\partial x} + \frac{\partial u}{\partial y}\right)\left(\frac{\partial u}{\partial x}\right) \\[2mm] +2A_{26}\left(\frac{\partial v}{\partial x} + \frac{\partial u}{\partial y}\right)\left(\frac{\partial v}{\partial y}\right) + A_{44}\left(\phi_y + \frac{\partial w}{\partial y}\right)^2 \\[2mm] +2A_{45}\left(\phi_y + \frac{\partial w}{\partial y}\right)\left(\phi_x + \frac{\partial w}{\partial x}\right) + A_{55}\left(\phi_x + \frac{\partial w}{\partial x}\right)^2 \end{array} \right\} dydx$$

$$+ \frac{1}{2} \int_0^a \int_0^b \left\{ \begin{array}{l} D_{11}\left(\frac{\partial \phi_x}{\partial x}\right)^2 + D_{22}\left(\frac{\partial \phi_y}{\partial y}\right)^2 + D_{66}\left(\frac{\partial \phi_x}{\partial y} + \frac{\partial \phi_y}{\partial x}\right)^2 \\[2mm] +2D_{12}\left(\frac{\partial \phi_y}{\partial y}\right)\left(\frac{\partial \phi_x}{\partial x}\right) + 2D_{16}\left(\frac{\partial \phi_x}{\partial y} + \frac{\partial \phi_y}{\partial x}\right) \\[2mm] \times\left(\frac{\partial \phi_x}{\partial x}\right) + 2D_{26}\left(\frac{\partial \phi_x}{\partial y} + \frac{\partial \phi_y}{\partial x}\right)\left(\frac{\partial \phi_y}{\partial y}\right) \end{array} \right\} dydx$$

$$+ \int_0^a \int_0^b \left\{ \begin{array}{l} B_{11}\left(\frac{\partial \phi_x}{\partial x}\right)\left(\frac{\partial v}{\partial y}\right) + B_{12}\left(\frac{\partial \phi_y}{\partial y}\right)\left(\frac{\partial u}{\partial x}\right) \\[2mm] +B_{16}\left(\frac{\partial \phi_x}{\partial y} + \frac{\partial \phi_y}{\partial x}\right)\left(\frac{\partial u}{\partial x}\right) + B_{12}\left(\frac{\partial \phi_x}{\partial x}\right)\left(\frac{\partial v}{\partial y}\right) \\[2mm] +B_{22}\left(\frac{\partial \phi_y}{\partial y}\right)\left(\frac{\partial v}{\partial y}\right) + B_{26}\left(\frac{\partial \phi_x}{\partial y} + \frac{\partial \phi_y}{\partial x}\right)\left(\frac{\partial v}{\partial y}\right) \\[2mm] +B_{16}\left(\frac{\partial \phi_x}{\partial x}\right)\left(\frac{\partial v}{\partial x} + \frac{\partial u}{\partial y}\right) + B_{26}\left(\frac{\partial \phi_y}{\partial y}\right)\left(\frac{\partial v}{\partial x} + \frac{\partial u}{\partial y}\right) \\[2mm] +B_{66}\left(\frac{\partial \phi_x}{\partial y} + \frac{\partial \phi_y}{\partial x}\right)\left(\frac{\partial v}{\partial x} + \frac{\partial u}{\partial y}\right) \end{array} \right\} dydx \qquad (2.84)$$

Substituting the displacement expressions of the plate (Eqs. 2.57 and 2.58) into the Lagrangian energy functional (Eq. 2.81) and minimizing the total expression of the Lagrangian energy functional by taking its derivatives with respect to each of the undetermined coefficients and making them equal to zero

$$\frac{\partial L}{\partial \Xi_{mn}} = 0, \quad \text{and} \begin{cases} \Xi = A, B, C, D, E \\ m = 0, 1, \ldots M; \quad n = 0, 1, \ldots N \end{cases}$$

$$\frac{\partial L}{\partial \Psi_{ln}} = 0, \quad \text{and} \begin{cases} \Psi = a, c, e, g, i \\ l = 1, 2; \quad n = 0, 1, \ldots N \end{cases} \qquad (2.85)$$

$$\frac{\partial L}{\partial \Upsilon_{lm}} = 0, \quad \text{and} \begin{cases} \Upsilon = b, d, f, h, j \\ l = 1, 2; \quad m = 0, 1, \ldots M \end{cases}$$

a total of $5*(M+1)*(N+1) + 10*(M+N+2)$ equations can be obtained. They are summed up in a matrix form as:

$$\left(\mathbf{K} - \omega^2 \mathbf{M}\right)\mathbf{G} = 0 \qquad (2.86)$$

where $\mathbf{K}$ is the stiffness matrix of the plate, and $\mathbf{M}$ is the mass matrix. Both of them are symmetric matrices and they can be expressed as

$$\mathbf{K} = \begin{bmatrix} \mathbf{K}_{uu} & \mathbf{K}_{uv} & \mathbf{K}_{uw} & \mathbf{K}_{ux} & \mathbf{K}_{uy} \\ \mathbf{K}_{uv}^T & \mathbf{K}_{vv} & \mathbf{K}_{vw} & \mathbf{K}_{vx} & \mathbf{K}_{vy} \\ \mathbf{K}_{uw}^T & \mathbf{K}_{vw}^T & \mathbf{K}_{ww} & \mathbf{K}_{wx} & \mathbf{K}_{wy} \\ \mathbf{K}_{ux}^T & \mathbf{K}_{vx}^T & \mathbf{K}_{wx}^T & \mathbf{K}_{xx} & \mathbf{K}_{xy} \\ \mathbf{K}_{uy}^T & \mathbf{K}_{vy}^T & \mathbf{K}_{wy}^T & \mathbf{K}_{xy}^T & \mathbf{K}_{yy} \end{bmatrix} \qquad (2.87a)$$

$$\mathbf{M} = \begin{bmatrix} \mathbf{M}_{uu} & \mathbf{0} & \mathbf{0} & \mathbf{M}_{ux} & \mathbf{0} \\ \mathbf{0} & \mathbf{M}_{vv} & \mathbf{0} & \mathbf{0} & \mathbf{M}_{vy} \\ \mathbf{0} & \mathbf{0} & \mathbf{M}_{ww} & \mathbf{0} & \mathbf{0} \\ \mathbf{M}_{ux}^T & \mathbf{0} & \mathbf{0} & \mathbf{M}_{xx} & \mathbf{0} \\ \mathbf{0} & \mathbf{M}_{vy}^T & \mathbf{0} & \mathbf{0} & \mathbf{M}_{yy} \end{bmatrix} \tag{2.87b}$$

The superscript T represents the transposition operator. The explicit forms of submatrices in the stiffness matrix $\mathbf{K}$ and mass matrix $\mathbf{M}$ are listed as follows

$$\mathbf{K}_{uu} = \int_0^a \int_0^b \left\{ A_{11} \frac{\partial \mathbf{H}^T}{\partial x} \frac{\partial \mathbf{H}}{\partial x} + A_{16} \frac{\partial \mathbf{H}^T}{\partial x} \frac{\partial \mathbf{H}}{\partial y} + A_{16} \frac{\partial \mathbf{H}^T}{\partial y} \frac{\partial \mathbf{H}}{\partial x} + A_{66} \frac{\partial \mathbf{H}^T}{\partial y} \frac{\partial \mathbf{H}}{\partial y} \right\} dydx$$
$$+ \int_0^b \left\{ k_{x0}^u \mathbf{H}^T \mathbf{H}|_{x=0} + k_{x1}^u \mathbf{H}^T \mathbf{H}|_{x=a} \right\} dy + \int_0^a \left\{ k_{y0}^u \mathbf{H}^T \mathbf{H}|_{y=0} + k_{y1}^u \mathbf{H}^T \mathbf{H}|_{y=b} \right\} dx$$

$$\mathbf{K}_{uv} = \int_0^a \int_0^b \left\{ A_{12} \frac{\partial \mathbf{H}^T}{\partial x} \frac{\partial \mathbf{H}}{\partial y} + A_{16} \frac{\partial \mathbf{H}^T}{\partial x} \frac{\partial \mathbf{H}}{\partial x} + A_{26} \frac{\partial \mathbf{H}^T}{\partial y} \frac{\partial \mathbf{H}}{\partial y} + A_{66} \frac{\partial \mathbf{H}^T}{\partial y} \frac{\partial \mathbf{H}}{\partial x} \right\} dydx$$

$$\mathbf{K}_{uw} = 0$$

$$\mathbf{K}_{ux} = \int_0^a \int_0^b \left\{ B_{11} \frac{\partial \mathbf{H}^T}{\partial x} \frac{\partial \mathbf{H}}{\partial x} + B_{16} \frac{\partial \mathbf{H}^T}{\partial x} \frac{\partial \mathbf{H}}{\partial y} + B_{16} \frac{\partial \mathbf{H}^T}{\partial y} \frac{\partial \mathbf{H}}{\partial x} + B_{66} \frac{\partial \mathbf{H}^T}{\partial y} \frac{\partial \mathbf{H}}{\partial y} \right\} dydx$$

$$\mathbf{K}_{uy} = \int_0^a \int_0^b \left\{ B_{12} \frac{\partial \mathbf{H}^T}{\partial x} \frac{\partial \mathbf{H}}{\partial y} + B_{16} \frac{\partial \mathbf{H}^T}{\partial x} \frac{\partial \mathbf{H}}{\partial x} + B_{26} \frac{\partial \mathbf{H}^T}{\partial y} \frac{\partial \mathbf{H}}{\partial y} + B_{66} \frac{\partial \mathbf{H}^T}{\partial y} \frac{\partial \mathbf{H}}{\partial x} \right\} dydx$$

$$\mathbf{K}_{vv} = \int_0^a \int_0^b \left\{ A_{22} \frac{\partial \mathbf{H}^T}{\partial y} \frac{\partial \mathbf{H}}{\partial y} + A_{26} \frac{\partial \mathbf{H}^T}{\partial x} \frac{\partial \mathbf{H}}{\partial y} + A_{26} \frac{\partial \mathbf{H}^T}{\partial y} \frac{\partial \mathbf{H}}{\partial x} + A_{66} \frac{\partial \mathbf{H}^T}{\partial x} \frac{\partial \mathbf{H}}{\partial x} \right\} dydx$$
$$+ \int_0^b \left\{ k_{x0}^v \mathbf{H}^T \mathbf{H}|_{x=0} + k_{x1}^v \mathbf{H}^T \mathbf{H}|_{x=a} \right\} dy + \int_0^a \left\{ k_{y0}^v \mathbf{H}^T \mathbf{H}|_{y=0} + k_{y1}^v \mathbf{H}^T \mathbf{H}|_{y=b} \right\} dx$$

$$\mathbf{K}_{vw} = 0$$

$$\mathbf{K}_{vx} = \int_0^a \int_0^b \left\{ B_{12} \frac{\partial \mathbf{H}^T}{\partial y} \frac{\partial \mathbf{H}}{\partial x} + B_{26} \frac{\partial \mathbf{H}^T}{\partial y} \frac{\partial \mathbf{H}}{\partial y} + B_{16} \frac{\partial \mathbf{H}^T}{\partial x} \frac{\partial \mathbf{H}}{\partial x} + B_{66} \frac{\partial \mathbf{H}^T}{\partial x} \frac{\partial \mathbf{H}}{\partial y} \right\} dydx$$

$$\mathbf{K}_{vy} = \int_0^a \int_0^b \left\{ B_{22} \frac{\partial \mathbf{H}^T}{\partial y} \frac{\partial \mathbf{H}}{\partial y} + B_{26} \frac{\partial \mathbf{H}^T}{\partial y} \frac{\partial \mathbf{H}}{\partial x} + B_{26} \frac{\partial \mathbf{H}^T}{\partial x} \frac{\partial \mathbf{H}}{\partial y} + B_{66} \frac{\partial \mathbf{H}^T}{\partial x} \frac{\partial \mathbf{H}}{\partial x} \right\} dydx$$

$$\mathbf{K}_{ww} = \int_0^a \int_0^b \left\{ A_{44} \frac{\partial \mathbf{H}^T}{\partial y} \frac{\partial \mathbf{H}}{\partial y} + A_{45} \frac{\partial \mathbf{H}^T}{\partial x} \frac{\partial \mathbf{H}}{\partial y} + A_{45} \frac{\partial \mathbf{H}^T}{\partial y} \frac{\partial \mathbf{H}}{\partial x} + A_{55} \frac{\partial \mathbf{H}^T}{\partial x} \frac{\partial \mathbf{H}}{\partial x} \right\} dydx$$
$$+ \int_0^b \left\{ k_{x0}^w \mathbf{H}^T \mathbf{H}|_{x=0} + k_{x1}^w \mathbf{H}^T \mathbf{H}|_{x=a} \right\} dy + \int_0^a \left\{ k_{y0}^w \mathbf{H}^T \mathbf{H}|_{y=0} + k_{y1}^w \mathbf{H}^T \mathbf{H}|_{y=b} \right\} dx \tag{2.88}$$

$$\mathbf{K}_{wx} = \int_0^a \int_0^b \left\{ A_{45} \frac{\partial \mathbf{H}^T}{\partial y} \mathbf{H} + A_{55} \frac{\partial \mathbf{H}^T}{\partial x} \mathbf{H} \right\} dydx$$

$$\mathbf{K}_{wy} = \int_0^a \int_0^b \left\{ A_{44} \frac{\partial \mathbf{H}^T}{\partial y} \mathbf{H} + A_{45} \frac{\partial \mathbf{H}^T}{\partial x} \mathbf{H} \right\} dydx$$

$$\mathbf{K}_{xx} = \int_0^a \int_0^b \left\{ \begin{array}{l} D_{11} \frac{\partial \mathbf{H}^T}{\partial x} \frac{\partial \mathbf{H}}{\partial x} + D_{16} \frac{\partial \mathbf{H}^T}{\partial x} \frac{\partial \mathbf{H}}{\partial y} + D_{16} \frac{\partial \mathbf{H}^T}{\partial y} \frac{\partial \mathbf{H}}{\partial x} \\ + D_{66} \frac{\partial \mathbf{H}^T}{\partial y} \frac{\partial \mathbf{H}}{\partial y} + A_{55} \mathbf{H}^T \mathbf{H} \end{array} \right\} dydx$$
$$+ \int_0^b \left\{ K_{x0}^x \mathbf{H}^T \mathbf{H}|_{x=0} + K_{x1}^x \mathbf{H}^T \mathbf{H}|_{x=a} \right\} dy + \int_0^a \left\{ K_{y0}^x \mathbf{H}^T \mathbf{H}|_{y=0} + K_{y1}^x \mathbf{H}^T \mathbf{H}|_{y=b} \right\} dx$$

$$\mathbf{K}_{xy} = \int_0^a \int_0^b \left\{ \begin{array}{l} D_{12} \frac{\partial \mathbf{H}^T}{\partial x} \frac{\partial \mathbf{H}}{\partial y} + D_{16} \frac{\partial \mathbf{H}^T}{\partial x} \frac{\partial \mathbf{H}}{\partial x} + D_{26} \frac{\partial \mathbf{H}^T}{\partial y} \frac{\partial \mathbf{H}}{\partial y} \\ + D_{66} \frac{\partial \mathbf{H}^T}{\partial y} \frac{\partial \mathbf{H}}{\partial x} + A_{45} \mathbf{H}^T \mathbf{H} \end{array} \right\} dydx$$

$$\mathbf{K}_{yy} = \int_0^a \int_0^b \left\{ \begin{array}{l} D_{22} \frac{\partial \mathbf{H}^T}{\partial y} \frac{\partial \mathbf{H}}{\partial y} + D_{26} \frac{\partial \mathbf{H}^T}{\partial y} \frac{\partial \mathbf{H}}{\partial x} + D_{26} \frac{\partial \mathbf{H}^T}{\partial x} \frac{\partial \mathbf{H}}{\partial y} \\ + D_{66} \frac{\partial \mathbf{H}^T}{\partial x} \frac{\partial \mathbf{H}}{\partial x} + A_{44} \mathbf{H}^T \mathbf{H} \end{array} \right\} dydx$$
$$+ \int_0^b \left\{ K_{x0}^y \mathbf{H}^T \mathbf{H}|_{x=0} + K_{x1}^y \mathbf{H}^T \mathbf{H}|_{x=a} \right\} dy + \int_0^a \left\{ K_{y0}^y \mathbf{H}^T \mathbf{H}|_{y=0} + K_{y1}^y \mathbf{H}^T \mathbf{H}|_{y=b} \right\} dx$$

$$\mathbf{M}_{uu} = \mathbf{M}_{vv} = \mathbf{M}_{ww} = \int_0^a \int_0^b I_0 \mathbf{H}^T \mathbf{H} dydx$$

$$\mathbf{M}_{ux} = \mathbf{M}_{vy} = \int_0^a \int_0^b I_1 \mathbf{H}^T \mathbf{H} dydx$$

$$\mathbf{M}_{xx} = \mathbf{M}_{xy} = \int_0^a \int_0^b I_2 \mathbf{H}^T \mathbf{H} dydx$$

G is a column vector which contains, in an appropriate order, the unknown expansion coefficients that appear in the series expansions, namely:

$$\mathbf{G} = [\mathbf{G}^u, \mathbf{G}^v, \mathbf{G}^w, \mathbf{G}^x, \mathbf{G}^y]^T \qquad (2.89)$$

where $\mathbf{G}^u$, $\mathbf{G}^v$, $\mathbf{G}^w$, $\mathbf{G}^x$ and $\mathbf{G}^y$ are given in Eq. (2.79). Obviously, the natural frequencies and eigenvectors can be easily obtained by solving a standard matrix eigenproblem. Once the coefficient eigenvector **G** is determined for a given frequency, the displacements of the plate can be determined by substituting the coefficients into Eqs. (2.57) and (2.58).

The modified Fourier series solution procedure derived herein is focused on rectangular plates, it can readily be used for other laminated structures, such as beams, cylindrical shells, conical shells, spherical shells and shallow shell, see Chaps. 3–8.

Chapter 3
Straight and Curved Beams

Beams, plates and shells are commonly utilized in engineering applications, and they are named according to their size or/and shape characteristics and different theories have been developed to study their structural behaviors. A beam is typically described as a structural component having one dimension relatively greater than the other dimensions. Specially, a beam can be referred to as a rod or bar when subjected to tension, a column when subjected to compression and a shaft when subjected to torsional loads (Qatu 2004). Beams are one of the most fundamental structural elements. Almost every machine contains one or more beam components, such as bridges, steel framed structures and building frames. In addition, many structures can be modeled at a preliminary level as beams. For example, spring boards, supports of a wind power generation can be treated as cantilever beams, and a span of an overhead viaduct or bridge can be viewed as a simply supported beam. In recent decades, laminated beams made from advanced composite materials are extensively used in many engineering applications where higher strength to weight ratio is desired, such as aircraft structures, space vehicles, turbo-machines, deep-sea equipment and other industrial applications. Researches on the vibration and dynamic analyses of laminated composite beams have been increasing rapidly in recent decades. A paper which reviewed most of the researches done in years (1989–2012) on the vibration analysis of composite beams by Hajianmaleki and Qatu (2013) showed that research articles on the subject during period 2000–2012 are more than twice than those of 1989–2000. Due to the great importance, this chapter considers the vibration of laminated beams in the framework of classical thin beam theory (CBT) and shear deformation beam theory (SDBT).

Beams can be straight or curved. Both straight and curved beams are considered in this chapter. Generally, a straight beam can be considered as a degenerated curved beam with infinite radius of curvature (zero curvature). This chapter is concerned with the development of the fundamental equations of laminated curved beams according to the CBT and SDBT. Equations for the straight beams can be derived by setting curvatures to zero in those of curved beams. Strain-displacement relations, force and moment resultants, energy functions, governing equations and boundary conditions are derived and shown for both theories. Natural frequencies and mode shapes are presented for straight and curved beams with different boundary

© Science Press, Beijing and Springer-Verlag Berlin Heidelberg 2015
G. Jin et al., *Structural Vibration*, DOI 10.1007/978-3-662-46364-2_3

conditions, lamination schemes and geometry parameters in both strong and weak forms of the proposed modified Fourier series method. The effects of boundary conditions, geometry parameters and material properties are studied as well.

3.1 Fundamental Equations of Thin Laminated Beams

Fundamental equations of laminated thin beams are presented in this section in the framework of classical beam theory. As shown in Fig. 3.1, a laminated curved beam with uniform thickness h, width b is selected as the model. The beam is characterized by its middle surface, in which R represents the mean radius of the beam and θ_0 denotes the included angle of the curved beam. To describe the beam clearly, we introduce the following coordinate system: the α-coordinate is taken along the length of the beam, and β- and z-coordinates are taken along the width and the thickness directions, respectively. u, v and w separately indicate the middle surface displacement variations of the beam in the α, β and z directions. It should be stressed that this chapter addresses vibrations of laminated beams in their plane of curvature, therefore, the fundamental equations derived for thin deep shells can be specialized to those for curved beams by further assuming that the displacement v is identical with zero and the displacements u and w along the coordinate system are only functions of the α- and z-coordinates.

3.1.1 Kinematic Relations

Letting $\alpha = \theta$, the Lamé parameters of laminated curved beams can be obtained as $A = R$. Introducing the Lamé parameters into Eq. (1.7), the middle surface strain and curvature change of thin beams are:

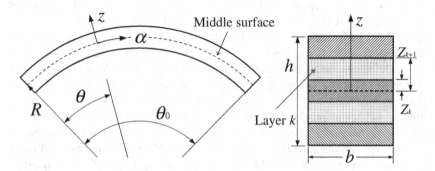

Fig. 3.1 Laminated curved beams

$$\varepsilon_\theta^0 = \frac{\partial u}{R\partial\theta} + \frac{w}{R} \quad \chi_\theta = \frac{\partial u}{R^2\partial\theta} - \frac{\partial^2 w}{R^2\partial\theta^2} \tag{3.1}$$

According to the thin beam assumption, the strain at an arbitrary point in the kth layer of thin laminated beams can be defined as:

$$\varepsilon_\theta = \varepsilon_\theta^0 + z\chi_\theta \tag{3.2}$$

where $Z_k < z < Z_{k+1}$. Z_{k+1} and Z_k denote the distances from the top surface and the bottom surface of the layer to the referenced middle surface, respectively.

3.1.2 Stress-Strain Relations and Stress Resultants

Suppose the laminated thin beam is composed of N composite layers which are bonded together rigidly. And the angle between the principal direction of the composite material in kth layer and the α axis is denoted by ϑ^k. According to Hooke's law, the corresponding stress-strain relations in the kth layer of the beam can be written as:

$$\{\sigma_\theta\}_k = \overline{Q_{11}^k}\{\varepsilon_\theta\}_k \tag{3.3}$$

where σ_θ is the normal stress in the θ direction. The constant $\overline{Q_{11}^k}$ is the elastic stiffness coefficient of this layer, which is found from following equations:

$$\overline{Q_{11}^k} = Q_{11}^k \cos^4\vartheta^k$$
$$Q_{11}^k = \frac{E_1}{1 - \mu_{12}\mu_{21}} \tag{3.4}$$

where E_1 is modulus of elasticity of the composite material in the principal direction. μ_{12} and μ_{21} are the Poisson's ratios. The subscript (11) in Eqs. (3.3) and (3.4) can be omitted but is maintained here for the direct use and comparison with the thin shell equations presented in Chap. 1. By carrying the integration of the normal stress over the cross-section results in

$$N_\theta = b \int_{-h/2}^{h/2} \sigma_\alpha dz \quad M_\theta = b \int_{-h/2}^{h/2} \sigma_\alpha z dz \tag{3.5}$$

where N_θ is the force resultant and M_θ is the moment resultant. The force and moment resultant relations to the strains in the middle surface and curvature change are defined as

$$\begin{bmatrix} N_\theta \\ M_\theta \end{bmatrix} = \begin{bmatrix} A_{11} & B_{11} \\ B_{11} & D_{11} \end{bmatrix} \begin{bmatrix} \varepsilon_\theta^0 \\ \chi_\theta \end{bmatrix} \tag{3.6}$$

where A_{11}, B_{11}, and D_{11} are the stiffness coefficients arising from the piecewise integration over the beam thickness:

$$\begin{aligned}
A_{11} &= b \sum_{k=1}^{N} \overline{Q_{11}^k} (z_{k+1} - z_k) \\
B_{11} &= \frac{b}{2} \sum_{k=1}^{N} \overline{Q_{11}^k} \left(z_{k+1}^2 - z_k^2 \right) \\
D_{11} &= \frac{b}{3} \sum_{k=1}^{N} \overline{Q_{11}^k} \left(z_{k+1}^3 - z_k^3 \right)
\end{aligned} \tag{3.7}$$

Notably, when the beam is laminated symmetrically with respect to its middle surface, the constants B_{ij} equal to zero. The above equations are valid for cylindrical bending of beams (Qatu 2004).

3.1.3 Energy Functions

The strain energy (U_s) of a thin beam during vibration is defined in terms of the middle surface strains and stress resultants as:

$$U_s = \frac{1}{2} \int_\theta \left\{ N_\theta \varepsilon_\theta^0 + M_\theta \chi_\theta \right\} R d\theta \tag{3.8}$$

Substituting Eqs. (3.1) and (3.6) into Eq. (3.8), the strain energy of the beam can be rewritten in terms of the displacements as:

$$U_s = \frac{1}{2} \int_\theta \left\{ \begin{aligned} & A_{11} \left(\frac{\partial u}{R \partial \theta} + \frac{w}{R} \right)^2 + D_{11} \left(\frac{\partial u}{R^2 \partial \theta} - \frac{\partial^2 w}{R^2 \partial \theta^2} \right)^2 \\ & + 2B_{11} \left(\frac{\partial u}{R^2 \partial \theta} - \frac{\partial^2 w}{R^2 \partial \theta^2} \right) \left(\frac{\partial u}{R \partial \theta} + \frac{w}{R} \right) \end{aligned} \right\} R d\theta \tag{3.9}$$

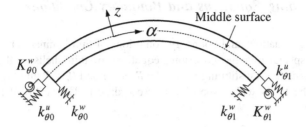

Fig. 3.2 Boundary conditions of thin laminated beams

The kinetic energy (T) of the beam is written as:

$$T = \frac{1}{2} \int_\theta I_0 \left\{ \left(\frac{\partial u}{\partial t} \right)^2 + \left(\frac{\partial w}{\partial t} \right)^2 \right\} R d\theta \qquad (3.10)$$

where the inertia term are:

$$I_0 = b \sum_{k=1}^{N} \int_{z_k}^{z_{k+1}} \rho^k dz \qquad (3.11)$$

ρ^k is the mass of the kth layer per unit middle surface area. The external work is expressed as:

$$W_e = \int_\theta \{ q_\theta u + q_z w \} R d\theta \qquad (3.12)$$

where q_θ and q_z are the external loads in the θ and z directions, respectively.

As described earlier, the general boundary conditions of a beam are implemented by using the artificial spring boundary technique, in which each end of the beam is assumed to be restrained by two groups of linear springs (k_u and k_w) and one group of rotational springs (K_w) to simulate the given or typical boundary conditions expressed in the form of boundary forces and the flexural moments, respectively (see Fig. 3.2). Specifically, symbols k_ψ^u, k_ψ^w and K_ψ^w ($\psi = \theta_0$ and θ_1) are used to indicate the stiffness of the boundary springs at the boundaries $\theta = 0$ and $\theta = \theta_0$, respectively. Therefore, the deformation strain energy (U_{sp}) stored in the boundary springs can be written as:

$$U_{sp} = \frac{1}{2} \left\{ \begin{array}{l} \left[k_{\theta 0}^u u^2 + k_{\theta 0}^w w^2 + K_{\theta 0}^w (\partial w / R \partial \theta)^2 \right]_{\theta = 0} \\ + \left[k_{\theta 1}^u u^2 + k_{\theta 1}^w w^2 + K_{\theta 1}^w (\partial w / R \partial \theta)^2 \right]_{\theta = \theta_0} \end{array} \right\} \qquad (3.13)$$

3.1.4 Governing Equations and Boundary Conditions

The governing equations and boundary conditions of thin laminated beams can be obtained by specializing the governing equations of thin shells to those of thin laminated beams (i.e. substituting $\alpha = \theta$, $A = R$, $B = 1$ and $R_\alpha = R$ into Eq. (1.28) and deleting all the terms with respect to β). According to Eq. (1.28), the governing equations are:

$$\frac{\partial N_\theta}{R\partial\theta} + \frac{Q_\theta}{R} + q_\theta = I_0 \frac{\partial^2 u}{\partial t^2}$$

$$-\frac{N_\theta}{R} + \frac{\partial Q_\theta}{R\partial\theta} + q_z = I_0 \frac{\partial^2 w}{\partial t^2} \tag{3.14}$$

where

$$Q_\theta = \frac{\partial M_\theta}{R\partial\theta} \tag{3.15}$$

And the general boundary conditions of thin laminated beams are

$$\theta = 0 : \begin{cases} N_\theta + \frac{M_\theta}{R} - k_{\theta0}^u u = 0 \\ Q_\theta - k_{\theta0}^w w = 0 \\ -M_\theta - K_{\theta0}^w \frac{\partial w}{R\partial\theta} = 0 \end{cases} \qquad \theta = \theta_0 : \begin{cases} N_\theta + \frac{M_\theta}{R} + k_{\theta1}^u u = 0 \\ Q_\theta + k_{\theta1}^w w = 0 \\ -M_\theta + K_{\theta1}^w \frac{\partial w}{R\partial\theta} = 0 \end{cases} \tag{3.16}$$

Alternately, the governing equations and boundary conditions of the considered beam can be obtained by applying Hamilton's principle in the same manner as following describe. The Lagrangian function (L) of thin laminated beams can be expressed in terms of strain energy, kinetic energy and external work as:

$$L = T - U_s - U_{sp} + W_e \tag{3.17}$$

Substituting Eqs. (3.9), (3.10), (3.12) and (3.13) into Eq. (3.17) and applying Hamilton's principle:

$$\delta \int_0^t \left(T - U_s - U_{sp} + W_e \right) dt = 0 \tag{3.18}$$

yields:

$$\int_0^t \int_\theta \left\{ I_0 \left(\frac{\partial u}{\partial t} \frac{\partial \delta u}{\partial t} + \frac{\partial w}{\partial t} \frac{\partial \delta w}{\partial t} \right) + q_\theta \delta u + q_z \right\} R d\theta dt$$

$$= \int_0^t \int_\theta \left\{ N_\theta \left(\frac{\partial \delta u}{\partial \theta} + \delta w \right) + M_\theta \left(\frac{\partial \delta u}{R \partial \theta} - \frac{\partial^2 \delta w}{R \partial \theta^2} \right) \right\} d\theta dt$$

$$+ \int_0^t \left\{ \begin{array}{l} \left[k_{\theta 0}^u u \delta u + k_{\theta 0}^w w \delta w + K_{\theta 0}^w \frac{\partial w}{R \partial \theta} \frac{\partial \delta w}{R \partial \theta} \right]_{\theta=0} \\ + \left[k_{\theta 1}^u u \delta u + k_{\theta 1}^w w \delta w + K_{\theta 1}^w \frac{\partial w}{R \partial \theta} \frac{\partial \delta w}{R \partial \theta} \right]_{\theta=\theta_0} \end{array} \right\} dt \quad (3.19)$$

Integrating by parts to relieve the virtual displacements δu and δw, we have

$$0 = \int_0^t \int_\theta \left\{ \frac{\partial N_\theta}{\partial \theta} + R \left(\frac{Q_\theta}{R} + q_\theta - I_0 \frac{\partial^2 u}{\partial t^2} \right) \right\} \delta u d\theta dt$$

$$+ \int_0^t \int_\theta \left\{ -N_\theta + \frac{\partial Q_\theta}{\partial \theta} + R \left(q_z - I_0 \frac{\partial^2 w}{\partial t^2} \right) \right\} \delta w d\theta dt$$

$$- \int_0^t \left\{ \begin{array}{l} \left(N_\theta + \frac{M_\theta}{R} + k_{\theta 1}^u u \right) \delta u|_{\theta_0} - \left(N_\theta + \frac{M_\theta}{R} - k_{\theta 0}^u u \right) \delta u|_0 \\ + \left(-M_\theta + K_{\theta 1}^w \frac{\partial w}{R \partial \theta} \right) \frac{\partial \delta w}{R \partial \theta}|_{\theta_0} + \left(Q_\theta + k_{\theta 1}^w w \right) \delta w|_{\theta_0} \\ + \left(M_\theta + K_{\theta 0}^w \frac{\partial w}{R \partial \theta} \right) \frac{\partial \delta w}{R \partial \theta}|_0 - \left(Q_\theta - k_{\theta 0}^w w \right) \delta w|_0 \end{array} \right\} dt \quad (3.20)$$

Since the virtual displacements δu and δw are arbitrary, the Eq. (3.20) can be satisfied only is the coefficients of the virtual displacements are zero. Thus, the governing equations and boundary conditions of thin laminated beams are obtained, which is the same as those presented in Eqs. (3.14) and (3.16). For general curved beams and asymmetrically laminated straight beams, each boundary can exist two possible combinations for each type of classical boundary conditions (free, simply-supported, and clamped). At each boundary of θ = constant, the possible combinations for each classical boundary condition are given in Table 3.1.

By using the artificial spring boundary technique, taking edge $\theta = 0$ for example, the F, S (simply-supported), SD (shear-diaphragm) and C (completely clamped) boundary conditions which are of particular interest can be readily realized by assigning the stiffness of the boundary springs at proper values as follows:

Table 3.1 Possible classical boundary conditions for thin laminated curved beams

Boundary type	Conditions
Free boundary conditions	
F	$N_\theta + \frac{M_\theta}{R_\theta} = Q_\theta = M_\theta = 0$
F2	$u = Q_\theta = M_\theta = 0$
Simply supported boundary conditions	
S	$u = w = M_\theta = 0$
SD	$N_\theta + \frac{M_\theta}{R_\theta} = w = M_\theta = 0$
Clamped boundary conditions	
C	$u = w = \frac{\partial w}{R\partial\theta} = 0$
C2	$N_\theta + \frac{M_\theta}{R_\theta} = w = \frac{\partial w}{R\partial\theta} = 0$

$$\begin{aligned}
\text{F}: \ &k_{\theta0}^u = k_{\theta0}^w = K_{\theta0}^w = 0 \\
\text{SD}: \ &k_{\theta0}^w = 10^7 D, \ k_{\theta0}^u = K_{\theta0}^w = 0 \\
\text{S}: \ &k_{\theta0}^u = k_{\theta0}^w = 10^7 D, \ K_{\theta0}^w = 0 \\
\text{C}: \ &k_{\theta0}^u = k_{\theta0}^w = K_{\theta0}^w = 10^7 D
\end{aligned} \tag{3.21}$$

where $D = E_1 h^3/12(1 - \mu_{12}\mu_{21})$ is the flexural stiffness of the beam.

Figure 3.3 shows the variations of the fundamental frequency parameters $\Delta\Omega$ (where $\Delta\Omega$ is defined as the difference of the fundamental frequency to that of the elastic restraint parameter $\Gamma = 10^{-3}$, namely, $\Delta\Omega = f(\Gamma) - f(10^{-3} D)$) versus restraint parameter Γ of a steel ($E = 210$ GPa, $\mu = 0.3$, $\rho = 7800$ kg/m^3) thin curved beam with different geometry parameters. The beam is clamped at boundary $\theta = \theta_0$ and elastically supported at boundary $\theta = 0$ (i.e., $k_{\theta0}^u = k_{\theta0}^w = K_{\theta0}^w = \Gamma D$). According to Fig. 3.3, we can see that the change of the restraint parameter Γ has little effect on

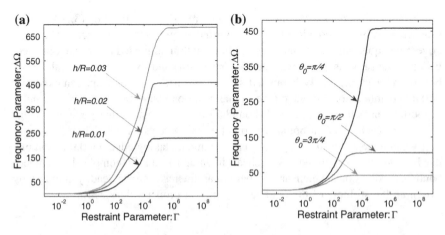

Fig. 3.3 Variation of fundamental frequency parameter $\Delta\Omega$ versus elastic restraint parameters Γ for a *thin* beam ($R = 1$) with different geometry parameters (a) $\theta_0 = \pi/4$; (b) $h/R = 0.02$

frequency parameter $\Delta\Omega$ when it is smaller than 10^0. However, when it is increased from 10^0 to 10^5, the frequency parameters increase rapidly. Then, the frequency parameters approach their utmost and remain unchanged when Γ approaches infinity. In such a case, the beam can be deemed as clamped in both ends. In conclusion, by assigning the stiffness of the boundary springs at $10^7 D$, the completely clamped boundary conditions of a beam can be realized.

3.2 Fundamental Equations of Thick Laminated Beams

In the CBT, the effects of shear deformation and rotary inertia are neglected. It is only applicable for thin beams. For beams with higher thickness ratios, the assumption that normals to the undeformed middle surface remain straight and normal to the deformed middle surface and suffer no extension of the classical beam theory should be relaxed and both shear deformation and rotary inertia effects should be included in the calculation. In this section, fundamental equations of laminated beams in the framework of shear deformation beam theory are developed and the deepness term $(1 + z/R_a)$ is considered in the formulation.

3.2.1 Kinematic Relations

Assuming that normals to the undeformed middle surface remain straight but not normal to the deformed middle surface, the displacement field in the beam space can be expressed in terms of middle surface displacements and rotation component as:

$$U(\theta, z) = u(\theta) + z\phi_\theta(\theta), \quad W(\theta, z) = w(\theta) \tag{3.22}$$

where u and w are the displacements at the middle surface in the θ and z directions. ϕ_θ represents the rotation of transverse normal, see Fig. 3.1. Letting $\alpha = \theta$, $A = R$ and $R_\alpha = R$ and specializing Eqs. (1.33) and (1.34) to those of beams, the normal and shear strains at any point of the beam space can be defined in terms of middle surface strains and curvature change as:

$$\varepsilon_\theta = \frac{1}{(1 + z/R)} \left(\varepsilon_\theta^0 + z\chi_\theta \right)$$
$$\gamma_{\theta z} = \frac{1}{(1 + z/R)} \gamma_{\theta z}^0 \tag{3.23}$$

where ε_θ^0 and $\gamma_{\theta z}^0$ denote the normal and shear strains in the reference surface. χ_θ is the curvature change. They are defined in terms of the middle surface displacements and rotation component as:

$$\varepsilon_\theta^0 = \frac{\partial u}{R\partial\theta} + \frac{w}{R}, \quad \chi_\theta = \frac{\partial\phi_\theta}{R\partial\theta}$$

$$\gamma_{\theta z}^0 = \frac{\partial w}{R\partial\theta} - \frac{u}{R} + \phi_\theta$$

(3.24)

3.2.2 Stress-Strain Relations and Stress Resultants

According to Hooke's law, the corresponding stress-strain relations in the kth layer of thick laminated curved beams can be written as:

$$\left\{\begin{matrix} \sigma_\theta \\ \tau_{\theta z} \end{matrix}\right\}_k = \begin{bmatrix} \overline{Q_{11}^k} & 0 \\ 0 & \overline{Q_{55}^k} \end{bmatrix} \left\{\begin{matrix} \varepsilon_\theta \\ \gamma_{\theta z} \end{matrix}\right\}_k$$

(3.25)

where σ_θ represents the normal stress, $\tau_{\theta z}$ is the shear stress. The elastic stiffness coefficients $\overline{Q_{ii}^k}$ ($i = 1, 5$) are defined by following equations:

$$\overline{Q_{11}^k} = Q_{11}^k \cos^4 \vartheta^k \text{ and } Q_{11}^k = \frac{E_1}{1 - \mu_{12}\mu_{21}}$$

$$\overline{Q_{55}^k} = Q_{55}^k \cos^2 \vartheta^k \text{ and } Q_{55}^k = G_{13}$$

(3.26)

where E_1 is the modulus of elasticity of the composite material in the principal direction. μ_{12} and μ_{21} are the Poisson's ratios. G_{13} is the shear modulus. ϑ^k represents the included angle between the principal direction of the layer and the θ-axis. By carrying the integration of the normal stress over the cross-section, the force and moment resultants can be obtained:

$$\begin{bmatrix} N_\theta \\ Q_\theta \end{bmatrix} = b \int_{-h/2}^{h/2} \begin{bmatrix} \sigma_\theta \\ \tau_{\theta z} \end{bmatrix} dz \quad M_\theta = b \int_{-h/2}^{h/2} \sigma_\theta z dz$$

(3.27)

where Q_θ represents the transverse shear force resultant. Performing the integration operation in Eq. (3.27), the force and moment resultants can be written in terms of the middle surface strains and curvature change as:

$$\begin{bmatrix} N_\theta \\ M_\theta \\ Q_\theta \end{bmatrix} = \begin{bmatrix} A_{11} & B_{11} & 0 \\ B_{11} & D_{11} & 0 \\ 0 & 0 & A_{55} \end{bmatrix} \begin{bmatrix} \varepsilon_\theta^0 \\ \chi_\theta \\ \gamma_{\theta z}^0 \end{bmatrix}$$

(3.28)

The stiffness coefficients A_{11}, A_{55}, B_{11} and D_{11} are defined as follows:

$$A_{11} = Rb \sum_{k=1}^{N} \overline{Q_{11}^k} \ln\left(\frac{R + z_{k+1}}{R + z_k}\right)$$

$$A_{55} = K_s Rb \sum_{k=1}^{N} \overline{Q_{55}^k} \ln\left(\frac{R + z_{k+1}}{R + z_k}\right)$$

$$B_{11} = Rb \sum_{k=1}^{N} \overline{Q_{11}^k} \left[(z_{k+1} - z_k) - R\ln\left(\frac{R + z_{k+1}}{R + z_k}\right)\right]$$

$$D_{11} = Rb \sum_{k=1}^{N} \overline{Q_{11}^k} \left[\frac{1}{2}(z_{k+1}^2 - z_k^2) - 2R(z_{k+1} - z_k) + R^2 \ln\left(\frac{R + z_{k+1}}{R + z_k}\right)\right]$$

(3.29)

in which K_s is the shear correction factor, typically taken at 5/6 (Qatu 2004; Reddy 2003).

3.2.3 Energy Functions

The strain energy (U_s) of thick laminated curved beams during vibration can be defined as

$$U_s = \frac{1}{2}\int_{\theta} \{N_\theta \varepsilon_\theta^0 + M_\theta \chi_\theta + Q_\theta \gamma_{\theta z}^0\} Rd\theta \qquad (3.30)$$

Substituting Eqs. (3.24) and (3.28) into Eq. (3.30), the strain energy function of thick curved beams can be rewritten as:

$$U_s = \frac{1}{2}\int_{\theta} \left\{ \begin{array}{l} A_{11}\left(\dfrac{\partial u}{R\partial\theta} + \dfrac{w}{R}\right)^2 + 2B_{11}\dfrac{\partial\phi_\theta}{R\partial\theta}\left(\dfrac{\partial u}{R\partial\theta} + \dfrac{w}{R}\right) \\ + D_{11}\left(\dfrac{\partial\phi_\theta}{R\partial\theta}\right)^2 + A_{55}\left(\dfrac{\partial w}{R\partial\theta} - \dfrac{u}{R} + \phi_\theta\right)^2 \end{array} \right\} Rd\theta \qquad (3.31)$$

The corresponding kinetic energy (T) function of the beams is written as:

$$T = \frac{1}{2}\int_{\theta} \left\{ \overline{I_0}\left(\frac{\partial u}{\partial t}\right)^2 + 2\overline{I_1}\frac{\partial u}{\partial t}\frac{\partial\phi_\theta}{\partial t} + \overline{I_2}\left(\frac{\partial\phi_\theta}{\partial t}\right)^2 + \overline{I_0}\left(\frac{\partial w}{\partial t}\right)^2 \right\} Rd\theta \qquad (3.32)$$

where the inertia terms are defined as:

$$\bar{I_0}=I_0 + \frac{I_1}{R}$$

$$\bar{I_1}=I_1 + \frac{I_2}{R}$$

$$\bar{I_2}=I_2 + \frac{I_3}{R} \qquad (3.33)$$

$$[I_0, I_1, I_2, I_3] = b \sum_{k=1}^{N} \int_{z_k}^{z_{k+1}} \rho^k [1, z, z^2, z^3]dz$$

in which ρ^k is the mass of the k'th layer per unit middle surface area. The external work is expressed as:

$$W_e = \int_{\theta} \{q_\theta u + q_z w + m_\theta \phi_\theta\}Rd\theta \qquad (3.34)$$

where q_θ and q_z denote the external loads. m_θ is the external couples in the middle surface of the beam. Using the artificial spring boundary technique similar to that described earlier, symbols k_ψ^u, k_ψ^w and K_ψ^θ ($\psi = \theta_0$ and θ_1) are used to indicate the rigidities (per unit length) of the boundary springs at the boundaries $\theta = 0$ and $\theta = \theta_0$, respectively, see Fig. 3.4. Therefore, the deformation strain energy (U_{sp}) stored in the boundary springs during vibration can be defined as:

$$U_{sp} = \frac{1}{2} \left\{ \left[k_{\theta 0}^u u^2 + k_{\theta 0}^w w^2 + K_{\theta 0}^\theta \phi_\theta^2 \right]_{\theta=0} + \left[k_{\theta 1}^u u^2 + k_{\theta 1}^w w^2 + K_{\theta 1}^\theta \phi_\theta^2 \right]_{\theta=\theta_0} \right\} \quad (3.35)$$

3.2.4 Governing Equations and Boundary Conditions

Specializing the governing equations and boundary conditions of the thick shell (Eq. 1.59) to those of thick beams, we have

$$\frac{\partial N_\theta}{R\partial\theta} + \frac{Q_\theta}{R} + q_\theta = \bar{I_0}\frac{\partial^2 u}{\partial t^2} + \bar{I_1}\frac{\partial^2 \phi_\theta}{\partial t^2}$$

$$-\frac{N_\theta}{R} + \frac{\partial Q_\theta}{R\partial\theta} + q_z = \bar{I_0}\frac{\partial^2 w}{\partial t^2} \qquad (3.36)$$

$$\frac{\partial M_\theta}{R\partial\theta} - Q_\theta + m_\theta = \bar{I_1}\frac{\partial^2 u}{\partial t^2} + \bar{I_2}\frac{\partial^2 \phi_\theta}{\partial t^2}$$

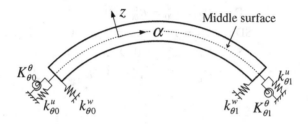

Fig. 3.4 Boundary conditions of a *thick* laminated beam

Similarly, according to Eq. (1.60), the general boundary conditions of thick laminated curved beams are written as:

$$\theta = 0: \begin{cases} N_\theta - k_{\theta 0}^u u = 0 \\ Q_\theta - k_{\theta 0}^w w = 0 \\ M_\theta - K_{\theta 0}^\theta \phi_\theta = 0 \end{cases}, \quad \theta = \theta_0: \begin{cases} N_\theta + k_{\theta 1}^u u = 0 \\ Q_\theta + k_{\theta 1}^w w = 0 \\ M_\theta + K_{\theta 1}^\theta \phi_\theta = 0 \end{cases} \quad (3.37)$$

Alternately, the governing equations and boundary conditions for the thick beams can be obtained by applying Hamilton's principle in the same manner as described in Sect. 3.1.4.

For thick curved beams or unsymmetrically laminated straight beams, each boundary exits two possible combinations for each classical boundary condition. At each boundary, the possible combinations for each classical boundary condition are given in Table 3.2.

In the framework of artificial spring boundary technique, taking edge $\theta = 0$ for example, the frequently encountered boundary conditions F, S, SD and C can be readily realized by assigning the stiffness of the boundary springs at proper values as follows:

Table 3.2 Possible classical boundary conditions for thick curved beams

Boundary type	Conditions
Free boundary conditions	
F	$N_\theta = Q_\theta = M_\theta = 0$
F2	$u = Q_\theta = M_\theta = 0$
Simply supported boundary conditions	
S	$u = w = M_\theta = 0$
SD	$N_\theta = w = M_\theta = 0$
Clamped boundary conditions	
C	$u = w = \phi_\theta = 0$
C2	$N_\theta = w = \phi_\theta = 0$

$$\text{F}: \quad k_{\theta 0}^u = k_{\theta 0}^w = K_{\theta 0}^\theta = 0$$

$$\text{SD}: \quad k_{\theta 0}^w = 10^7 D, \ k_{\theta 0}^u = K_{\theta 0}^\theta = 0$$

$$\text{S}: \quad k_{\theta 0}^u = k_{\theta 0}^w = 10^7 D, \ K_{\theta 0}^\theta = 0 \tag{3.38}$$

$$\text{C}: \quad k_{\theta 0}^u = k_{\theta 0}^w = K_{\theta 0}^\theta = 10^7 D$$

3.3 Solution Procedures

With above fundamental equations and modified Fourier series method developed in Chap. 2, both strong and weak form solution procedures of laminated beams with general boundary conditions are presented in this section. To fully illustrate the modified Fourier series solution procedure, we only consider the free vibration analysis of laminated beams with general boundary conditions based on the SDBT. For other beam theories, the corresponding solution procedures can be obtained in the same manner.

Combining Eqs. (3.24), (3.28) and (3.36), it is obvious that the displacements and rotation components of thick laminated curved beams are required to have up to the second derivative. Therefore, regardless of boundary conditions, each displacements and rotation component of a laminated beam can be expanded as a one-dimensional modified Fourier series as

$$u(\theta) = \sum_{m=0}^M A_m \cos \lambda_m \theta + a_1 P_1(\theta) + a_2 P_2(\theta)$$

$$w(\theta) = \sum_{m=0}^M B_m \cos \lambda_m \theta + b_1 P_1(\theta) + b_2 P_2(\theta) \tag{3.39}$$

$$\phi_\theta(\theta) = \sum_{m=0}^M C_m \cos \lambda_m \theta + c_1 P_1(\theta) + c_2 P_2(\theta)$$

where $\lambda_m = m\pi/\theta_0$. $P_1(\theta)$ and $P_2(\theta)$ denote the auxiliary polynomial functions introduced to remove all the discontinuities potentially associated with the first-order derivatives at the boundaries then ensure and accelerate the convergence of the series expansion of the beam displacements and rotation component. M is the truncation number. A_m, B_m and C_m are the expansion coefficients of standard cosine Fourier series. a_1, a_2, b_1, b_2, c_1 and c_2 represent the corresponding expansion coefficients of auxiliary functions $P_1(\theta)$ and $P_2(\theta)$. These two auxiliary functions are defined as

$$P_1(\theta) = \theta\left(\frac{\theta}{\theta_0} - 1\right)^2 \quad P_2(\theta) = \frac{\theta^2}{\theta_0}\left(\frac{\theta}{\theta_0} - 1\right) \tag{3.40}$$

It should be stressed that in the CBT cases, the displacements of a laminated beam are required to have up to the fourth-order derivatives. In such case, the following four auxiliary polynomial functions are introduced to remove all the discontinuities potentially associated with the first-order and third-order derivatives at the boundaries (Jin et al. 2013a).

$$
\begin{aligned}
P_1(\theta) &= \frac{9\theta_0}{4\pi}\sin\left(\frac{\pi\theta}{2\theta_0}\right) - \frac{\theta_0}{12\pi}\sin\left(\frac{3\pi\theta}{2\theta_0}\right) \\
P_2(\theta) &= -\frac{9\theta_0}{4\pi}\cos\left(\frac{\pi\theta}{2\theta_0}\right) - \frac{\theta_0}{12\pi}\cos\left(\frac{3\pi\theta}{2\theta_0}\right) \\
P_3(\theta) &= \frac{\theta_0^3}{\pi^3}\sin\left(\frac{\pi\theta}{2\theta_0}\right) - \frac{\theta_0^3}{3\pi^3}\sin\left(\frac{3\pi\theta}{2\theta_0}\right) \\
P_4(\theta) &= -\frac{\theta_0^3}{\pi^3}\cos\left(\frac{\pi\theta}{2\theta_0}\right) - \frac{\theta_0^3}{3\pi^3}\cos\left(\frac{3\pi\theta}{2\theta_0}\right)
\end{aligned}
\tag{3.41}
$$

3.3.1 Strong Form Solution Procedure

Substituting Eqs. (3.24) and (3.28) into Eq. (3.36) the governing equations of laminated curved beams including shear deformation and rotary inertia effects can be rewritten as:

$$
\left(
\begin{bmatrix}
L_{11} & L_{12} & L_{13} \\
L_{21} & L_{22} & L_{23} \\
L_{31} & L_{32} & L_{33}
\end{bmatrix}
- \omega^2
\begin{bmatrix}
M_{11} & 0 & M_{13} \\
0 & M_{22} & 0 \\
M_{31} & 0 & M_{33}
\end{bmatrix}
\right)
\begin{bmatrix} u \\ w \\ \phi_\theta \end{bmatrix}
=
\begin{bmatrix} 0 \\ 0 \\ 0 \end{bmatrix}
\tag{3.42}
$$

where the coefficients of the linear operator ($M_{ij} = M_{ji}$) are given below:

$$
\begin{aligned}
L_{11} &= \frac{A_{11}}{R^2}\frac{\partial^2}{\partial\theta^2} - \frac{A_{55}}{R^2}, & L_{12} &= \frac{A_{11}}{R^2}\frac{\partial}{\partial\theta} + \frac{A_{55}}{R^2}\frac{\partial}{\partial\theta} \\
L_{13} &= \frac{B_{11}}{R^2}\frac{\partial^2}{\partial\theta^2} + \frac{A_{55}}{R}, & L_{21} &= -\frac{A_{11}}{R^2}\frac{\partial}{\partial\theta} - \frac{A_{55}}{R^2}\frac{\partial}{\partial\theta} \\
L_{22} &= -\frac{A_{11}}{R^2} + \frac{A_{55}}{R^2}\frac{\partial^2}{\partial\theta^2}, & L_{23} &= -\frac{B_{11}}{R^2}\frac{\partial}{\partial\theta} + \frac{A_{55}}{R}\frac{\partial}{\partial\theta} \\
L_{31} &= \frac{B_{11}}{R^2}\frac{\partial^2}{\partial\theta^2} + \frac{A_{55}}{R}, & L_{32} &= \frac{B_{11}}{R^2}\frac{\partial}{\partial\theta} - \frac{A_{55}}{R}\frac{\partial}{\partial\theta} \\
L_{33} &= \frac{D_{11}}{R^2}\frac{\partial^2}{\partial\theta^2} - A_{55}, & M_{11} &= M_{22} = -\bar{I}_0 \\
M_{13} &= M_{31} = -\bar{I}_1, & M_{33} &= -\bar{I}_2
\end{aligned}
\tag{3.43}
$$

Rewriting the modified Fourier series expressions (Eq. 3.39) in the matrix form as:

$$u(\theta) = \mathbf{H}_{sc}\mathbf{A} + \mathbf{H}_{af}\mathbf{a}$$
$$w(\theta) = \mathbf{H}_{sc}\mathbf{B} + \mathbf{H}_{af}\mathbf{b} \qquad (3.44)$$
$$\phi_\theta(\theta) = \mathbf{H}_{sc}\mathbf{C} + \mathbf{H}_{af}\mathbf{c}$$

where

$$\mathbf{H}_{sc} = [\cos \lambda_0\theta, \ldots, \cos \lambda_m\theta, \ldots, \cos \lambda_M\theta]$$
$$\mathbf{H}_{af} = [P_1(\theta), P_2(\theta)]$$
$$\mathbf{A} = [A_0, \ldots, A_m, \ldots, A_M]^T \quad \mathbf{a} = [a_1, a_2]^T \qquad (3.45)$$
$$\mathbf{B} = [B_0, \ldots, B_m, \ldots, B_M]^T \quad \mathbf{b} = [b_1, b_2]^T$$
$$\mathbf{C} = [C_0, \ldots, C_m, \ldots, C_M]^T \quad \mathbf{c} = [c_1, c_2]^T$$

where superscript T represents the transposition operator. Substituting Eq. (3.44) into Eq. (3.42) results in

$$\mathbf{L}_{sc}\begin{bmatrix}\mathbf{A}\\\mathbf{B}\\\mathbf{C}\end{bmatrix} + \mathbf{L}_{af}\begin{bmatrix}\mathbf{a}\\\mathbf{b}\\\mathbf{c}\end{bmatrix} - \omega^2\left(\mathbf{M}_{sc}\begin{bmatrix}\mathbf{A}\\\mathbf{B}\\\mathbf{C}\end{bmatrix} + \mathbf{M}_{af}\begin{bmatrix}\mathbf{a}\\\mathbf{b}\\\mathbf{c}\end{bmatrix}\right) = 0 \qquad (3.46)$$

where

$$\mathbf{L}_i = \begin{bmatrix} L_{11}\mathbf{H}_i & L_{12}\mathbf{H}_i & L_{13}\mathbf{H}_i \\ L_{21}\mathbf{H}_i & L_{22}\mathbf{H}_i & L_{23}\mathbf{H}_i \\ L_{31}\mathbf{H}_i & L_{32}\mathbf{H}_i & L_{33}\mathbf{H}_i \end{bmatrix} (i = sc, af)$$
$$\qquad (3.47)$$
$$\mathbf{M}_i = \begin{bmatrix} M_{11}\mathbf{H}_i & 0 & M_{13}\mathbf{H}_i \\ 0 & M_{22}\mathbf{H}_i & 0 \\ M_{31}\mathbf{H}_i & 0 & M_{33}\mathbf{H}_i \end{bmatrix} (i = sc, af)$$

Similarly, substituting Eq. (3.44) into Eq. (3.37), the boundary conditions of thick laminated curved beams can be rewritten as

$$\begin{bmatrix} \mathbf{L}_{sc}^{\theta 0} \\ \mathbf{L}_{\theta 1} \\ \mathbf{L}_{sc}^{\theta 1} \end{bmatrix} \begin{bmatrix} \mathbf{A} \\ \mathbf{B} \\ \mathbf{C} \end{bmatrix} + \begin{bmatrix} \mathbf{L}_{af}^{\theta 0} \\ \mathbf{L}_{\theta 1} \\ \mathbf{L}_{af} \end{bmatrix} \begin{bmatrix} \mathbf{a} \\ \mathbf{b} \\ \mathbf{c} \end{bmatrix} = 0 \qquad (3.48)$$

where

$$\mathbf{L}_i^{\theta 0} = \begin{bmatrix} \dfrac{A_{11}}{R}\dfrac{\partial \mathbf{H}_i}{\partial \theta} - k_{\theta 0}^u \mathbf{H}_i & \dfrac{A_{11}}{R}\mathbf{H}_i & \dfrac{B_{11}}{R}\dfrac{\partial \mathbf{H}_i}{\partial \theta} \\[3mm] -\dfrac{A_{55}}{R}\mathbf{H}_i & \dfrac{A_{55}}{R}\dfrac{\partial \mathbf{H}_i}{\partial \theta} - k_{\theta 0}^w \mathbf{H}_i & A_{55}\mathbf{H}_i \\[3mm] \dfrac{B_{11}}{R}\dfrac{\partial \mathbf{H}_i}{\partial \theta} & \dfrac{B_{11}}{R}\mathbf{H}_i & \dfrac{D_{11}}{R}\dfrac{\partial \mathbf{H}_i}{\partial \theta} - K_{\theta 0}^\theta \mathbf{H}_i \end{bmatrix}_{\theta=0} \quad (i=sc,af)$$

$$\mathbf{L}_i^{\theta 1} = \begin{bmatrix} \dfrac{A_{11}}{R}\dfrac{\partial \mathbf{H}_i}{\partial \theta} + k_{\theta 1}^u \mathbf{H}_i & \dfrac{A_{11}}{R}\mathbf{H}_i & \dfrac{B_{11}}{R}\dfrac{\partial \mathbf{H}_i}{\partial \theta} \\[3mm] -\dfrac{A_{55}}{R}\mathbf{H}_i & \dfrac{A_{55}}{R}\dfrac{\partial \mathbf{H}_i}{\partial \theta} + k_{\theta 1}^w \mathbf{H}_i & A_{55}\mathbf{H}_i \\[3mm] \dfrac{B_{11}}{R}\dfrac{\partial \mathbf{H}_i}{\partial \theta} & \dfrac{B_{11}}{R}\mathbf{H}_i & \dfrac{D_{11}}{R}\dfrac{\partial \mathbf{H}_i}{\partial \theta} + K_{\theta 1}^\theta \mathbf{H}_i \end{bmatrix}_{\theta=\theta_0} \quad (i=sc,af)$$

$$(3.49)$$

Thus, the relation between the expansion coefficients of standard cosine Fourier series ($\mathbf{A}$, $\mathbf{B}$ and $\mathbf{C}$) and those of corresponding auxiliary functions ($\mathbf{a}$, $\mathbf{b}$ and $\mathbf{c}$) can be determined by the following equation:

$$\begin{bmatrix} \mathbf{a} \\ \mathbf{b} \\ \mathbf{c} \end{bmatrix} = - \begin{bmatrix} \mathbf{L}_{af}^{\theta 0} \\ \mathbf{L}_{af}^{\theta 1} \end{bmatrix}^{-1} \begin{bmatrix} \mathbf{L}_{sc}^{\theta 0} \\ \mathbf{L}_{sc}^{\theta 1} \end{bmatrix} \begin{bmatrix} \mathbf{A} \\ \mathbf{B} \\ \mathbf{C} \end{bmatrix} \qquad (3.50)$$

In order to derive the constraint equations for the unknown expansion coefficients, all the sine terms, the auxiliary polynomial functions and their derivatives in Eq. (3.46) are expanded into Fourier cosine series then collecting the similar terms, i.e., multiplying Eq. (3.46) with $\mathbf{H}_e$ in the left side and integrating it from 0 to θ_0 with respect to θ, we have

$$\overline{\mathbf{L}}_{sc}\begin{bmatrix} \mathbf{A} \\ \mathbf{B} \\ \mathbf{C} \end{bmatrix} + \overline{\mathbf{L}}_{af}\begin{bmatrix} \mathbf{a} \\ \mathbf{b} \\ \mathbf{c} \end{bmatrix} - \omega^2 \left(\overline{\mathbf{M}}_{sc}\begin{bmatrix} \mathbf{A} \\ \mathbf{B} \\ \mathbf{C} \end{bmatrix} + \overline{\mathbf{M}}_{af}\begin{bmatrix} \mathbf{a} \\ \mathbf{b} \\ \mathbf{c} \end{bmatrix} \right) = 0 \qquad (3.51)$$

where

$$\overline{\mathbf{L}}_{sc} = \int_0^{\theta 0} \mathbf{H}_e \mathbf{L}_{sc} d\theta$$

$$\overline{\mathbf{L}}_{af} = \int_0^{\theta 0} \mathbf{H}_e \mathbf{L}_{af} d\theta$$

$$\overline{\mathbf{M}}_{sc} = \int_0^{\theta 0} \mathbf{H}_e \mathbf{M}_{sc} d\theta \tag{3.52}$$

$$\overline{\mathbf{M}}_{af} = \int_0^{\theta 0} \mathbf{H}_e \mathbf{M}_{af} d\theta$$

$$\mathbf{H}_e = \mathbf{H}_{sc}^T = [\cos \lambda_0 \theta, \ldots, \cos \lambda_m \theta, \ldots, \cos \lambda_M \theta]^T$$

Finally, combining Eqs. (3.50) and (3.51) results in

$$(\mathbf{K} - \omega^2 \mathbf{M}) [\mathbf{A} \quad \mathbf{B} \quad \mathbf{C}]^T = \mathbf{0} \tag{3.53}$$

where $\mathbf{K}$ is the stiffness matrix and $\mathbf{M}$ is the mass matrix. They are defined as

$$\mathbf{K} = \overline{\mathbf{L}}_{sc} - \overline{\mathbf{L}}_{af} \begin{bmatrix} \mathbf{L}_{af}^{\theta 0} \\ \mathbf{L}_{af}^{\theta 1} \end{bmatrix}^{-1} \begin{bmatrix} \mathbf{L}_{sc}^{\theta 0} \\ \mathbf{L}_{sc}^{\theta 1} \end{bmatrix}$$

$$\mathbf{M} = \overline{\mathbf{M}}_{sc} - \overline{\mathbf{M}}_{af} \begin{bmatrix} \mathbf{L}_{af}^{\theta 0} \\ \mathbf{L}_{af}^{\theta 1} \end{bmatrix}^{-1} \begin{bmatrix} \mathbf{L}_{sc}^{\theta 0} \\ \mathbf{L}_{sc}^{\theta 1} \end{bmatrix} \tag{3.54}$$

Thus, the natural frequencies and modes of the beams under consideration can be determined easily by solving the standard characteristic equation.

3.3.2 Weak Form Solution Procedure (Rayleigh-Ritz Procedure)

Instead of seeking a solution in strong form as described in Sect. 3.3.1, all the expansion coefficients can be treated equally and independently as the generalized coordinates and solved directly from the Rayleigh–Ritz technique, which is the focus of the current section.

For free vibration analysis, the Lagrangian energy functional (L) of beams can be defined in terms of the strain energy and kinetic energy functions as:

$$L = T - U_s - U_{sp} \tag{3.55}$$

Substituting Eqs. (3.31), (3.32) (3.35) and (3.39) into Eq. (3.55) and taking its derivatives with respect to each of the undetermined coefficients and making them equal to zero

$$\frac{\partial L}{\partial \Xi} = 0 \quad \text{and} \quad \begin{cases} \Xi = A_m, B_m, C_m \\ m = 0, 1, 2, \ldots M \end{cases}$$

$$\frac{\partial L}{\partial \Psi} = 0 \quad \text{and} \quad \Psi = a_1, a_2, b_1, b_2, c_1, c_2 \tag{3.56}$$

a total of $3*(M + 3)$ equations can be obtained and they can be summed up in a matrix form as:

$$(\mathbf{K} - \omega^2 \mathbf{M})\mathbf{G} = \mathbf{0} \tag{3.57}$$

where $\mathbf{K}$ is the stiffness matrix for the beam, and $\mathbf{M}$ is the mass matrix. They are defined as

$$\mathbf{K} = \begin{bmatrix} \mathbf{K}_{uu} & \mathbf{K}_{uw} & \mathbf{K}_{u\theta} \\ \mathbf{K}_{uw}^T & \mathbf{K}_{ww} & \mathbf{K}_{w\theta} \\ \mathbf{K}_{u\theta}^T & \mathbf{K}_{w\theta}^T & \mathbf{K}_{\theta\theta} \end{bmatrix}$$

$$\mathbf{M} = \begin{bmatrix} \mathbf{M}_{uu} & \mathbf{0} & \mathbf{M}_{u\theta} \\ \mathbf{0} & \mathbf{M}_{ww} & \mathbf{0} \\ \mathbf{M}_{u\theta}^T & \mathbf{0} & \mathbf{M}_{\theta\theta} \end{bmatrix} \tag{3.58}$$

The explicit forms of submatrices $\mathbf{K}_{ij}$ and $\mathbf{M}_{ij}$ in the stiffness and mass matrices are listed in follow

$$\mathbf{K}_{uu} = \int\limits_0^{\theta 0} \left\{ \frac{A_{11}}{R} \frac{\partial \mathbf{H}^T}{\partial \theta} \frac{\partial \mathbf{H}}{\partial \theta} + \frac{A_{55}}{R} \mathbf{H}^T \mathbf{H} \right\} d\theta + k_{\theta 0}^u \mathbf{H}^T \mathbf{H}|_{\theta=0} + k_{\theta 1}^u \mathbf{H}^T \mathbf{H}|_{\theta=\theta 0}$$

$$\mathbf{K}_{uw} = \int\limits_0^{\theta 0} \left\{ \frac{A_{11}}{R} \frac{\partial \mathbf{H}^T}{\partial \theta} \mathbf{H} - \frac{A_{55}}{R} \mathbf{H}^T \frac{\partial \mathbf{H}}{\partial \theta} \right\} d\theta$$

$$\mathbf{K}_{u\theta} = \int\limits_0^{\theta 0} \left\{ \frac{B_{11}}{R} \frac{\partial \mathbf{H}^T}{\partial \theta} \frac{\partial \mathbf{H}}{\partial \theta} - A_{55} \mathbf{H}^T \mathbf{H} \right\} d\theta$$

$$\mathbf{K}_{ww} = \int\limits_0^{\theta 0} \left\{ \frac{A_{11}}{R} \mathbf{H}^T \mathbf{H} + \frac{A_{55}}{R} \frac{\partial \mathbf{H}^T}{\partial \theta} \frac{\partial \mathbf{H}}{\partial \theta} \right\} d\theta + k_{\theta 0}^w \mathbf{H}^T \mathbf{H}|_{\theta=0} + k_{\theta 1}^w \mathbf{H}^T \mathbf{H}|_{\theta=\theta 0}$$

$$\mathbf{K}_{w\theta} = \int\limits_0^{\theta 0} \left\{ \frac{B_{11}}{R} \mathbf{H}^T \frac{\partial \mathbf{H}}{\partial \theta} + A_{55} \frac{\partial \mathbf{H}^T}{\partial \theta} \mathbf{H} \right\} d\theta$$

$$\mathbf{K}_{\theta\theta} = \int\limits_0^{\theta 0} \left\{ \frac{D_{11}}{R} \frac{\partial \mathbf{H}^T}{\partial \theta} \frac{\partial \mathbf{H}}{\partial \theta} + A_{55} R \mathbf{H}^T \mathbf{H} \right\} d\theta + K_{\theta 0}^\theta \mathbf{H}^T \mathbf{H}|_{\theta=0} + K_{\theta 1}^u \mathbf{H}^T \mathbf{H}|_{\theta=\theta 0}$$

$$\mathbf{M}_{uu} = \mathbf{M}_{ww} = \int\limits_0^{\theta 0} \overline{I}_0 R \mathbf{H}^T \mathbf{H} d\theta$$

$$\mathbf{M}_{u\theta} = \int\limits_0^{\theta 0} \overline{I}_1 R \mathbf{H}^T \mathbf{H} d\theta$$

$$\mathbf{M}_{\theta\theta} = \int\limits_0^{\theta 0} \overline{I}_2 R \mathbf{H}^T \mathbf{H} d\theta$$

$$(3.59)$$

where

$$\mathbf{H} = [\,\mathbf{H}_{sc} \quad \mathbf{H}_{af}\,] = [\cos \lambda_0 \theta, \ldots, \cos \lambda_m \theta, \ldots, \cos \lambda_M \theta, P_1(\theta), P_2(\theta)] \quad (3.60)$$

G is a column vector which contains, in an appropriate order, the unknown expansion coefficients:

$$\mathbf{G} = [\mathbf{A} \quad \mathbf{a} \quad \mathbf{B} \quad \mathbf{b} \quad \mathbf{C} \quad \mathbf{c}]^T \quad (3.61)$$

Obviously, the vibration results can now be easily obtained by solving a standard matrix eigenproblem.

3.4 Laminated Beams with General Boundary Conditions

Vibration results of laminated straight and curved beams with general boundary conditions are given in this section. The isotropic beams are treated as special cases of laminated beams in the presentation. Natural frequencies and mode shapes for straight and curved beams with different boundary conditions, lamination schemes and geometry parameters are presented using both strong form and weak form solution procedures. The convergence of the solutions is studied and the effects of shear deformation and rotary inertia, deepness term $(1 + z/R)$ and beam parameters (boundary conditions, lamination schemes, geometry parameters and material properties) are investigated as well.

For the sake of simplicity, character strings CBT_w, CBT_s, $SDBT_w$ and $SDBT_s$ are introduced to represent the beam theories and methods used in the calculation (where subscripts w and s denote weak form and strong form solution procedures, respectively). In addition, a two-letter string is applied to indicate the end conditions of a beam, such as C-F denotes a beam with C and F boundary conditions at the boundaries $\theta = 0$ and $\theta = \theta_0$, respectively. Unless otherwise stated, the natural frequencies of the considered beams are expressed in the non-dimensional parameters as $\Omega = \omega L_\theta^2 \sqrt{12\rho/E_1 h^2}$ and the material properties of the beams are given as: $\mu_{12} = 0.25$, $G_{13} = 0.5E_2$ (where L_θ represents the span length of a curved beam, i.e., $L_\theta = R\theta_0$).

3.4.1 Convergence Studies and Result Verification

Table 3.3 shows the convergence studies made for the first six natural frequencies (Hz) of a moderately thick, two-layered laminated ($[0°/90°]$) curved beam with completely free (F-F) and clamped (C-C) boundary conditions. The geometry and material constants of the beam are given as: $R = 1$ m, $\theta_0 = 1$, $h/R = 0.1$, $E_2 = 10$ GPa, $E_1/E_2 = 10$, $\mu_{12} = 0.27$, $G_{13} = 5.5$ GPa, $\rho = 1,700$ kg/m^3. Both the $SDBT_w$ and $SDBT_s$ solutions for truncation schemes $M = 8, 9, 14, 15$ are included in studies. It is obvious that the modified Fourier series solution has an excellent convergence, and is sufficiently accurate even when only a small number of terms are included in the series expressions. In addition, from the table, we can see that the $SDBT_w$ solutions converge faster than the $SDBT_s$ ones. Unless otherwise stated, the truncation number (M) of the displacement expressions will be uniformly selected as $M = 15$ in the following calculation and the weak form solution procedure will be adopted in the calculation.

In Table 3.4, comparisons of the frequency parameters Ω for a two-layered, unsymmetrically laminated ($[90°/0°]$) curved beam with SD-SD boundary conditions are presented. The geometric properties of the layers of the beam are the same as those used in Table 3.3 except that the thickness-to-length ratio is given as

Table 3.3 Convergence of the first six natural frequencies (Hz) for a [0°/90°] laminated curved beam with F-F and C-C boundary conditions ($R = 1$ m, $\theta_0 = 1$, $h/R = 0.1$)

Boundary conditions	M	Mode number					
		1	2	3	4	5	6
F-F	SDBT$_w$						
	8	123.88	314.88	549.83	813.20	895.85	1099.3
	9	123.88	314.84	549.78	812.30	895.63	1098.3
	14	123.88	314.81	549.63	811.86	895.58	1096.5
	15	123.88	314.81	549.63	811.84	895.57	1096.5
	SDBT$_s$						
	8	124.60	316.99	553.41	821.66	897.25	1109.4
	9	124.59	316.20	552.42	817.25	896.32	1106.3
	14	124.01	315.15	550.23	813.00	895.80	1098.0
	15	124.01	315.09	550.11	812.80	895.72	1097.7
C-C	SDBT$_w$						
	8	262.54	279.29	514.27	723.46	894.04	990.46
	9	262.53	279.27	514.07	723.31	893.95	987.46
	14	262.52	279.21	513.96	722.71	893.93	986.40
	15	262.52	279.21	513.95	722.71	893.93	986.34
	SDBT$_s$						
	8	262.84	280.27	517.05	729.92	895.50	1005.6
	9	262.74	279.86	516.23	727.91	894.55	996.79
	14	262.57	279.36	514.43	723.68	894.11	988.60
	15	262.56	279.33	514.34	723.58	894.04	987.94

Table 3.4 Comparison of the frequency parameters Ω for a [90°/0°] laminated curved beam with SD-SD boundary conditions ($R = 1$ m, $\theta_0 = 1$)

h/L_θ	Qatu (1993)			SDBT$_w$		
	1	2	3	1	2	3
0.01	4.0094	18.000	41.286	4.0179	18.042	41.388
0.02	3.9885	17.839	40.681	3.9969	17.878	40.770
0.05	3.9109	17.089	37.667	3.9190	17.124	37.738
0.10	3.7419	15.329	31.300	3.7496	15.357	31.349
0.20	3.3312	11.808	21.481	3.3387	11.833	21.523

$h/L_\theta = 0.01$, 0.02, 0.05, 0.1 and 0.2 and the elementally material parameters of the layers are: $E_1/E_2 = 15$. From the table, we can see that the present solutions agree very well with exact solutions published by Qatu (1993). The differences between the two results are very small, and do not exceed 0.27 % for the worst case. To further prove the validity of the present method, Table 3.5 lists comparisons of the frequency parameters Ω for a two-layered, unsymmetrically laminated ([90°/0°]) curved beam with various boundary conditions. The layers of the beam are thought

Table 3.5 Comparison of the frequency parameters Ω for a [90°/0°] laminated curved beam with different boundary conditions ($h/L_\theta = 0.01$, $R/L_\theta = 2$, $E_1/E_2 = 15$)

B.C.	Theory	Mode number				
		1	2	3	4	5
S-S	Qatu and Elsharkawy (1993)	18.434	37.935	74.549	99.888	143.10
	SDBT$_w$	18.448	37.910	74.303	99.435	142.01
	CBT$_w$	18.473	38.009	74.703	99.905	142.85
SD-SD	Qatu and Elsharkawy (1993)	4.5173	18.593	42.050	74.884	117.92
	SDBT$_w$	4.5249	18.606	42.009	74.632	116.37
	CBT$_w$	4.5268	18.631	42.137	75.037	117.32
C-C	Qatu and Elsharkawy (1993)	29.015	51.348	94.637	111.64	149.21
	SDBT$_w$	28.973	51.167	93.982	111.02	147.71
	CBT$_w$	29.076	51.436	94.833	111.35	149.17
C2-C2	Qatu and Elsharkawy (1993)	10.445	29.145	57.291	94.837	144.65
	SDBT$_w$	10.255	29.028	56.995	94.103	140.25
	CBT$_w$	10.467	29.205	57.406	95.008	141.98

to be of equal thickness and made from material with following properties: $h/L_\theta = 0.01$, $R/L_\theta = 2$, $E_1/E_2 = 15$. The Ritz solutions obtained by Qatu and Elsharkawy (1993) by using the classical beam theory are selected as the benchmark solutions. A consistent agreement between the present results and the referential data is seen from the table. Furthermore, comparing the sixth frequencies of the beam, we can see that the CBT solutions are less accurate in the higher modes. The maximum difference (in the sixth frequency parameters) between the SDBT$_w$ and CBT$_w$ solutions can be 1.67 % for the worst case.

3.4.2 Effects of Shear Deformation and Rotary Inertia

In this section, effects of the shear deformation and rotary inertia which are neglected in the CBT will be investigated. Shear deformation was first applied in the analysis of beams by Timoshenko (1921). This effect is higher in composite materials since the longitudinal to shear modulus ratio is much higher in composites than metallic materials (Hajianmaleki and Qatu 2012b). Figures 3.5 and 3.6 show the differences between the lowest three frequency parameters Ω obtained by CBT$_w$ and SDBT$_w$ for an isotropic curved beam and a two-layered, [0°/90°] laminated curved beam with different boundary conditions and thickness-to-span length ratios, respectively. Both F-F and C-C boundary conditions are shown in each figure. The geometric and material constants of the layers of the two beams are: $R/L_\theta = 1$, $E_1/E_2 = 1$(Fig. 3.5)

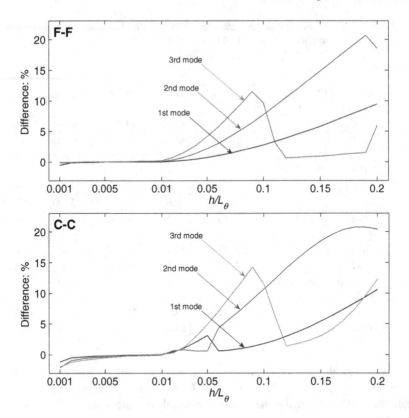

Fig. 3.5 Differences between the lowest three frequency parameters Ω obtained by CBT_w and $SDBT_w$ for an isotropic curved beam ($R/L_\theta = 1$, $E_1/E_2 = 1$)

and $E_1/E_2 = 15$ (Fig. 3.6). The thickness-to-span length ratio h/L_θ is varied from 0.001 to 0.2, corresponding to very thin to thick beams. From the figures, we can see that the effects of the shear deformation and rotary inertia increase as the orthotropy ratio (E_1/E_2) increases. Furthermore, when the thickness-to-span length ratio h/L_θ is less than 0.02, the maximum difference between the frequency parameters Ω obtained by CBT_w and $SDBT_w$ is less than 4 %. However, when the thickness-to-span length ratio h/L_θ is equal to 0.1, this difference can be as many as 11.5 % for the isotropic curved beam and 32.3 % for the [0°/90°] laminated one. As expected, it can be seen that the difference between the CBT_w and $SDBT_w$ solutions increases with thickness-to-span length ratio increases. Figure 3.6 also shows that the maximum difference between these two results can be as many as 50.4 % for a thickness-to-length ratio of 0.2. In such case, the CBT_w results are utterly inaccurate. This investigation shows that the CBT only applicable for thin beams. For beams with higher thickness ratios, both shear deformation and rotary inertia effects should be included in the calculation. These results can be used in establishing the limits of classical shell and plate theories as well.

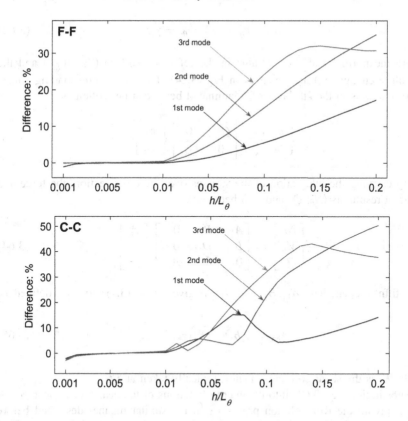

Fig. 3.6 Differences between the lowest three frequency parameters Ω obtained by CBT_w and $SDBT_w$ for a $[0°/90°]$ laminated curved beam $(R/L_\theta = 1, E_1/E_2 = 15)$

3.4.3 Effects of the Deepness Term $(1 + z/R)$

Considering Eq. (3.23), the deepness term $(1 + z/R)$ introduces curvature complexity in the kinematic relations. When the thickness of the beam, h, is small compared to its radius of curvature R, i.e., $h/R < <1$ and $|z/R| < <1$, then deepness term $(1 + z/R)$ approximately equals to 1. In such case, the shear deformation shallow beam theories (SDSBT) can be obtained from the general SDBT (Khdeir and Reddy 1997; Qatu 1992). Qatu (2004) pointed out that this term should not be neglected in the analysis especially when the span length-to-radius ratio is more than 1/2. In this section, effects of the deepness term $(1 + z/R)$ will be investigated.

Neglecting the deepness term and including the effects of shear deformation and rotary inertia, the normal and shear strains at any point of a beam can be rewritten as:

$$\varepsilon_\theta = \varepsilon_\theta^0 + z\chi_\theta \quad \gamma_{\theta z} = \gamma_{\theta z}^0 \tag{3.62}$$

where the normal and shear strains in the reference surface (ε_θ^0 and $\gamma_{\theta z}^0$) and the curvature change (χ_θ) are given as in Eq. (3.24). Thus, the corresponding stress-strain relations in the kth layer of a laminated beam can be written as:

$$\left\{ \begin{array}{c} \sigma_\theta \\ \tau_{\theta z} \end{array} \right\}_k = \left[\begin{array}{cc} \overline{Q_{11}^k} & 0 \\ 0 & \overline{Q_{55}^k} \end{array} \right] \left\{ \begin{array}{c} \varepsilon_\theta \\ \gamma_{\theta z} \end{array} \right\}_k \tag{3.63}$$

By carrying the integration of the stresses over the cross-section, the force and moment resultants (N_θ, Q_θ and M_θ) become:

$$\left[\begin{array}{c} N_\theta \\ M_\theta \\ Q_\theta \end{array} \right] = \left[\begin{array}{ccc} A_{11} & B_{11} & 0 \\ B_{11} & D_{11} & 0 \\ 0 & 0 & A_{55} \end{array} \right] \left[\begin{array}{c} \varepsilon_\theta^0 \\ \chi_\theta \\ \gamma_{\theta z} \end{array} \right] \tag{3.64}$$

The stiffness coefficients A_{11}, B_{11} and D_{11} are given in Eq. (3.7) and A_{55} is defined as:

$$A_{55} = K_s b \sum_{k=1}^{N} \overline{Q_{11}^k}(z_{k+1} - z_k) \tag{3.65}$$

where K_s is the shear correction factor, typically taken at 5/6.

Substituting Eq. (3.64) into the energy functions of the beams (see, Sect. 3.2.3) and applying the Ritz solution procedure in the similar manner described before (see, Sect. 3.3.2), the vibration solutions of laminated beams in the framework of SDSBT can be obtained (represent by SDSBT$_w$). It should be stressed that the inertia terms of the beam in the SDSBT are defined as:

$$\overline{I_0} = I_0$$
$$\overline{I_1} = I_1$$
$$\overline{I_2} = I_2 \tag{3.66}$$
$$[I_0, I_1, I_2] = b \sum_{k=1}^{N} \int_{z_k}^{z_{k+1}} \rho^k [1, z, z^2] dz$$

Figures 3.7 and 3.8 show the differences between the lowest three non-dimensional frequency parameters Ω obtained by SDBT$_w$ and SDSBT$_w$ for a two layered, unsymmetrically laminated [0°/90°] curved beam and an isotropic curved beam with different span length-to-radius and thickness-to-radius ratios, respectively. The 'difference' is defined as: difference = (SDSBT$_w$ − SDBT$_w$)/SDSBT$_w$*100 %. The geometric and material constants of the layers of the beam are: $R = 1$, $E_1/E_2 = 1$ or

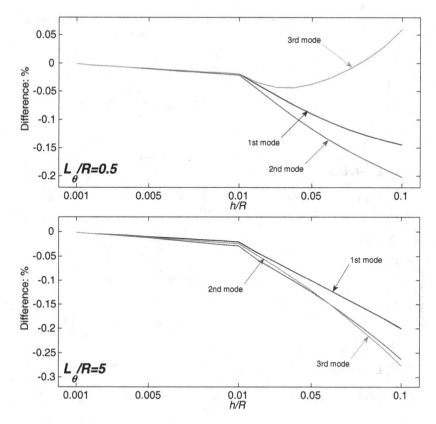

Fig. 3.7 Differences between the lowest three frequency parameters Ω obtained by SDBT$_w$ and SDSBT$_w$ for a $[0°/90°]$ laminated curved beam ($R = 1$, $E_1/E_2 = 15$)

15 (laminated). The beams are assumed to be C-F supported. Two span length-to-radius ratios, i.e., $L_\theta/R = 0.5$ and 5, corresponding to shallow and deep curved beams are shown in each figure. The thickness-to-radius ratio h/R is varied from 0.001 to 0.1. As expected, the effects of the deepness term increase as the thickness-to-radius ratio increases. The included angle θ_0 of the curved beam is the span length-to-radius ratio. This means that the span length-to-radius ratio of 5 indicates a very deep curved beam with an included angle of 286.48°, which are more than three quarters of the closed circle. In this case, the difference between the frequency parameters Ω obtained by SDSBT$_w$ and SDBT$_w$ is very small and the maximum difference is less than 0.41 % for the worst case. In addition, it is clear from the figures that the effects of the deepness term vary with mode number and span length-to-radius ratio and orthotropy ratio (E_1/E_2). In the case of span length-to-radius ratio $L_\theta/R = 5$, the SDBT$_w$ results are generally higher than those of SDSBT$_w$. These results can be used in establishing the limits of shell theories and shell equations.

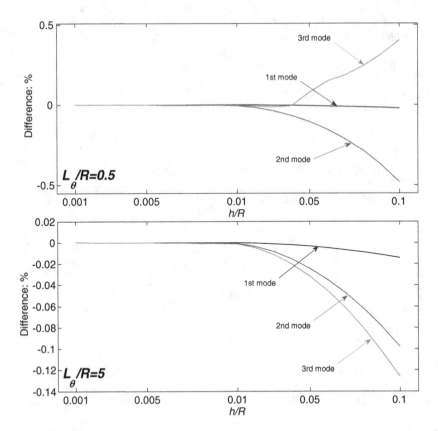

Fig. 3.8 Differences between the lowest three frequency parameters Ω obtained by $SDBT_w$ and $SDSBT_w$ for an isotropic curved beam ($R = 1$, $E_1/E_2 = 1$)

3.4.4 Isotropic and Laminated Beams with General Boundary Conditions

In this section, isotropic and laminated beams with various boundary conditions including the classical restraints and the elastic ones will be studied. Only the free vibration solutions (natural frequencies and mode shapes) based on SDBT and weak form solution procedure are considered in this section.

Table 3.6 shows the frequency parameters Ω obtained for a two-layered, [0°/90°] laminated curved beam with different thickness-to-radius ratios and various classical boundary conditions. The geometry and material parameters used in the study are: $R = 1$, $\theta_0 = 2\pi/3$, $E_1/E_2 = 15$. And the thickness-to-radius ratios performed in the study are $h/R = 0.01$, 0.05 and 0.1, corresponding to thin to moderately thick curved beams. The first observation is that the frequency parameters for curved beams with F-SD boundary conditions (no constraints on the in-plane displacement) are higher than those of curved beam with F-S boundary conditions.

Table 3.6 Frequency parameters Ω for a [0°/90°] laminated curved beam with different thickness-to-radius ratios and boundary conditions ($R = 1$, $\theta_0 = 2\pi/3$)

h/R	Mode	Boundary conditions							
		F-F	F-S	F-SD	F-C	S-S	SD-SD	S-C	C-C
0.01	1	9.5195	5.1928	5.6583	1.8539	14.614	2.1861	19.448	25.039
	2	27.392	20.523	21.455	7.7666	37.008	15.969	43.325	49.995
	3	55.452	46.358	47.226	25.569	71.295	39.538	80.689	90.812
	4	93.083	81.757	82.608	53.771	113.13	72.651	124.27	135.87
	5	140.16	126.70	127.44	91.461	166.27	115.27	180.51	195.52
0.05	1	9.5681	5.2531	5.6975	1.8445	14.878	2.2102	19.435	24.690
	2	27.523	20.756	21.581	7.7620	37.546	16.093	42.965	48.220
	3	55.053	46.447	47.003	25.389	70.686	39.438	78.179	86.441
	4	90.949	80.485	80.920	52.785	110.31	71.371	117.75	124.06
	5	134.28	122.32	122.47	87.966	157.68	111.09	165.51	176.82
0.1	1	9.6198	5.3307	5.7443	1.8487	14.964	2.2341	18.948	23.492
	2	27.102	20.648	21.357	7.7249	36.701	16.008	40.390	43.401
	3	52.379	44.743	45.037	24.451	65.996	38.040	70.061	75.703
	4	83.207	74.347	74.565	49.195	92.469	66.257	94.377	94.698
	5	117.69	108.20	108.23	78.109	99.117	98.930	108.30	121.93

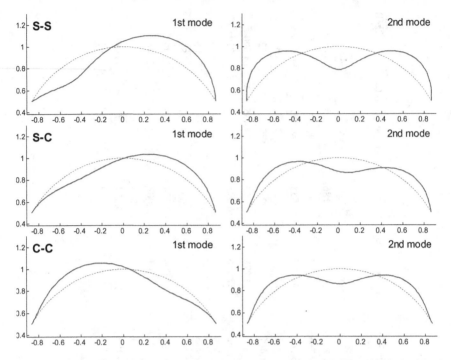

Fig. 3.9 The lowest three mode shapes for a moderately *thick* ($h/R = 0.05$), $[0°/90°]$ laminated curved beam with different boundary conditions ($R = 1$, $\theta_0 = 2\pi/3$)

The second observation is that the fundamental frequency parameter of the beam increases with thickness-to-radius ratio increases when subjected to F-F, F-S, F-SD, S-S and SD-SD boundary conditions. In other cases, the beam frequency parameters decrease with thickness-to-radius ratio increases. In Fig. 3.9, the lowest two mode shapes for the beam with thickness-to-radius ratio $h/R = 0.05$ are presented for S-S, S-C and C-C boundary conditions, respectively. These mode shapes are determined by substituting the related eigenvectors into the assumed displacement expansions.

Table 3.7 studies the effects of the orthotropy ratio (E_1/E_2) on the frequency parameters Ω for a two-layered, $[0°/90°]$ laminated curved beam with different boundary conditions. The geometry parameters used in the investigation are: $R = 1$, $\theta_0 = 2\pi/3$, $h/R = 0.05$. Four different orthotropy ratios considered in the investigation are: $E_1/E_2 = 1, 10, 20$ and 40. From the table, we can see that the increment in the orthotropy ratio results in decreases of the frequency parameters.

Since the effects of both shear deformation and rotary inertia and the deepness term are included in the SDBT, therefore, it can be applied to predict vibration characteristics of moderately thick curved beams with arbitrary included angles. Table 3.8 shows the frequency parameters Ω obtained for a two-layered, $[0°/90°]$

Table 3.7 Frequency parameters Ω for a [0°/90°] laminated curved beam with different orthotropy ratios and boundary conditions ($R = 1$, $\theta_0 = 2\pi/3$, $h/R = 0.05$)

E_1/E_2	Mode	Boundary conditions							
		F-F	F-S	F-SD	F-C	S-S	SD-SD	S-C	C-C
1	1	20.400	11.119	12.142	3.969	31.287	4.7104	41.488	53.239
	2	58.737	43.932	46.030	16.555	78.374	34.295	91.007	103.85
	3	118.35	98.751	100.92	54.567	150.95	84.576	169.72	189.98
	4	197.48	172.94	175.47	113.83	231.46	154.54	245.55	254.89
	5	295.26	265.45	268.86	191.89	310.10	243.48	315.43	328.61
10	1	10.504	5.7639	6.2542	2.0255	16.332	2.4261	21.377	27.203
	2	30.248	22.797	23.714	8.527	41.263	17.681	47.317	53.219
	3	60.628	51.111	51.750	27.941	77.959	43.411	86.432	95.797
	4	100.42	88.810	89.326	58.211	121.95	78.764	130.29	137.23
	5	148.74	135.41	135.62	97.288	175.12	122.98	183.35	197.31
20	1	9.0415	4.9652	5.3841	1.7430	14.052	2.0888	18.327	23.246
	2	25.977	19.597	20.372	7.329	35.409	15.194	40.445	45.305
	3	51.862	43.773	44.287	23.931	66.471	37.166	73.360	80.937
	4	85.465	75.661	76.056	49.652	103.46	67.098	110.27	116.08
	5	125.82	114.65	114.78	82.521	147.35	104.14	154.61	164.50
40	1	8.1490	4.4770	4.8532	1.5721	12.623	1.8833	16.375	20.657
	2	23.301	17.591	18.282	6.584	31.598	13.642	35.852	39.892
	3	46.183	39.014	39.462	21.360	58.749	33.140	64.359	70.464
	4	75.416	66.816	67.161	43.956	90.500	59.296	95.798	100.33
	5	109.88	100.19	100.31	72.335	127.43	91.082	133.01	140.01

Table 3.8 Frequency parameters Ω for a $[0°/90°]$ laminated curved beam with different included angles and boundary conditions ($R = 1$, $h/R = 0.05$)

θ_0	Mode	Boundary conditions								
		F-F	F-S	F-SD	F-C	S-S	SD-SD	S-C	C-C
$\pi/2$	1	10.005	5.9054	6.3409	1.7722	16.561	3.2096	20.938	26.105
	2	28.178	21.953	22.468	8.586	38.863	17.248	43.534	47.549
	3	55.053	47.152	47.364	26.473	70.658	40.139	76.711	84.959
	4	89.297	79.705	79.831	53.088	106.03	70.809	106.34	106.36
	5	129.63	118.68	118.70	86.151	108.37	108.12	121.43	136.93
π	1	8.7706	4.3969	4.5494	2.0633	10.931	12.841	15.543	20.777
	2	25.365	17.744	18.975	6.569	33.468	36.216	39.449	45.584
	3	52.896	43.201	44.337	22.381	67.002	69.019	75.214	83.912
	4	89.877	78.130	79.207	49.875	108.64	110.69	118.03	127.44
	5	135.61	122.09	122.91	86.475	159.48	160.71	170.57	182.12
$3\pi/2$	1	8.2428	3.9844	4.1003	2.6458	5.1157	3.2921	9.8808	14.992
	2	21.531	13.115	14.420	5.8982	25.651	6.6963	32.026	38.680
	3	47.428	36.490	38.078	17.548	57.966	28.905	66.578	75.602
	4	84.055	70.815	72.626	43.413	100.26	61.559	110.72	121.60
	5	130.50	115.27	116.99	79.703	152.25	103.92	164.70	177.47

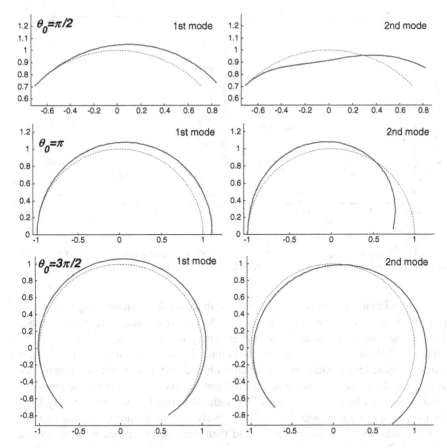

Fig. 3.10 The lowest two mode shapes for a [0°/90°] laminated curved beam with F-C boundary conditions and different included angles ($R = 1$, $h/R = 0.05$)

laminated curved beam with different included angles and various combinations of classical boundary conditions. The geometry and material parameters of the beam are assumed to be: $R = 1$, $h/R = 0.05$, $E_1/E_2 = 15$. It can be seen from the table that in all the boundary condition cases, all the frequency parameters of the beam decrease with included angle increases. It may be attributed to the stiffness of the curved beam reduces when the included angle increases. In order to enhance our understanding of the effects of the included angle, Fig. 3.10 presents the lowest two mode shapes of the beam with F-C boundary conditions. From the figure, we can see that the influence of the included angle on the mode shapes of the beam varies with mode sequence.

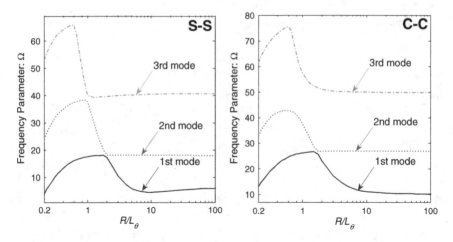

Fig. 3.11 Variation of frequency parameters Ω versus radius-to-span length ratio (R/L_θ) for a [0°/90] laminated curved beam with S-S and C-C boundary conditions ($h/L_\theta = 0.05$)

Then, influence of the radius-to-span length ratio (R/L_θ) on the frequencies of a [0°/90°] laminated curved beam ($E_1/E_2 = 15$) is investigated. The beam is made from composite layers with following geometry constants: $L_\theta = 1$, $h/L_\theta = 0.05$. Figure 3.11 shows variations of the lowest three frequency parameters Ω versus the radius-to-span length ratio for the beam with S-S and C-C boundary conditions. Increasing the radius-to-span length ratio from 0.2 (very deep curved beam) to 10 (shallow curved beam), the effects of the radius-to-span length ratio are very large for the frequency parameters. We can see clearly that the frequency parameter trace of the fundamental modes climb up and then decline, and reach its crest around $R/L_\theta = 2$. The same tendency can be seen in the second and third modes as well while the second and third ones reach their crests around $R/L_\theta = 1$ and $R/L_\theta = 0.8$. The similar characteristics can be found in the right subfigure. When the radius-to-span length ratio is increased from 10 to 100, increasing the radius-to-span length ratio has very limited influence on the frequency parameters due to the curved beam is very shallow and its vibration behaviors approximate to a straight beam.

Finally, Table 3.9 shows the first three frequency parameters Ω for a [−45°/45°] laminated curved beam ($L_\theta = 1$, $h/L_\theta = 0.05$, $E_1/E_2 = 15$) with different radius-to-span length ratios and elastic boundary conditions. The radius-to-span length ratios included in the calculation are $1/\pi$, $4/\pi$ and ∞. The radius-to-span length ratio of $1/\pi$ and $4/\pi$ correspond to curved beam with an included angle of π and $\pi/4$, respectively. The last radius-to-span length ratio of ∞ corresponds to a straight

Table 3.9 The first three frequency parameters Ω for a $[-45°/45°]$ laminated curved beam with different radius-to-span length ratios and S-Elastic boundary conditions ($L_\theta = 1$, $h/L_\theta = 0.05$)

R/L_θ	Γ	$k^u_{\theta1}=\Gamma, k^w_{\theta1}=10^7 D, K^\theta_{\theta1}=0$			$k^u_{\theta1}=0, k^w_{\theta1}=10^7 D, K^\theta_{\theta1}=\Gamma$		
		1	2	3	1	2	3
$1/\pi$	$10^{-1}D$	0.4465	12.851	36.277	0.6846	12.974	36.380
	$10^0 D$	1.4104	12.863	36.280	1.8096	13.816	37.125
	$10^1 D$	4.4087	12.992	36.314	2.8424	15.779	39.292
	$10^2 D$	10.805	16.473	36.709	3.0734	16.507	40.291
	$10^3 D$	11.828	32.677	49.626	3.0996	16.600	40.426
	$10^4 D$	11.887	34.153	64.561	3.1022	16.610	40.440
$4/\pi$	$10^{-1}D$	4.6699	20.169	43.834	4.7857	20.328	43.957
	$10^0 D$	4.7105	20.169	43.835	5.4645	21.404	44.860
	$10^1 D$	5.0975	20.169	43.844	6.5092	23.900	47.523
	$10^2 D$	7.8777	20.170	43.932	6.7873	24.804	48.738
	$10^3 D$	17.166	20.189	44.503	6.8198	24.918	48.902
	$10^4 D$	20.147	25.119	45.374	6.8231	24.929	48.918
∞ (straight)	$10^{-1}D$	5.4235	20.949	44.711	5.5870	21.101	44.843
	$10^0 D$	5.4235	20.949	44.711	6.5033	22.146	45.818
	$10^1 D$	5.4235	20.949	44.711	7.8962	24.625	48.748
	$10^2 D$	5.4234	20.949	44.711	8.2619	25.535	50.111
	$10^3 D$	5.4236	20.949	44.711	8.3044	25.649	50.295
	$10^4 D$	5.4234	20.949	44.711	8.3087	25.661	50.314

beam. The beam is simply-supported at the edge of $\theta = 0$ and elastically restrained at the other edge (S-Elastic). Table 3.10 shows similar studies for the C-Elastic boundary conditions. The tables show that increasing the axial restrained rigidity has very limited effects on the straight beam. It is attributed to the lower frequency parameters of a straight beam are dominated by its transverse vibration.

In conclusion, vibration of isotropic and laminated straight and curved beams is studied in this chapter. The effects of shear deformation and rotary inertia, deepness term $(1 + z/R)$ and other beam parameters are clearly outlined. It is shown that the difference (for the lowest three frequency parameters) between the CBT and SDBT solutions increases with thickness-to-length ratio increases. For a laminated curved beam ($R/L_\theta = 1$, $E_1/E_2 = 15$, $[0°/90°]$ lamination scheme) with thickness-to-length ratios of 0.05, 0.1, 0.15 and 0.2, the maximum differences can be as many as 8.8, 32.3, 42 and 50.4 % for the worst case, respectively. The effects of the deepness term $(1 + z/R)$ on the frequency parameters is very small, the maximum difference between the frequency parameters Ω obtained by SDSBT (neglect the deepness term) and SDBT is less than 0.41 % for the worse case of a curved beam of span length-to-radius ratio $L_\theta/R = 5$ and thickness-to-radius ratio $h/R = 0.1$. A variety of

Table 3.10 The first three frequency parameters Ω for a $[-45°/45°]$ laminated curved beam with different radius-to-span length ratios and C-Elastic boundary conditions ($L_\theta = 1$, $h/L_\theta = 0.05$)

R/L_θ	Spring rigidity	$k^u_{\theta 1}=\Gamma, k^w_{\theta 1}=10^7D, K^\theta_{\theta 1}=0$			$k^u_{\theta 1}=0, k^w_{\theta 1}=10^7D, K^\theta_{\theta 1}=\Gamma$		
		1	2	3	1	2	3
$1/\pi$	$10^{-1}D$	3.0784	17.040	40.988	3.0752	17.181	41.080
	10^0D	3.3805	17.045	40.991	3.2670	18.139	41.756
	10^1D	5.5452	17.094	41.022	3.5970	20.410	43.784
	10^2D	13.775	18.337	41.374	3.6929	21.272	44.749
	10^3D	16.321	35.923	51.778	3.7044	21.382	44.882
	10^4D	16.397	38.427	69.273	3.7055	21.394	44.895
$4/\pi$	$10^{-1}D$	7.5129	24.694	48.910	7.6416	24.848	49.019
	10^0D	7.5373	24.694	48.912	8.4238	25.915	49.821
	10^1D	7.7754	24.695	48.929	9.7649	28.553	52.224
	10^2D	9.7491	24.703	49.084	10.149	29.564	53.331
	10^3D	17.734	24.797	49.989	10.195	29.693	53.480
	10^4D	23.782	26.081	51.044	10.200	29.706	53.495
∞ (straight)	$10^{-1}D$	8.3092	25.662	50.316	8.4796	25.807	50.442
	10^0D	8.3092	25.662	50.316	9.4924	26.822	51.368
	10^1D	8.3092	25.662	50.316	11.219	29.395	54.250
	10^2D	8.3092	25.662	50.316	11.710	30.398	55.641
	10^3D	8.3092	25.662	50.316	11.768	30.526	55.832
	10^4D	8.3092	25.662	50.316	11.774	30.539	55.851

new vibration results including frequencies and mode shapes for straight and curved laminated beams with classical and elastic restraints as well as different geometric and material parameters are given, which may serve as benchmark solutions for the future researches.

Chapter 4
Plates

Plates are one of the most fundamental structural elements which are widely used in a variety of engineering applications. A plate can be defined as a solid body bounded by two parallel flat surfaces having two dimensions relatively greater than the other one (thickness). It can also be viewed as a special case of shells with zero curvature (infinite radii of curvature). The vibration of plates is an old topic and a lot of books, papers and reports have already been published in the past decades. In 1969, A.W. Leissa published an excellent monograph titled *Vibration of Plates*, in which theoretical and experimental results of approximately 500 research papers and reports were presented. Most 90 % of this book considers the homogeneous thin plates. A plate is typically considered to be thin when the ratio of its thickness to representative lateral dimension is less than 1/20 (Leissa 1969; Qatu 2004). As pointed by Leissa (Liew et al. 1998), the classical thin plate theory (CPT) permits one to obtain a fundamental frequency with good accuracy. However, the higher frequencies of a plate with thickness ratio of 1/20, determined by thin plate theory, will not be accurate. It will be somewhat too high. The inaccuracies can be largely eliminated by use of the shear deformation plate theories (SDPTs) for they include both shear deformation and rotary inertia due to rotations. Liew et al. (1998) presented a book deal with thick isotropic plates using the *p*-version Ritz method.

Plates can be rectangular, circular, annular, sectorial, elliptical, triangular, trapezoidal and of other shapes. Chakraverty (2009) analyzed vibrations of plates of various shapes and classical boundary conditions by using the boundary characteristic orthogonal polynomials with the Ritz method. The homogeneous thin plates are the subject of Chakraverty's work as well. With the development of new industries and modern processes, laminated plates composed of composite laminas are extensively used in many fields of modern engineering practices such as space vehicles, civil constructions and deep-sea engineering equipments. Laminated plates are treated by various studies and books. The development of research on this subject has been well documented in several monographs respectively by Qatu (2004), Reddy (2003), Carrera et al. (2011), Ye (2003), and review or survey articles (Carrera 2002, 2003; Liew et al. 2011).

© Science Press, Beijing and Springer-Verlag Berlin Heidelberg 2015
G. Jin et al., *Structural Vibration*, DOI 10.1007/978-3-662-46364-2_4

This chapter considers vibrations of laminated plates with various shapes and general boundary conditions. The fundamental equations of shells in the framework of CST and SDST described in Chap. 1 will be specialized to those of plates by setting the curvatures to zero. We will begin with the fundamental equations of rectangular plates, followed by vibration results of laminated rectangular plates with general boundary conditions. Strain-displacement relations, force and moment resultants, energy functions, governing equations and boundary conditions are derived and shown for both theories. On the basis of SDPT, numerous natural frequencies and mode shapes are presented for laminated rectangular plates with different boundary conditions, lamination schemes and geometry parameters by using the modified Fourier series and weak form solution procedure (see, Chap. 2) because previous studies showed that convergence of solutions with weak form solution procedure is faster (Table 3.3). Effects of boundary conditions, geometry parameters and material properties are studied as well. Vibration of sectorial, annular and circular plates will then be treated in the later sections of this chapter.

4.1 Fundamental Equations of Thin Laminated Rectangular Plates

As shown in Fig. 4.1, a rectangular laminated plate with length a, width b and total thickness of h is selected as the analysis model. To describe the plate clearly, we introduce the following coordinate system: the x-coordinate is taken along the length of the plate, and y- and z-coordinates are taken along the width and the thickness directions, respectively. The middle surface displacements of the plate in the x, y and z directions are denoted by u, v and w, respectively. The laminated rectangular plate is assumed to be composed of N_L composite layers. Consider the laminated rectangular plate and its rectangular coordinate system in Fig. 4.1, the coordinates, characteristics of the Lamé parameters and radii of curvatures are: $\alpha = x$, $\beta = y$, $A = 1$, $B = 1$, $R_\alpha = \infty$, $R_\beta = \infty$.

Fundamental equations of thin laminated rectangular plates are presented in this section by substituting their geometry parameters into those of general thin laminated shell equations.

4.1.1 Kinematic Relations

Based on the assumptions of Kirchhoff, the displacement field of thin rectangular plates is restricted to the following linear relationships (see Fig. 4.2, Reddy (2003)):

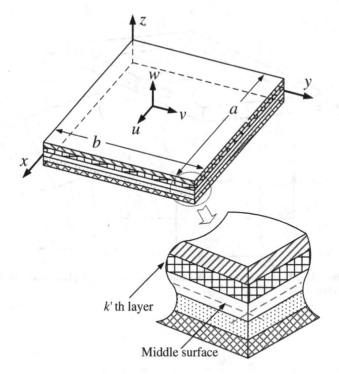

Fig. 4.1 Laminated rectangular plates

$$U(x, y, z) = u(x, y) - z\frac{\partial w(x, y)}{\partial x}$$

$$V(x, y, z) = v(x, y) - z\frac{\partial w(x, y)}{\partial y} \tag{4.1}$$

$$W(x, y, z) = w(x, y)$$

where u, v and w are the middle surface displacements. Letting $\alpha = x$, $\beta = y$, $A = B = 1$ and $R_\alpha = R_\beta = \infty$, the strain-displacement relations of rectangular plates can be obtained from Eqs. (1.6) and (1.7) as

$$\varepsilon_x = \varepsilon_x^0 + z\chi_x$$

$$\varepsilon_y = \varepsilon_y^0 + z\chi_y \tag{4.2}$$

$$\gamma_{xy} = \gamma_{xy}^0 + z\chi_{xy}$$

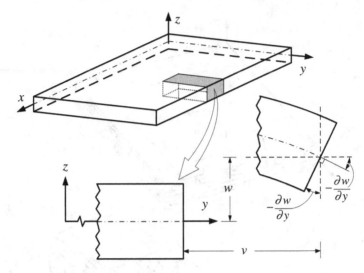

Fig. 4.2 Undeformed and deformed geometries of an edge of a plate under the Kirchhoff assumption (Reddy 2003)

where

$$
\begin{aligned}
\varepsilon_x^0 &= \frac{\partial u}{\partial x}, & \chi_x &= -\frac{\partial^2 w}{\partial x^2} \\
\varepsilon_y^0 &= \frac{\partial v}{\partial y}, & \chi_y &= -\frac{\partial^2 w}{\partial y^2} \\
\gamma_{xy}^0 &= \frac{\partial v}{\partial x} + \frac{\partial u}{\partial y}, & \chi_{xy} &= -2\frac{\partial^2 w}{\partial x \partial y}
\end{aligned}
\tag{4.3}
$$

4.1.2 Stress-Strain Relations and Stress Resultants

According to Hooke's law, the corresponding stresses in the k'th layer of the plate are written as:

$$
\left\{ \begin{array}{c} \sigma_x \\ \sigma_y \\ \tau_{yz} \end{array} \right\}_k =
\begin{bmatrix} \overline{Q}_{11}^k & \overline{Q}_{12}^k & \overline{Q}_{16}^k \\ \overline{Q}_{12}^k & \overline{Q}_{22}^k & \overline{Q}_{26}^k \\ \overline{Q}_{16}^k & \overline{Q}_{26}^k & \overline{Q}_{66}^k \end{bmatrix}
\left\{ \begin{array}{c} \varepsilon_x \\ \varepsilon_y \\ \gamma_{xy} \end{array} \right\}_k
\tag{4.4}
$$

The lamina stiffness coefficients $\overline{Q}_{ij}^k(i, j = 1, 2, 6)$ can be written as in Eq. (1.9).

Then, the force and moment resultants are obtained by carrying the integration of stresses over the cross-section:

$$[N_x \quad N_y \quad N_{xy}] = \int_{-h/2}^{h/2} [\sigma_x \quad \sigma_y \quad \tau_{xy}]dz$$

$$[M_x \quad M_y \quad M_{xy}] = \int_{-h/2}^{h/2} [\sigma_x \quad \sigma_y \quad \tau_{xy}]zdz$$

(4.5)

Performing the integration operation in Eq. (4.5) yields

$$\begin{bmatrix} N_x \\ N_y \\ N_{xy} \\ M_x \\ M_y \\ M_{xy} \end{bmatrix} = \begin{bmatrix} A_{11} & A_{12} & A_{16} & B_{11} & B_{12} & B_{16} \\ A_{12} & A_{22} & A_{26} & B_{12} & B_{22} & B_{26} \\ A_{16} & A_{26} & A_{66} & B_{16} & B_{26} & B_{66} \\ B_{11} & B_{12} & B_{16} & D_{11} & D_{12} & D_{16} \\ B_{12} & B_{22} & B_{26} & D_{12} & D_{22} & D_{26} \\ B_{16} & B_{26} & B_{66} & D_{16} & D_{26} & D_{66} \end{bmatrix} \begin{bmatrix} \varepsilon_x^0 \\ \varepsilon_y^0 \\ \gamma_{xy}^0 \\ \chi_x \\ \chi_y \\ \chi_{xy} \end{bmatrix}$$

(4.6)

where N_x, N_y and N_{xy} are the normal and shear force resultants and M_x, M_y, M_{xy} denote the bending and twisting moment resultants. A_{ij}, B_{ij}, and D_{ij} are the stiffness coefficients arising from the piecewise integration over the plate thickness, they can be written as in Eq. (1.15).

In comparing to homogeneous thin plates, there exist some fundamental differences in the equations of generally thin laminated plates. The most important difference is that the transverse vibration is coupled with in-plane vibration for generally laminated plates. When a laminated plate is symmetrically laminated with respect to the middle surface, the constants B_{ij} equal to zero and the in-plane vibration is then decoupled from the transverse vibration, which will sufficiently reduce the complexity of the stress-strain relations, energy functions, governing equations and boundary condition equations of the plate.

4.1.3 Energy Functions

The strain energy (U_s) of thin laminated rectangular plates during vibration can be written as:

$$U_s = \frac{1}{2} \int_0^a \int_0^b \left\{ \begin{array}{l} N_x \varepsilon_x^0 + N_y \varepsilon_y^0 + N_{xy} \gamma_{xy}^0 \\ +M_x \chi_x + M_y \chi_y + M_{xy} \chi_{xy} \end{array} \right\} dxdy$$

(4.7)

Substituting Eqs. (4.3) and (4.6) into Eq. (4.7), the strain energy of the thin laminated plates can be rewritten in terms of displacements as:

$$
U_s = \frac{1}{2} \int_0^a \int_0^b \left\{ \begin{array}{l} A_{11}\left(\dfrac{\partial u}{\partial x}\right)^2 + 2A_{12}\left(\dfrac{\partial v}{\partial y}\right)\left(\dfrac{\partial u}{\partial x}\right) + 2A_{16}\left(\dfrac{\partial v}{\partial x} + \dfrac{\partial u}{\partial y}\right)\left(\dfrac{\partial u}{\partial x}\right) \\[3mm] + A_{22}\left(\dfrac{\partial v}{\partial y}\right)^2 + 2A_{26}\left(\dfrac{\partial v}{\partial x} + \dfrac{\partial u}{\partial y}\right)\left(\dfrac{\partial v}{\partial y}\right) + A_{66}\left(\dfrac{\partial v}{\partial x} + \dfrac{\partial u}{\partial y}\right)^2 \end{array} \right\} dy dx
$$

$$
- \int_0^a \int_0^b \left\{ \begin{array}{l} B_{11}\left(\dfrac{\partial^2 w}{\partial x^2}\right)\left(\dfrac{\partial u}{\partial x}\right) + B_{12}\left(\dfrac{\partial^2 w}{\partial y^2}\right)\left(\dfrac{\partial u}{\partial x}\right) + 2B_{16}\left(\dfrac{\partial^2 w}{\partial x \partial y}\right)\left(\dfrac{\partial u}{\partial x}\right) \\[3mm] + B_{12}\left(\dfrac{\partial^2 w}{\partial x^2}\right)\left(\dfrac{\partial v}{\partial y}\right) + B_{22}\left(\dfrac{\partial^2 w}{\partial y^2}\right)\left(\dfrac{\partial v}{\partial y}\right) + 2B_{26}\left(\dfrac{\partial^2 w}{\partial x \partial y}\right)\left(\dfrac{\partial v}{\partial y}\right) \\[3mm] + \left(B_{16}\left(\dfrac{\partial^2 w}{\partial x^2}\right) + B_{26}\left(\dfrac{\partial^2 w}{\partial y^2}\right) + 2B_{66}\left(\dfrac{\partial^2 w}{\partial x \partial y}\right)\right)\left(\dfrac{\partial v}{\partial x} + \dfrac{\partial u}{\partial y}\right) \end{array} \right\} dy dx
$$

$$
+ \frac{1}{2} \int_0^a \int_0^b \left\{ \begin{array}{l} D_{11}\left(\dfrac{\partial^2 w}{\partial x^2}\right)^2 + 2D_{12}\left(\dfrac{\partial^2 w}{\partial y^2}\right)\left(\dfrac{\partial^2 w}{\partial x^2}\right) + 4D_{16}\left(\dfrac{\partial^2 w}{\partial x \partial y}\right)\left(\dfrac{\partial^2 w}{\partial x^2}\right) \\[3mm] + D_{22}\left(\dfrac{\partial^2 w}{\partial y^2}\right)^2 + 4D_{26}\left(\dfrac{\partial^2 w}{\partial x \partial y}\right)\left(\dfrac{\partial^2 w}{\partial y^2}\right) + 4D_{66}\left(\dfrac{\partial^2 w}{\partial x \partial y}\right)^2 \end{array} \right\} dy dx
$$

$$(4.8)$$

The corresponding kinetic energy (T) is:

$$
T = \frac{1}{2} \int_0^a \int_0^b I_0 \left\{ \left(\frac{\partial u}{\partial t}\right)^2 + \left(\frac{\partial v}{\partial t}\right)^2 + \left(\frac{\partial w}{\partial t}\right)^2 \right\} dx dy \qquad (4.9)
$$

where the inertia term I_0 is given as in Eq. (1.19). The external work (W_e) is expressed as:

$$
W_e = \int_0^a \int_0^b \{ q_x u + q_y v + q_z w \} dx dy \qquad (4.10)
$$

where q_x, q_y and q_z are the external loads in the x, y and z directions, respectively.

As described in Sect. 1.2.3, the general boundary conditions of a plate are implemented by using the artificial spring boundary technique. Specifically, symbols k_ψ^u, k_ψ^v, k_ψ^w and K_ψ^w ($\psi = x0, y0, x1$ and $y1$) are used to indicate the stiffness of the boundary springs at the boundaries $x = 0$, $y = 0$, $x = a$ and $y = b$, respectively. Therefore, the deformation strain energy about the boundary springs (U_{sp}) is defined as (Fig. 4.3):

Fig. 4.3 Boundary conditions of thin laminated rectangular plates

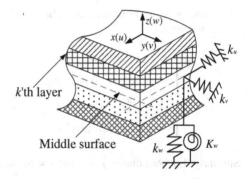

$$U_{sp} = \frac{1}{2} \int_0^b \left\{ \begin{array}{l} \left[k_{x0}^u u^2 + k_{x0}^v v^2 + k_{x0}^w w^2 + K_{x0}^w (\partial w / \partial x) \right] |_{x=0} \\ + \left[k_{x1}^u u^2 + k_{x1}^v v^2 + k_{x1}^w w^2 + K_{x1}^w (\partial w / \partial x) \right] |_{x=a} \end{array} \right\} dy$$

$$+ \frac{1}{2} \int_0^a \left\{ \begin{array}{l} \left[k_{y0}^u u^2 + k_{y0}^v v^2 + k_{y0}^w w^2 + K_{y0}^w (\partial w / \partial y) \right] |_{y=0} \\ + \left[k_{y1}^u u^2 + k_{y1}^v v^2 + k_{y1}^w w^2 + K_{y1}^w (\partial w / \partial y) \right] |_{y=b} \end{array} \right\} dx$$

(4.11)

4.1.4 Governing Equations and Boundary Conditions

The governing equations and boundary conditions of thin rectangular plates can be obtained by specializing those of thin shells or applying the Hamilton's principle in the same manner as described in Sect. 1.2.4. Substituting $\alpha = x$, $\beta = y$, $A = B = 1$ and $R_\alpha = R_\beta = \infty$ into Eq. (1.28) yields following governing equations:

$$\frac{\partial N_x}{\partial x} + \frac{\partial N_{xy}}{\partial y} + q_x = I_0 \frac{\partial^2 u}{\partial t^2}$$

$$\frac{\partial N_{xy}}{\partial x} + \frac{\partial N_y}{\partial y} + q_y = I_0 \frac{\partial^2 v}{\partial t^2}$$

$$\frac{\partial^2 M_x}{\partial x^2} + 2 \frac{\partial^2 M_{xy}}{\partial x \partial y} + \frac{\partial^2 M_y}{\partial y^2} + q_z = I_0 \frac{\partial^2 w}{\partial t^2}$$

(4.12)

The corresponding boundary conditions of thin laminated plates at boundaries of $x =$ constant are:

$$
x = 0 : \begin{cases} N_x - k_{x0}^u u = 0 \\ N_{xy} - k_{x0}^v v = 0 \\ Q_x + \frac{\partial M_{xy}}{\partial y} - k_{x0}^w w = 0 \\ -M_x - K_{x0}^w \frac{\partial w}{\partial x} = 0 \end{cases} \qquad x = a : \begin{cases} N_x + k_{x1}^u u = 0 \\ N_{xy} + k_{x1}^v v = 0 \\ Q_x + \frac{\partial M_{xy}}{\partial y} + k_{x1}^w w = 0 \\ -M_x + K_{x1}^w \frac{\partial w}{\partial x} = 0 \end{cases} \quad (4.13)
$$

Similarly, for boundaries $y = 0$ and $y = b$, the boundary conditions are obtained as:

$$
y = 0 : \begin{cases} N_{xy} - k_{y0}^u u = 0 \\ N_y - k_{y0}^v v = 0 \\ Q_y + \frac{\partial M_{xy}}{\partial x} - k_{y0}^w w = 0 \\ -M_y - K_{y0}^w \frac{\partial w}{\partial y} = 0 \end{cases} \qquad y = b : \begin{cases} N_{xy} + k_{y1}^u u = 0 \\ N_y + k_{y1}^v v = 0 \\ Q_y + \frac{\partial M_{xy}}{\partial x} + k_{y1}^w w = 0 \\ -M_y + K_{y1}^w \frac{\partial w}{\partial y} = 0 \end{cases} \quad (4.14)
$$

For generally laminated thin plates, each boundary can exist 12 possible classical boundary conditions. Taking boundaries $x =$ constant for example, the possible combinations for each classical boundary condition are given in Table 4.1.

Table 4.1 Possible classical boundary conditions for generally laminated thin rectangular plates at each boundary of $x =$ constant

Boundary type	Conditions
Free boundary conditions	
F	$N_x = N_{xy} = Q_x + \dfrac{\partial M_{xy}}{\partial y} = M_x = 0$
F2	$u = N_{xy} = Q_x + \dfrac{\partial M_{xy}}{\partial y} = M_x = 0$
F3	$N_x = v = Q_x + \dfrac{\partial M_{xy}}{\partial y} = M_x = 0$
F4	$u = v = Q_x + \dfrac{\partial M_{xy}}{\partial y} = M_x = 0$
Simply supported boundary conditions	
S	$u = v = w = M_x = 0$
SD	$N_x = v = w = M_x = 0$
S3	$u = N_{xy} = w = M_x = 0$
S4	$N_x = N_{xy} = w = M_x = 0$
Clamped boundary conditions	
C	$u = v = w = \dfrac{\partial w}{\partial x} = 0$
C2	$N_x = v = w = \dfrac{\partial w}{\partial x} = 0$
C3	$u = N_{xy} = w = \dfrac{\partial w}{\partial x} = 0$
C4	$N_x = N_{xy} = w = \dfrac{\partial w}{\partial x} = 0$

In the framework of artificial spring boundary technique, taking edge $x = 0$ for example, the frequently encountered boundary conditions F, SD, S and C can be readily realized by assigning the stiffness of the boundary springs at proper values as follows:

$$
\begin{aligned}
&\text{F: } k_{x0}^u = k_{x0}^v = k_{x0}^w = K_{x0}^w = 0 \\
&\text{SD: } k_{x0}^v = k_{x0}^w = 10^7 D, \ k_{x0}^u = K_{x0}^w = 0 \\
&\text{S: } k_{x0}^u = k_{x0}^v = k_{x0}^w = 10^7 D, \ K_{x0}^w = 0 \\
&\text{C: } k_{x0}^u = k_{x0}^v = k_{x0}^w = K_{x0}^w = 10^7 D
\end{aligned}
\tag{4.15}
$$

where $D = E_1 h^3/12(1-\mu_{12}\,\mu_{21})$ is the flexural stiffness of the plate.

4.2 Fundamental Equations of Thick Laminated Rectangular Plates

This section presents fundamental equations that can be used for thick laminated plates. The treatment that follows is a specialization of shear deformation shell theory (SDST) to laminated plates. For thick plates, the Kirchhoff hypothesis is relaxed by assuming that normals to the undeformed middle surface remain straight but do not normal to the deformed middle surface, see Fig. 4.4 (Reddy 2003). The following equations are referred to as shear deformation plate theory (SDPT).

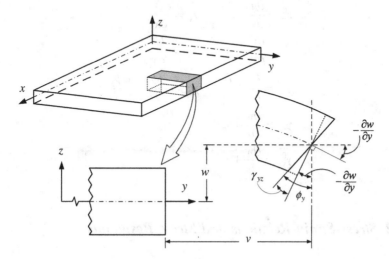

Fig. 4.4 Undeformed and deformed geometries of an edge of a plate including shear deformation

4.2.1 Kinematic Relations

Based on the plate model presented in Fig. 4.1 and the assumptions of the first-order shear deformation theory, the displacement field of thick laminated rectangular plates is of the form

$$
\begin{aligned}
U(x,y,z) &= u(x,y) + z\phi_x \\
V(x,y,z) &= v(x,y) + z\phi_y \\
W(x,y,z) &= w(x,y)
\end{aligned}
\tag{4.16}
$$

where u, v and w are the middle surface displacements of the plate in the x, y and z directions, respectively; ϕ_x and ϕ_y represent the transverse normal rotations of the reference surface respect to the y and x axes. Specializing Eqs. (1.33) and (1.34) to those of plates, the normal and shear strains at any point of the plate space can be defined in terms of the middle surface strains and curvature changes as:

$$
\begin{aligned}
\varepsilon_x &= \varepsilon_x^0 + z\chi_x, & \gamma_{xz} &= \gamma_{xz}^0 \\
\varepsilon_y &= \varepsilon_y^0 + z\chi_y, & \gamma_{yz} &= \gamma_{yz}^0 \\
\gamma_{xy} &= \gamma_{xy}^0 + z\chi_{xy}
\end{aligned}
\tag{4.17}
$$

where γ_{xz}^0 and γ_{yz}^0 indicate the transverse shear strains, it is assumed to be constant through the thickness. The middle surface strains and curvature changes are written in terms of middle surface displacements and rotation components as:

$$
\begin{aligned}
\varepsilon_x^0 &= \frac{\partial u}{\partial x}, & \chi_x &= \frac{\partial \phi_x}{\partial x} \\
\varepsilon_y^0 &= \frac{\partial v}{\partial y}, & \chi_y &= \frac{\partial \phi_y}{\partial y} \\
\gamma_{xy}^0 &= \frac{\partial v}{\partial x} + \frac{\partial u}{\partial y}, & \chi_{xy} &= \frac{\partial \phi_x}{\partial y} + \frac{\partial \phi_y}{\partial x} \\
\gamma_{xz}^0 &= \frac{\partial w}{\partial x} + \phi_x \\
\gamma_{yz}^0 &= \frac{\partial w}{\partial y} + \phi_y
\end{aligned}
\tag{4.18}
$$

4.2.2 Stress-Strain Relations and Stress Resultants

According to Eq. (1.35), the corresponding stress-strain relations in the layer k of a thick laminated rectangular plate are:

$$\begin{Bmatrix} \sigma_x \\ \sigma_y \\ \tau_{yz} \\ \tau_{xz} \\ \tau_{xy} \end{Bmatrix}_k = \begin{bmatrix} \overline{Q}_{11}^k & \overline{Q}_{12}^k & 0 & 0 & \overline{Q}_{16}^k \\ \overline{Q}_{12}^k & \overline{Q}_{22}^k & 0 & 0 & \overline{Q}_{26}^k \\ 0 & 0 & \overline{Q}_{44}^k & \overline{Q}_{45}^k & 0 \\ 0 & 0 & \overline{Q}_{45}^k & \overline{Q}_{55}^k & 0 \\ \overline{Q}_{16}^k & \overline{Q}_{26}^k & 0 & 0 & \overline{Q}_{66}^k \end{bmatrix} \begin{Bmatrix} \varepsilon_x \\ \varepsilon_y \\ \gamma_{yz} \\ \gamma_{xz} \\ \gamma_{xy} \end{Bmatrix}_k \tag{4.19}$$

The force and moment resultants are obtained by integrating the stresses over the plate thickness

$$[N_x \ \ N_y \ \ N_{xy} \ \ Q_x \ \ Q_y] = \int_{-h/2}^{h/2} [\sigma_x \ \ \sigma_y \ \ \tau_{xy} \ \ \tau_{xz} \ \ \tau_{yz}] dz$$

$$[M_x \ \ M_y \ \ M_{xy}] = \int_{-h/2}^{h/2} [\sigma_x \ \ \sigma_y \ \ \tau_{xy}] z dz \tag{4.20}$$

where Q_y and Q_x are the transverse shear force resultants. Carrying out the integration over the thickness, from layer to layer, yields

$$\begin{bmatrix} N_x \\ N_y \\ N_{xy} \\ M_x \\ M_y \\ M_{xy} \end{bmatrix} = \begin{bmatrix} A_{11} & A_{12} & A_{16} & B_{11} & B_{12} & B_{16} \\ A_{12} & A_{22} & A_{26} & B_{12} & B_{22} & B_{26} \\ A_{16} & A_{26} & A_{66} & B_{16} & B_{26} & B_{66} \\ B_{11} & B_{12} & B_{16} & D_{11} & D_{12} & D_{16} \\ B_{12} & B_{22} & B_{26} & D_{12} & D_{22} & D_{26} \\ B_{16} & B_{26} & B_{66} & D_{16} & D_{26} & D_{66} \end{bmatrix} \begin{bmatrix} \varepsilon_x^0 \\ \varepsilon_y^0 \\ \gamma_{xy}^0 \\ \chi_x \\ \chi_y \\ \chi_{xy} \end{bmatrix} \tag{4.21}$$

$$\begin{bmatrix} Q_y \\ Q_x \end{bmatrix} = \begin{bmatrix} A_{44} & A_{45} \\ A_{45} & A_{55} \end{bmatrix} \begin{bmatrix} \gamma_{yz}^0 \\ \gamma_{xz}^0 \end{bmatrix} \tag{4.22}$$

The stiffness coefficients A_{ij}, B_{ij} and D_{ij} are defined as in Eq. (1.43). When a thick rectangular plate is symmetrically laminated with respect to the middle surface, the constants B_{ij} equal to zero, however, the in-plane vibration will not be decoupled from the bending vibration due to the shear deformation.

4.2.3 Energy Functions

The strain energy (U_s) of thick laminated rectangular plates during vibration can be defined in terms of the middle surface strains and curvature changes and stress resultants as

$$U_s = \frac{1}{2} \int_0^a \int_0^b \left\{ \begin{array}{l} N_x \varepsilon_x^0 + N_y \varepsilon_y^0 + N_{xy} \varepsilon_{xy}^0 + M_x \chi_x \\ +M_y \chi_y + M_{xy} \chi_{xy} + Q_y \gamma_{yz} + Q_x \gamma_{xz} \end{array} \right\} dydx \qquad (4.23)$$

Substituting Eqs. (4.18) and (4.21) into Eq. (4.23), the strain energy of the laminated plate can be expressed in terms of displacements (u, v, w) and rotation components (ϕ_x, ϕ_y) as in Eq. (2.84).

The corresponding kinetic energy (T) function is:

$$T = \frac{1}{2} \int_0^a \int_0^b \left\{ \begin{array}{l} I_0 \left(\dfrac{\partial u}{\partial t}\right)^2 + 2I_1 \dfrac{\partial u}{\partial t}\dfrac{\partial \phi_x}{\partial t} + I_2 \left(\dfrac{\partial \phi_x}{\partial t}\right)^2 + I_0 \left(\dfrac{\partial v}{\partial t}\right)^2 \\ +2I_1 \dfrac{\partial v}{\partial t}\dfrac{\partial \phi_y}{\partial t} + I_2 \left(\dfrac{\partial \phi_y}{\partial t}\right)^2 + I_0 \left(\dfrac{\partial w}{\partial t}\right)^2 \end{array} \right\} dydx \qquad (4.24)$$

The inertia terms are written as in Eq. (1.52). Assuming the distributed external forces q_x, q_y and q_z are in the x, y and z directions, respectively and m_x and m_y represent the external couples in the middle surface, thus, the work done by the external forces and moments is

$$W_e = \int_0^a \int_0^b \left\{ q_x u + q_y v + q_z w + m_x \phi_x + m_y \phi_y \right\} dydx \qquad (4.25)$$

Using the artificial spring boundary technique similar to that described earlier, symbols $k_\psi^u, k_\psi^v, k_\psi^w, K_\psi^x$ and K_ψ^y ($\psi = x0, y0, x1$ and $y1$) are used to indicate the rigidities (per unit length) of the boundary springs at the boundaries $x = 0$, $y = 0$, $x = a$ and $y = b$, respectively, see Fig. 4.5. Therefore, the deformation strain energy (U_{sp}) stored in the boundary springs during vibration is defined as:

Fig. 4.5 Boundary conditions of a thick laminated rectangular plate

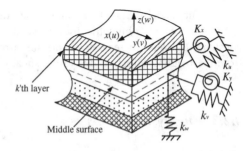

$$U_{sp} = \frac{1}{2} \int_0^b \left\{ \begin{array}{l} \left[k_{x0}^u u^2 + k_{x0}^v v^2 + k_{x0}^w w^2 + K_{x0}^x \phi_x^2 + K_{x0}^y \phi_y^2 \right]|_{x=0} \\ + \left[k_{x1}^u u^2 + k_{x1}^v v^2 + k_{x1}^w w^2 + K_{x1}^x \phi_x^2 + K_{x1}^y \phi_y^2 \right]|_{x=a} \end{array} \right\} dy$$
$$+ \frac{1}{2} \int_0^a \left\{ \begin{array}{l} \left[k_{y0}^u u^2 + k_{y0}^v v^2 + k_{y0}^w w^2 + K_{y0}^x \phi_x^2 + K_{y0}^y \phi_y^2 \right]|_{y=0} \\ + \left[k_{y1}^u u^2 + k_{y1}^v v^2 + k_{y1}^w w^2 + K_{y0}^x \phi_x^2 + K_{y0}^y \phi_y^2 \right]|_{y=b} \end{array} \right\} dx \qquad (4.26)$$

4.2.4 Governing Equations and Boundary Conditions

Specializing the governing equations and boundary conditions of thick general shells (Eq. (1.59)) to those of thick rectangular plates, we have

$$\frac{\partial N_x}{\partial x} + \frac{\partial N_{xy}}{\partial y} + q_x = I_0 \frac{\partial^2 u}{\partial t^2} + I_1 \frac{\partial^2 \phi_x}{\partial t^2}$$

$$\frac{\partial N_{xy}}{\partial x} + \frac{\partial N_y}{\partial y} + q_y = I_0 \frac{\partial^2 v}{\partial t^2} + I_1 \frac{\partial^2 \phi_y}{\partial t^2}$$

$$\frac{\partial Q_x}{\partial x} + \frac{\partial Q_y}{\partial y} + q_z = I_0 \frac{\partial^2 w}{\partial t^2} \qquad (4.27)$$

$$\frac{\partial M_x}{\partial x} + \frac{\partial M_{xy}}{\partial y} - Q_x + m_x = I_1 \frac{\partial^2 u}{\partial t^2} + I_2 \frac{\partial^2 \phi_x}{\partial t^2}$$

$$\frac{\partial M_{xy}}{\partial x} + \frac{\partial M_y}{\partial y} - Q_y + m_y = I_1 \frac{\partial^2 v}{\partial t^2} + I_2 \frac{\partial^2 \phi_y}{\partial t^2}$$

Substituting Eqs. (4.17), (4.18), (4.21) and (4.22) into above equations, the governing equations can be written in terms of displacements. These equations are proved useful when exact solutions are desired. These equations can be written as

$$\left(\begin{bmatrix} L_{11} & L_{12} & L_{13} & L_{14} & L_{15} \\ L_{21} & L_{22} & L_{23} & L_{24} & L_{25} \\ L_{31} & L_{32} & L_{33} & L_{34} & L_{35} \\ L_{41} & L_{42} & L_{43} & L_{44} & L_{45} \\ L_{51} & L_{52} & L_{53} & L_{54} & L_{55} \end{bmatrix} - \omega^2 \begin{bmatrix} M_{11} & 0 & 0 & M_{14} & 0 \\ 0 & M_{22} & 0 & 0 & M_{25} \\ 0 & 0 & M_{33} & 0 & 0 \\ M_{41} & 0 & 0 & M_{44} & 0 \\ 0 & M_{52} & 0 & 0 & M_{55} \end{bmatrix} \right) \begin{bmatrix} u \\ v \\ w \\ \phi_x \\ \phi_y \end{bmatrix} = \begin{bmatrix} -p_x \\ -p_y \\ -p_z \\ -m_x \\ -m_y \end{bmatrix}$$
$$(4.28)$$

The coefficients of the linear operator L_{ij} and M_{ij} are given as in Eq. (2.49). The general boundary conditions of the plate are:

$$x=0: \begin{cases} N_x - k_{x0}^u u = 0 \\ N_{xy} - k_{x0}^v v = 0 \\ Q_x - k_{x0}^w w = 0 \\ M_x - K_{x0}^x \phi_x = 0 \\ M_{xy} - K_{x0}^y \phi_y = 0 \end{cases} \quad x=a: \begin{cases} N_x + k_{x1}^u u = 0 \\ N_{xy} + k_{x1}^v v = 0 \\ Q_x + k_{x1}^w w = 0 \\ M_x + K_{x1}^x \phi_x = 0 \\ M_{xy} + K_{x1}^y \phi_y = 0 \end{cases}$$

$$y=0: \begin{cases} N_{xy} - k_{y0}^u u = 0 \\ N_y - k_{y0}^v v = 0 \\ Q_y - k_{y0}^w w = 0 \\ M_{xy} - K_{y0}^x \phi_x = 0 \\ M_y - K_{y0}^y \phi_y = 0 \end{cases} \quad y=b: \begin{cases} N_{xy} + k_{y1}^u u = 0 \\ N_y + k_{y1}^v v = 0 \\ Q_y + k_{y1}^w w = 0 \\ M_{xy} + K_{y1}^x \phi_x = 0 \\ M_y + K_{y1}^y \phi_y = 0 \end{cases}$$

$$(4.29)$$

The classification of the classical boundary conditions shown in Table 1.3 for thick laminated shells is applicable for plates. Taking boundaries x = constant for example, the possible combinations for each classical boundary conditions are given in Table 4.2.

Similarly, in the framework of artificial spring boundary technique, the aforementioned classical boundary conditions can be readily realized by assigning the stiffness of the boundary springs at proper values. Taking edge $x = 0$ for example, the frequently encountered boundary conditions F, SD (shear-diaphragm), S (simply-supported) and C can be defined in terms of boundary spring rigidities as follows:

$$
\begin{aligned}
&\text{F: } k_{x0}^u = k_{x0}^v = k_{x0}^w = K_{x0}^x = K_{x0}^y = 0 \\
&\text{SD: } k_{x0}^v = k_{x0}^w = K_{x0}^y = 10^7 D, \ k_{x0}^u = K_{x0}^x = 0 \\
&\text{S: } k_{x0}^u = k_{x0}^v = k_{x0}^w = K_{x0}^y = 10^7 D, \ K_{x0}^x = 0 \\
&\text{C: } k_{x0}^u = k_{x0}^v = k_{x0}^w = K_{x0}^x = K_{x0}^y = 10^7 D
\end{aligned}
$$

$$(4.30)$$

where $D = E_1 h^3/12(1-\mu_{12}\mu_{21})$ is the flexural stiffness of the plate. Figure 4.6 (Ye et al. 2014a) shows the variations of the first three frequency parameters $\Delta\Omega$ versus restraint parameters Γ_λ ($\lambda = u, v, w, x$ and y) of a [0°/90°/0°] laminated plate ($a/b = 1$, $h/a = 0.1$, $E_1/E_2 = 40$, $\mu_{12} = 0.25$, $G_{12} = 0.6E_2$, $G_{13} = 0.6E_2$ and $G_{23} = 0.5E_2$), where $\Delta\Omega$ is defined as the difference of the frequency parameter $\Omega = \omega b^2/\pi^2 \sqrt{\rho h/D}$ to that of the elastic restraint parameter $\Gamma_\lambda = 10^{-1}$, namely, $\Delta\Omega = \Omega(\Gamma_\lambda) - \Omega(10^{-1}D)$. The plates under consideration are completely free at boundaries $y = 0$, $y = b$, and completely clamped at boundary $x = 0$, while at edge $x = a$, the plates are elastically supported by only one group of spring components with stiffness assigned as $\Gamma \times D$. According to Fig. 4.6, we can see that when the restraint parameter Γ is increased from 10^0 to 10^5, the frequency parameters increase rapidly and approach their utmost. Then they remain unchanged when Γ approaches infinity. In such a case, the plate can be deemed as clamped in both ends. Thus, by assigning the stiffness of the entire boundary springs at $10^7 D$, the completely clamped boundary conditions of a plate can be realized.

Table 4.2 Possible classical boundary conditions for thick rectangular plates at each boundary of $x = $ constant

Boundary type	Conditions
Free boundary conditions	
F	$N_x = N_{xy} = Q_x = M_x = M_{xy} = 0$
F2	$u = N_{xy} = Q_x = M_x = M_{xy} = 0$
F3	$N_x = v = Q_x = M_x = M_{xy} = 0$
F4	$u = v = Q_x = M_x = M_{xy} = 0$
F5	$N_x = N_{xy} = Q_x = M_x = \phi_y = 0$
F6	$u = N_{xy} = Q_x = M_x = \phi_y = 0$
F7	$N_x = v = Q_x = M_x = \phi_y = 0$
F8	$u = v = Q_x = M_x = \phi_y = 0$
Simply supported boundary conditions	
S	$u = v = w = M_x = \phi_y = 0$
SD	$N_x = v = w = M_x = \phi_y = 0$
S3	$u = N_{xy} = w = M_x = \phi_y = 0$
S4	$N_x = N_{xy} = w = M_x = \phi_y = 0$
S5	$u = v = w = M_x = M_{xy} = 0$
S6	$N_x = v = w = M_x = M_{xy} = 0$
S7	$u = N_{xy} = w = M_x = M_{xy} = 0$
S8	$N_x = N_{xy} = w = M_x = M_{xy} = 0$
Clamped boundary conditions	
C	$u = v = w = \phi_x = \phi_y = 0$
C2	$N_x = v = w = \phi_x = \phi_y = 0$
C3	$u = N_{xy} = w = \phi_x = \phi_y = 0$
C4	$N_x = N_{xy} = w = \phi_x = \phi_y = 0$
C5	$u = v = w = \phi_x = M_{xy} = 0$
C6	$N_x = v = w = \phi_x = M_{xy} = 0$
C7	$u = N_{xy} = w = \phi_x = M_{xy} = 0$
C8	$N_x = N_{xy} = w = \phi_x = M_{xy} = 0$

4.3 Vibration of Laminated Rectangular Plates

In this section, we consider free vibration laminated rectangular plates with general boundary conditions. The homogeneous rectangular plates are treated as a special case of the laminated ones in the study. Only solutions in the framework of SDPT are considered in this section. Natural frequencies and mode shapes for thin and thick laminated rectangular plates with different boundary conditions, lamination schemes and geometry parameters are presented. For the sake of simplicity, a four-letter string is employed to represent the boundary condition of a plate, such as FCSSD identify a rectangular plate with edges $x = 0$, $y = 0$, $x = a$, $y = b$ having F, C, S and SD boundary conditions, respectively.

Combining Eqs. (2.49) and (4.28), it is obvious that the displacements and rotation components of laminated rectangular plates in the framework of SDPT are

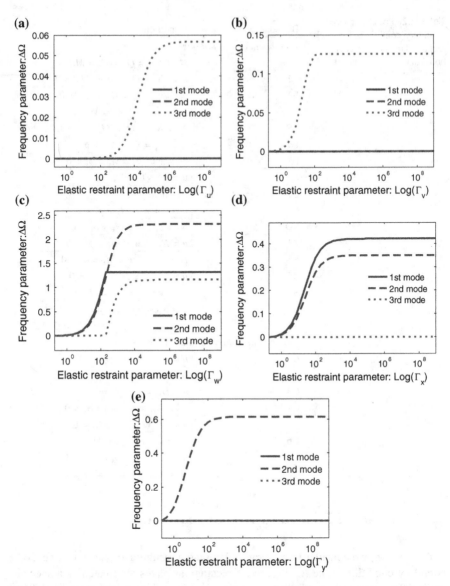

Fig. 4.6 Variation of the frequency parameters $\Delta\Omega$ versus the elastic restraint parameters Γ_λ for a three-layered, [0°/90°/0°] rectangular plate

required to have up to the second derivatives. Therefore, regardless of boundary conditions, each displacement and rotation component of a laminated plate is expanded as a two-dimensional modified Fourier series as:

$$u(x,y) = \sum_{m=0}^{M}\sum_{n=0}^{N} A_{mn} \cos \lambda_m x \, \cos \lambda_n y + \sum_{l=1}^{2}\sum_{n=0}^{N} a_{ln} P_l(x) \cos \lambda_n y$$

$$+ \sum_{l=1}^{2}\sum_{m=0}^{M} b_{lm} P_l(y) \cos \lambda_m x$$

$$v(x,y) = \sum_{m=0}^{M}\sum_{n=0}^{N} B_{mn} \cos \lambda_m x \, \cos \lambda_n y + \sum_{l=1}^{2}\sum_{n=0}^{N} c_{ln} P_l(x) \cos \lambda_n y$$

$$+ \sum_{l=1}^{2}\sum_{m=0}^{M} d_{lm} P_l(y) \cos \lambda_m x$$

$$w(x,y) = \sum_{m=0}^{M}\sum_{n=0}^{N} C_{mn} \cos \lambda_m x \, \cos \lambda_n y + \sum_{l=1}^{2}\sum_{n=0}^{N} e_{ln} P_l(x) \cos \lambda_n y$$

$$+ \sum_{l=1}^{2}\sum_{m=0}^{M} f_{lm} P_l(y) \cos \lambda_m x$$

$$\phi_x(x,y) = \sum_{m=0}^{M}\sum_{n=0}^{N} D_{mn} \cos \lambda_m x \cos \lambda_n y + \sum_{l=1}^{2}\sum_{n=0}^{N} g_{ln} P_l(x) \cos \lambda_n y$$

$$+ \sum_{l=1}^{2}\sum_{m=0}^{M} h_{lm} P_l(y) \cos \lambda_m x$$

$$\phi_y(x,y) = \sum_{m=0}^{M}\sum_{n=0}^{N} E_{mn} \cos \lambda_m x \cos \lambda_n y + \sum_{l=1}^{2}\sum_{n=0}^{N} i_{ln} P_l(x) \cos \lambda_n y$$

$$+ \sum_{l=1}^{2}\sum_{m=0}^{M} j_{lm} P_l(y) \cos \lambda_m x$$

$$(4.31)$$

where $\lambda_m = m\pi/a$ and $\lambda_n = n\pi/b$. A_{mn}, B_{mn}, C_{mn}, D_{mn} and E_{mn} are expansion coefficients of the standard cosine Fourier series. a_{ln}, b_{lm}, c_{ln}, d_{lm}, e_{ln}, f_{lm} g_{ln}, h_{lm}, i_{ln} and j_{lm} are the corresponding supplement coefficients. M and N denote the truncation numbers with respect to variables x and y, respectively. $P_l(x)$ and $P_l(y)$ denote the auxiliary polynomial functions introduced to remove all the discontinuities potentially associated with the first-order derivatives at the boundaries. These auxiliary functions are defined as

$$P_1(x) = x\left(\frac{x}{a} - 1\right)^2 \quad P_2(x) = \frac{x^2}{a}\left(\frac{x}{a} - 1\right) \tag{4.32}$$

$$P_1(y) = y\left(\frac{y}{b} - 1\right)^2 \quad P_2(y) = \frac{y^2}{b}\left(\frac{y}{b} - 1\right) \tag{4.33}$$

The solutions of laminated rectangular plates under consideration can be sought in the strong form solution procedure as described in Sect. 2.1.2. Alternately, all the expansion coefficients in Eq. (4.31) can be treated equally and independently as the generalized coordinates and solved directly from the Rayleigh–Ritz technique in a fashion similar to that done in Sect. 2.2. Table 3.3 shows that the solutions obtained by the weak form solution procedure (i.e., Ritz technique) converge faster than those of strong form solution procedure, therefore, the weak form solution procedure will be adopted in the following calculation. Unless otherwise stated, the natural frequencies of the considered plates are expressed in the non-dimensional parameters as $\Omega = \omega a^2 \sqrt{\rho/E_2 h^2}$ and the material properties of the layers are given as: $E_2 = 10$ GPa, $E_1/E_2 =$ open, $\mu_{12} = 0.25$, $G_{12} = 0.6E_2$, $G_{13} = 0.6E_2$, $G_{23} = 0.5E_2$ and $\rho = 1{,}450$ kg/m^3.

4.3.1 Convergence Studies and Result Verification

A moderately thick, symmetrically laminated rectangular plate with completely free boundary condition has been selected to demonstrate the convergence and accuracy of the current method. The material properties and geometrical dimensions of the plate are given as follows: $a/b = 3/2$, $h/a = 0.1$, $E_1/E_2 = 20$, $\mu_{12} = 0.25$, $G_{12} = 0.5E_2$, $G_{13} = 0.5E_2$ and $G_{23} = 0.33E_2$. In Table 4.3 (Ye et al. 2014a), the first six frequency parameters $\Omega = \omega a^2 \sqrt{\rho/E_1 h^2}$ for the plate with [30°/−30°/−30°/30°] lamination scheme are examined. Results without considering the supplementary terms in the admissible functions are also included in the table. The table shows the proposed method has fast convergence behavior. The maximum discrepancy in the worst case between the 11×11 truncated configuration and the 13×13 one is less than 0.0013 %. In order to check the model, the present results are also compared with

Table 4.3 Comparison and convergence of the first six frequency parameters $\Omega = \omega a^2 \sqrt{\rho/E_1 h^2}$ for a completely free, [30°/−30°/−30°/30°] laminated rectangular plate ($a/b = 3/2$, $h/a = 0.1$)

Methods	$M \times N$	Mode number					
		1	2	3	4	5	6
Present	11 × 11	2.3449	2.8767	4.9502	5.1610	5.6962	8.4310
	12 × 12	2.3449	2.8767	4.9502	5.1610	5.6962	8.4310
	13 × 13	2.3449	2.8767	4.9502	5.1610	5.6962	8.4310
Present *	11 × 11	0.7499	1.0034	2.2191	2.8797	3.7857	4.2837
	12 × 12	0.7141	0.9565	2.1952	2.7305	3.7630	4.2406
	13 × 13	0.7051	0.9472	2.1472	2.6935	3.6792	4.1875
Frederiksen (1995)	16 × 16	2.3223	2.8745	4.9152	5.0674	5.6589	8.2288
	18 × 18	2.3223	2.8745	4.9152	5.0674	5.6589	8.2288
Messina and Soldatos (1999a)		2.3251	2.8777	4.9458	5.0910	5.6871	8.3580

*Present** Results without considering the supplementary terms in Eq. (4.31)

Table 4.4 Comparison of the frequency parameters $\Omega = \omega b^2/\pi^2 \sqrt{\rho/E_2 h^2}$ for a thick, [45°/−45°/45°/−45°/45°] laminated plate ($a/b = 1$, $h/a = 0.1$, $E_1/E_2 = 40$)

Boundary conditions	Method	Mode number					
		1	2	3	4	5	6
SSSS	Present	1.8803	3.3763	3.6923	4.9665	5.4801	5.5985
	Karami et al. (2006)	1.8788	3.3776	3.6921	4.9680	5.4834	5.6000
	Wang (1997)	1.8792	3.3776	3.6924	4.9682	5.4835	5.6002
	Liew et al. (2005)	1.8466	3.3774	3.6425	4.9661	5.4348	5.5187
CCCC	Present	2.2855	3.7363	3.9792	5.1777	5.6966	5.8416
	Karami et al. (2006)	2.2857	3.7392	3.9813	5.1799	5.7019	5.8454
	Wang (1997)	2.2857	3.7392	3.9813	5.1800	5.7019	5.8455
	Liew et al. (2005)	2.2785	3.7383	3.9583	5.1836	5.6808	5.8066

data published by Frederiksen (1995) who used Ritz method, Messina and Soldatos (1999a) based on HSDT formulation. From the table, one can see that the present solutions agree very well with the referential results.

To further validate the accuracy and reliability of current solutions, Table 4.4 (Ye et al. 2014a) shows the comparison of the first six frequency parameters $\Omega = \omega b^2/\pi^2 \sqrt{\rho/E_2 h^2}$ of a thick, [45°/−45°/45°/−45°/45°] laminated plate which was studied by Karami et al. (2006), Liew et al. (2005) and Wang (1997). The material properties and geometrical dimensions used in the investigation are: $a/b = 1$, $h/a = 0.1$, $E_1/E_2 = 40$. The SSSS and CCCC boundary conditions are performed in the comparison. It is observed that the comparison is very good. The discrepancies are negligible and the worst one is less than 1.83 %.

4.3.2 Laminated Rectangular Plates with Arbitrary Classical Boundary Conditions

The excellent agreement between the current solutions and those provided by other researchers observed from Tables 4.3 and 4.4 indicate that the proposed method is sufficiently accurate to deal with laminated rectangular plates with arbitrary boundary conditions. It also verified that the definition of the four types of classical boundaries in Eq. (4.30) is appropriate. In this section, laminated rectangular plates with various boundary conditions including the classical restraints and the elastic ones will be studied.

In Table 4.5, the first four non-dimension frequency parameters Ω of a two-layered, angle-ply [45°/−45°] laminated square plate ($E_1/E_2 = 40$) subjected to five possible combinations of boundary conditions are presented. Four different thickness-length ratios, i.e. $h/a = 0.01, 0.05, 0.1$ and 0.15, corresponding to thin to thick

Table 4.5 The first four frequency parameters Ω for a [45°/−45°] laminated square plate with various boundary conditions and thickness–length ratios

h/a	Mode number	Boundary conditions				
		FFFF	FSFS	FCFC	SSSS	CCCC
0.01	1	7.9118	6.2094	12.320	18.457	23.325
	2	11.467	12.849	17.489	37.712	46.788
	3	14.782	21.385	31.281	37.712	46.788
	4	24.979	28.160	34.054	63.436	73.851
	5	24.979	32.609	41.532	64.007	80.563
0.05	1	7.8140	6.0044	11.703	17.628	21.405
	2	10.964	12.234	16.111	34.803	41.031
	3	14.391	20.220	27.942	34.803	41.031
	4	22.971	26.027	30.872	56.213	61.328
	5	22.971	29.825	36.684	56.379	67.024
0.10	1	7.5763	5.6488	10.480	15.621	17.910
	2	10.176	11.216	13.871	28.893	31.961
	3	13.515	18.065	23.029	28.893	31.961
	4	19.946	22.357	25.207	43.258	44.986
	5	19.946	25.281	29.347	44.388	48.908
0.15	1	7.2469	5.2481	9.1388	13.451	14.794
	2	9.3174	10.121	11.816	23.591	25.081
	3	12.437	15.834	19.078	23.591	25.081
	4	17.088	18.897	20.312	33.556	34.211
	5	17.088	21.168	23.548	35.125	37.055

plates are performed in the calculation. It can be seen from the table that the augmentation of the thickness-length ratio leads to the decrease of the frequency parameters. Then let us consider moderately thick (h/a = 0.05), [0°/90°] and [0°/ 90°/0°] laminated rectangular (b/a = 2) plates with various anisotropic degrees. In Table 4.6, the first three frequency parameters Ω for the plates with four types of boundary conditions and five different anisotropic degrees, i.e. E_1/E_2 = 1, 5, 10, 20 and 40 are listed, respectively. It can be seen from the table that the frequency parameters increase in general as the anisotropic ratio increases.

In a composite lamina, fibers are the principal load carrying members (Reddy 2003). By appropriately arranging the fiber directions in the layers of a laminated plate, special functional requirements can be satisfied. The influence of fiber orientations on the vibration characteristics of composite laminated plates is investigated. In Fig. 4.7 (Ye et al. 2014a), variation of the lowest four frequency parameters Ω of a three-layered, [0°/ϑ/0°] composite plate with CFFF and CFCF boundary conditions against the fiber direction angle ϑ are depicted, respectively. The geometric and material properties of the layers of the plate are: a/b = 1,

Table 4.6 The first three frequency parameters Ω for [0°/90°] and [0°/90°/0°] laminated rectangular plates with various anisotropic degrees ($b/a = 2$, $h/a = 0.05$)

Boundary conditions	E_1/E_2	[0°/90°]			[0°/90°/0°]		
		1	2	3	1	2	3
FFFF	1	1.6186	2.4368	4.4783	1.6186	2.4363	4.4779
	5	2.2796	2.4714	5.4939	1.7283	2.4512	4.7494
	10	2.4895	2.6649	5.6947	1.8618	2.4598	5.1119
	20	2.5122	3.2378	6.0173	2.1036	2.4729	5.4672
	40	2.5380	4.1190	6.5613	2.4927	2.5179	5.6761
SDSDSDSD	1	3.8245	6.2386	10.007	3.8244	6.2378	10.005
	5	4.7235	7.2379	11.887	6.7027	8.3403	11.545
	10	5.3177	7.9484	13.174	9.0302	10.352	13.267
	20	6.2521	9.1169	15.236	12.270	13.351	16.049
	40	7.7320	11.025	18.515	16.547	17.487	20.152
SSSS	1	3.8837	6.3851	10.207	3.8839	6.3851	10.206
	5	5.4240	7.9582	12.629	6.7356	8.4471	11.711
	10	6.7036	9.2081	14.405	9.0538	10.435	13.407
	20	8.5373	11.043	17.090	12.287	13.413	16.158
	40	11.135	13.755	21.157	16.559	17.530	20.232
CCCC	1	7.3712	9.7520	13.697	7.3733	9.7535	13.697
	5	9.6590	12.147	16.906	13.969	15.344	18.173
	10	11.049	13.668	18.954	18.547	19.622	22.041
	20	13.112	15.991	22.086	23.992	24.886	27.090
	40	16.171	19.520	26.832	29.592	30.499	32.762

$h/a = 0.1$, $E_1/E_2 = 40$. Many interesting characteristics can be observed from the figures. Firstly, all the figures are symmetrical about $\vartheta = 90°$. From Fig. 4.7a, we can see that there is little variation in the 1st and 2nd mode frequency parameters when ϑ is increased from 0 to 90°. However, for the 3rd mode, the frequency parameter traces climb up and then decline, and may reach its crest around $\vartheta = 75°$. Figure 4.7b shows that the 1st and 2nd mode frequency parameters of the plate have decreased slightly with the fiber direction angle ϑ increased (from 0 to 90°). And there is little variation in frequency parameters of the 3rd modes. Comparing with the lowest three modes, the 4th mode frequency curves in all subfigures have more significant changes which also climb up and then decline.

The effects of the fiber direction angle on frequency parameters and mode shapes of single-layered composite rectangular plates are further reported. In Tables 4.7 and 4.8, the lowest three frequency parameters Ω of a single-layered composite rectangular plate with various boundary conditions and fiber directions are presented. The aspect ratio is chosen to be $b/a = 2/3$. The thickness-to-width ratio $h/b = 0.1$ is used in the calculation. The fiber direction angle ϑ is varied from 0° to 90° with an increment of 15°. It can be noticed that increasing ϑ from 0° to 90°

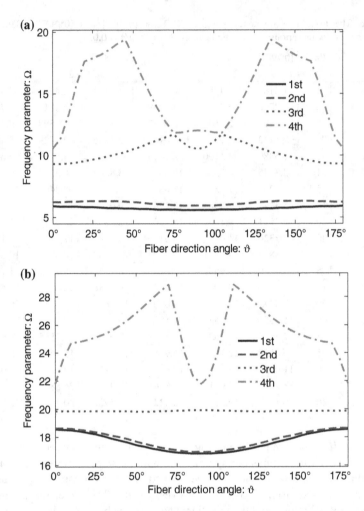

Fig. 4.7 Variation of the frequency parameters Ω versus the fiber direction angle ϑ for a $[0°/\vartheta/0°]$ laminated plate with different boundary conditions: **a** CFFF; **b** CFCF

increases the fundamental frequency parameters for plates with FSDSDSD, FCCC and FFFC boundary conditions. For the plates with orthotropy ratio $E_1/E_2 = 25$ and FFSDSD or FFFSD boundary conditions, when ϑ is varied from 0° to 45°, the fundamental frequency parameters increased as well. However, increasing the fiber direction angles from 45° to 90° decreases these frequency parameters. Furthermore, it can be seen from the table that the frequency parameters increase in general as the orthotropy ratio increases. Contour plots of the mode shapes for the plate with CCCC, FCCC and FFCC boundary conditions and orthotropy ratio $E_1/E_2 = 25$

Table 4.7 The first three frequency parameters Ω for a single-layered rectangular plate with various boundary conditions and fiber direction angles ($b/a = 2/3$, $h/b = 0.1$)

ϑ	Mode	$E_1/E_2 = 5$			$E_1/E_2 = 25$		
		FSDSDSD	FFSDSD	FFFSD	FSDSDSD	FFSDSD	FFFSD
0°	1	7.1214	1.8181	3.5800	7.1137	1.8435	3.5894
	2	15.243	10.782	9.8028	18.251	11.130	9.7959
	3	18.251	11.359	14.387	23.004	20.706	14.450
15°	1	7.4299	1.7988	3.8005	8.0005	1.9288	4.2139
	2	15.531	9.8858	10.017	20.939	10.880	10.023
	3	19.368	12.610	13.413	22.275	20.818	14.875
30°	1	8.2871	1.8809	4.1752	9.7757	2.1294	5.0365
	2	15.979	8.8869	10.716	21.598	9.8323	11.117
	3	21.867	14.214	12.576	27.293	20.197	15.100
45°	1	9.7045	1.9282	4.2298	12.792	2.2254	4.9192
	2	16.202	8.0423	10.768	21.975	8.6321	11.569
	3	23.053	16.088	13.950	30.772	18.577	18.381
60°	1	11.627	1.8796	3.9882	17.759	2.0992	4.3109
	2	16.507	7.4340	10.219	24.223	7.8529	10.792
	3	21.324	16.829	16.952	24.768	17.493	21.032
75°	1	13.407	1.7920	3.7406	20.110	1.8774	3.8508
	2	17.030	7.0324	9.7324	23.942	7.2353	9.9781
	3	19.144	16.453	19.485	26.949	16.736	20.346
90°	1	14.131	1.7831	3.6445	18.251	1.7886	3.6870
	2	17.333	6.9315	9.5416	27.073	6.9836	9.6477
	3	18.251	16.318	19.854	28.512	16.368	19.985

are given in Figs. 4.8, 4.9 and 4.10, respectively. From Figs. 4.8 and 4.10, we can see that the node lines (i.e., lines with zero displacements) of the second modes are paralleled to the fiber orientation.

Figure 4.11 shows the lowest three frequency parameters Ω and mode shapes for a FFFF, single-layered square composite plate with different fiber direction angles. The plate is assumed to be thin ($h/a = 0.01$) and made of material with orthotropy ratio $E_1/E_2 = 25$. Since the plate symmetrize about the line $x = y$, the frequency parameters for the plate with $\vartheta = 0°$, $15°$ and $30°$ are the same as those with fiber direction angles $\vartheta = 90°$, $75°$ and $60°$, respectively. Similarly, mode shapes for $\vartheta = 0°$, $15°$ and $30°$ are similar to those given for $\vartheta = 90°$, $75°$ and $60°$. In addition, it can be observed that the increment of the fiber direction angle from $0°$ to $45°$ results in increases of the lowest two frequency parameters of the plate.

Table 4.8 The first three frequency parameters Ω for a single-layered rectangular plate with various boundary conditions and fiber direction angles ($b/a = 2/3$, $h/b = 0.1$)

ϑ	Mode	$E_1/E_2 = 5$			$E_1/E_2 = 25$		
		FCCC	FFCC	FFFC	FCCC	FFCC	FFFC
0°	1	14.643	4.1554	2.2781	15.20202	5.9673	2.2701
	2	22.421	14.664	4.8398	30.26969	15.835	4.8543
	3	36.706	15.571	13.701	36.83315	25.755	13.704
15°	1	15.083	4.5393	2.3376	16.05186	6.9774	2.3549
	2	22.761	14.194	5.0506	30.07846	16.832	5.3525
	3	37.749	16.991	13.868	39.00083	26.250	14.089
30°	1	16.606	4.8977	2.5768	19.28065	7.5381	2.8620
	2	23.696	13.336	5.5623	30.81169	17.481	6.6963
	3	37.543	19.019	13.278	45.52704	27.112	15.349
45°	1	19.407	5.2668	3.0716	25.0631	8.4171	3.9445
	2	25.138	12.410	6.0661	33.3452	16.559	8.3679
	3	36.231	21.782	12.658	44.98416	28.239	15.340
60°	1	23.010	5.6496	3.7998	31.95584	9.6762	5.5627
	2	27.055	11.613	6.3853	37.2734	15.532	9.7606
	3	35.664	22.128	12.250	44.64113	25.914	15.519
75°	1	26.273	5.8850	4.6070	38.42538	10.791	8.0540
	2	28.954	10.957	6.5741	40.75042	14.863	10.835
	3	35.693	20.858	11.783	45.28074	23.461	15.109
90°	1	27.580	5.6969	5.0260	41.36358	10.730	10.436
	2	29.797	10.453	6.6394	42.27473	13.663	11.169
	3	35.804	20.259	11.530	45.79356	21.932	14.438

4.3.3 Laminated Rectangular Plates with Elastic Boundary Conditions

In order to prove the validity of the present method for the free vibration analysis of laminated plates with elastic boundary conditions, Table 4.9 shows the comparison of the first five frequency parameters $\Omega = \omega b^2/\pi^2 \sqrt{\rho h/D}$ for $[0°/90°/0°...]_9$ laminated plates with edges elastically restrained against rotation and translation (i.e. at edges $x = 0$ and a: $N_x = 0$, $N_{xy} = 0$, $k_w = \Gamma^* D_{22}/b^3$, $K_x = \Gamma^* D_{22}/b$, $M_{xy} = 0$ and at edges $y = 0$ and b: $N_{xy} = 0$, $N_y = 0$, $k_w = \Gamma^* D_{22}/a^3$, $M_{xy} = 0$, $K_y = \Gamma^* D_{22}/a$). The material constants and geometry parameters of the considered plates are: $a/b = 1$, $E_1/E_2 = 40$. The theoretical results reported by Liew et al. (1997) by using Pb2-Ritz method and Karami et al. (2006) based on DQM are included in the table. The results for thin ($h/a = 0.001$) and moderately thick plates ($h/a = 0.2$) with two different values of stiffness of elastic edges are found and are shown in the comparison. It is obvious that the proposed modified Fourier solution is sufficient to yield the solutions in good agreements with those of Pb2-Ritz method and DQM.

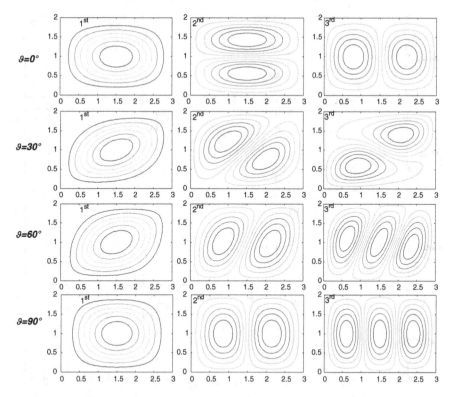

Fig. 4.8 Mode shapes for CCCC, single-layered rectangular plates (b/a = 2/3, h/b = 0.1, E_1/E_2 = 25)

In the last example in this section, vibration frequencies of composite laminated plates with some types of elastic restraints will be presented. Although we can obtain accurate solutions for composite laminated plates with arbitrary uniform and un-uniform elastic restraints, in this work, we choose three typical uniform elastic restraint conditions that are defined as follows (at $x = 0$):

E^1: the normal direction is elastically restrained ($u \neq 0$, $N_{xy} = Q_x = M_x = M_{xy} = 0$).

E^2: the transverse direction is elastically restrained ($w \neq 0$, $N_x = N_{xy} = M_x = M_{xy} = 0$).

E^3: the rotation is elastically restrained ($\phi_x \neq 0$, $N_x = N_{xy} = Q_x = M_{xy} = 0$).

Table 4.10 shows the lowest three frequency parameters Ω of laminated plates ($E_1/E_2 = 20$, $a/b = 3/2$) with different lamination schemes and thickness ratios. Two different lamination schemes, i.e. [0°/90°] and [0°/90°/0°], corresponding to symmetrically and unsymmetrically laminated plates are performed in the calculation. The thickness ratios used are $h/a = 0.01$, 0.05 and 0.1. The plates under consideration are clamped at the edge of $y = b$, free at edges $y = 0$, $x = a$ and with E^1 boundary conditions at the edge of $x = 0$ (E^1FFC). The table shows that increasing

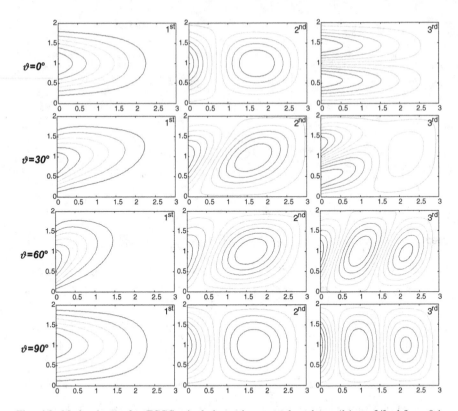

Fig. 4.9 Mode shapes for FCCC, single-layered rectangular plates (b/a = 2/3, h/b = 0.1, E_1/E_2 = 25)

the normal restrained rigidity has very limited effects on the frequency parameters of both symmetrically and unsymmetrically laminated plates. When the normal restrained rigidity is varied from $10^{-1}*D$ to 10^4*D, the maximum increment in the table are less than 1.83 % for all cases. It may be attributed to the lower frequency parameters of a laminated plate with lower thickness ratios are dominated by the transverse vibration. The further observation from the table is that the effects for the [0°/90°] plate are larger than the [0°/90°/0°] one.

Tables 4.11 and 4.12 show similar studies for the E^2FFC and E^3FFC boundary conditions, respectively. Table 4.11 reveals that the transverse restrained rigidity has a large effect on the frequency parameters of both [0°/90°] and [0°/90°/0°] laminated plates, especially the second mode. When the transverse restrained rigidity is varied from $10^{-1}*D$ to 10^4*D, the increments of the second mode can be 292.62, 286.48 and 263.75 % for the [0°/90°/0°] plate with thickness ratios of 0.01, 0.05 and 0.1, respectively. Table 4.12 shows that the rotation restrained rigidity has very limited effects on the fundamental frequency parameters of the plates, with the maximum increment less than 0.02 % in all cases when the rotation restrained rigidity is varied from $10^{-1}*D$ to 10^4*D. However, for the second mode, increasing

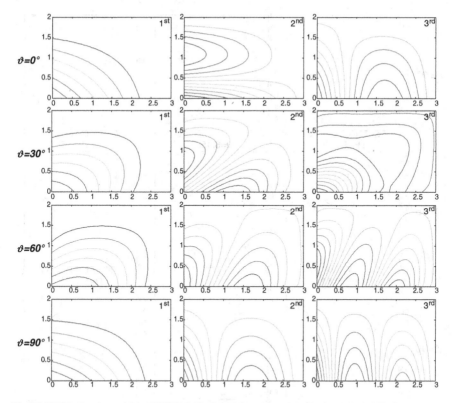

Fig. 4.10 Mode shapes for FFCC, single-layered rectangular plates ($b/a = 2/3$, $h/b = 0.1$, $E_1/E_2 = 25$)

the rotation restrained rigidity from $10^{-1}*D$ to 10^4*D increases the frequency parameters by almost 8 and 50 % in each plate configuration. From these tables, it is obvious that the effects of elastic restraint rigidity on the frequency parameters of composite laminated plates varies with mode sequences, lamination schemes and spring components. These results may serve as benchmark solutions for further researchers.

4.3.4 Laminated Rectangular Plates with Internal Line Supports

In the engineering practices, laminated plates are often restrained by internal line supports to reduce the magnitude of dynamic and static stresses and displacements of the structure or satisfy special architectural and functional requirements. The study of the vibrations of laminated plates with internal line supports is an important aspect in the successful applications of these structures. Thus, in this part,

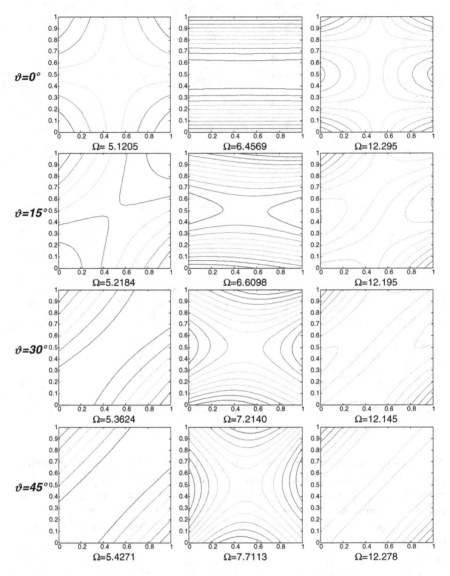

Fig. 4.11 Frequency parameters Ω and mode shapes for FFFF, thin single-layered square plates ($h/a = 0.01$, $E_1/E_2 = 25$)

the present method is applied to investigate the free vibration behaviors of laminated rectangular plates with internal line supports and arbitrary boundary conditions. As shown in Fig. 4.12, a laminated rectangular plate restrained by arbitrary internal line supports is considered. x_i and y_j represent the position of the ith and jth line supports along the y- and x-axes, respectively. The displacement fields in the position of the line support satisfy $w\,(x_i,\, y) = 0$ and $w\,(x,\, y_j) = 0$ (Cheung and Zhou

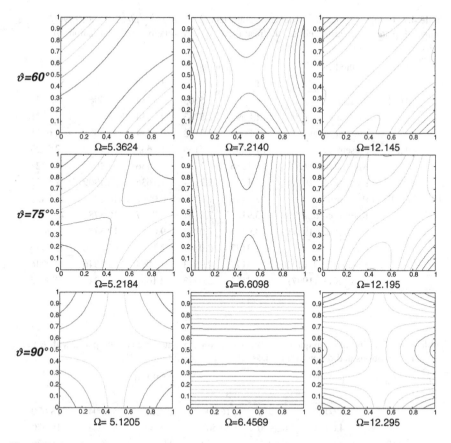

Fig. 4.11 (continued)

2001a, b). This condition can be readily obtained by introducing a group of continuously distributed linear springs at the location of each line support and setting the stiffness of these springs equal to infinite (which is represented by a very large number, 10^7*D). Thus, the potential energy (P_{ils}) stored in these springs is:

$$P_{ils} = \frac{1}{2} \int_0^b \left\{ \sum_{i=1}^{N_i} k_{xi}^i w(x_i, y)^2 \right\} dy + \frac{1}{2} \int_0^a \left\{ \sum_{j=1}^{M_j} k_{yj}^j w(x, y_j)^2 \right\} dx \qquad (4.34)$$

where N_i and M_j are the amount of line supports in the y and x directions. k_{xi}^i, k_{yj}^j denote the corresponding line supported springs distributed at $x = x_i$ and $y = y_j$. By adding the potential energy P_{ils} functions in the Lagrangian energy functional and applying the weak form solution procedure, the characteristic equation for a plate with arbitrary boundary conditions and internal line supports is readily obtained.

Table 4.9 Comparison of the frequency parameters $\Omega = \omega b^2/\pi^2\sqrt{\rho h/D}$ for a $[0°/90°/0°...]_9$ laminated plate with edges elastically restrained against rotation and translation ($a/b = 1$, $E_1/E_2 = 40$)

Γ	h/a	Method	Mode number				
			1	2	3	4	5
10^2	0.001	Present	1.8096	2.5703	2.7234	3.3061	4.6721
		Karami et al. (2006)	1.8095	2.5703	2.7234	3.3061	4.6721
		Liew et al. (1997)	1.8096	2.5703	2.7234	3.3061	4.6721
	0.2	Present	1.1236	1.7134	1.7635	2.1909	2.5323
		Karami et al. (2006)	1.1236	1.7134	1.7636	2.1909	2.5323
		Liew et al. (1997)	1.1236	1.7134	1.7636	2.1909	2.5323
10^8	0.001	Present	3.9126	7.0767	9.0087	10.878	12.743
		Karami et al. (2006)	3.9126	7.0767	9.0087	10.878	12.743
		Liew et al. (1997)	3.9126	7.0743	9.0088	10.878	12.744
	0.2	Present	1.3177	2.0543	2.1064	2.6346	2.9945
		Karami et al. (2006)	1.3177	2.0543	2.1064	2.6346	2.9946
		Liew et al. (1997)	1.3177	2.0543	2.1064	2.6346	2.9945

Table 4.10 The first three frequency parameters Ω for laminated rectangular plates with E^1FFC boundary conditions and different thickness ratios ($E_1/E_2 = 20$, $a/b = 3/2$)

h/a	k_{x0}^u/D	$[0°/90°]$			$[0°/90°/0°]$		
		1	2	3	1	2	3
0.01	10^{-1}	4.6013	6.5918	16.506	2.9838	5.5004	18.670
	10^0	4.6013	6.5920	16.507	2.9838	5.5004	18.670
	10^1	4.6013	6.5939	16.512	2.9838	5.5004	18.670
	10^2	4.6013	6.6070	16.554	2.9838	5.5004	18.670
	10^3	4.6014	6.6536	16.707	2.9838	5.5004	18.670
	10^4	4.6014	6.6846	16.809	2.9838	5.5004	18.670
0.05	10^{-1}	4.5518	6.4066	15.935	2.9686	5.3613	18.045
	10^0	4.5518	6.4124	15.954	2.9686	5.3613	18.047
	10^1	4.5518	6.4356	16.012	2.9686	5.3613	18.045
	10^2	4.5518	6.4874	16.165	2.9686	5.3613	18.048
	10^3	4.5518	6.5046	16.197	2.9686	5.3613	18.048
	10^4	4.5518	6.5158	16.205	2.9686	5.3613	18.044
0.10	10^{-1}	4.4076	6.0284	14.736	2.9232	5.1045	16.449
	10^0	4.4076	6.0435	14.795	2.9232	5.1045	16.449
	10^1	4.4076	6.0917	14.812	2.9232	5.1045	16.449
	10^2	4.4076	6.1232	14.693	2.9232	5.1045	16.449
	10^3	4.4076	6.1250	14.827	2.9232	5.1045	16.449
	10^4	4.4076	6.1250	14.825	2.9232	5.1045	16.449

Table 4.11 The first three frequency parameters Ω for laminated rectangular plates with E^2FFC boundary conditions and different thickness ratios ($E_1/E_2 = 20$, $a/b = 3/2$)

h/a	k_{x0}^w/D	$[0°/90°]$			$[0°/90°/0°]$		
		1	2	3	1	2	3
0.01	10^{-1}	4.8727	7.5360	17.057	3.3908	6.7176	18.785
	10^0	5.1045	10.714	21.873	3.7219	12.806	19.279
	10^1	5.1603	12.039	29.342	3.8062	19.314	19.938
	10^2	5.1724	12.198	29.386	3.8233	19.632	20.806
	10^3	5.1751	12.217	29.393	3.8268	19.652	20.911
	10^4	5.1760	12.221	29.395	3.8280	19.657	20.922
0.05	10^{-1}	4.8173	7.3853	16.448	3.3651	6.5993	18.162
	10^0	5.0194	10.468	21.285	3.6688	12.578	18.618
	10^1	5.0688	11.701	27.306	3.7430	18.529	19.135
	10^2	5.0781	11.841	27.343	3.7574	18.869	19.801
	10^3	5.0848	11.816	27.346	3.7626	18.893	19.889
	10^4	5.0820	11.846	27.353	3.7635	18.906	19.893
0.10	10^{-1}	4.6552	7.0274	15.301	3.3015	6.3737	16.568
	10^0	4.8229	9.9146	17.015	3.5636	11.976	16.810
	10^1	4.8608	10.922	17.018	3.6276	16.595	16.808
	10^2	4.8681	11.037	17.018	3.6370	16.810	17.028
	10^3	4.8690	11.051	17.013	3.6387	16.809	17.054
	10^4	4.8695	11.045	17.019	3.6354	16.810	17.051

In order to prove the validity of the present method in dealing with vibration of laminated composite plate with internal line supports, Table 4.13 presents the comparison of the first nine frequency parameters $\Omega = \omega ab\sqrt{\rho h/D}$ for a square cross-ply $[0°/90°]$ laminated plate with a central line support in each direction. The material used is graphite-epoxy with the following proprieties: $E_1/E_2 = 40$, $\mu_{12} = 0.25$, $G_{12} = 0.5E_2$, $G_{13} = 0.5E_2$, $G_{23} = 0.2E_2$. Three different classical boundary conditions, i.e. FFFF, SDSDSDSD and CCCC are considered in the comparison. The benchmark solutions provided by Cheung and Zhou (2001a) based on CPT are referenced. From the table a consistent agreement of present results and referential date is seen. The discrepancy is very small and does not exceeds 0.52 % for the worst case although different trial functions are used in the literature. In addition, the table shows that the line supports increase the frequencies of the plate.

Influence of the locations of line supports on the frequency parameters of a three-layered, cross-ply $[0°/90°/0°]$ plate is investigated as well. The plate is assumed to be thick ($h/b = 0.2$) and with similar material parameters as those used in Table 4.13. For the sake of brevity, only a line support along y direction (x_1) is considered in the analysis. In Fig. 4.13, variations of the lowest three mode frequency parameters $\Omega = \omega ab\sqrt{\rho h/D}$ of the considered plate with high length–width ratio ($a/b = 5$)

Table 4.12 The first three frequency parameters Ω for laminated rectangular plates with E^3FFC boundary conditions and different thickness ratios ($E_1/E_2 = 20$, $a/b = 3/2$)

h/a	K_{x0}^x/D	[0°/90°]			[0°/90°/0°]		
		1	2	3	1	2	3
0.01	10^{-1}	4.6014	6.8477	17.434	2.9838	5.9550	18.671
	10^0	4.6014	7.2371	19.370	2.9840	7.5831	18.675
	10^1	4.6014	7.3529	20.096	2.9842	8.7787	18.678
	10^2	4.6014	7.3669	20.188	2.9843	8.9800	18.679
	10^3	4.6014	7.3682	20.197	2.9843	9.0014	18.679
	10^4	4.6014	7.3684	20.198	2.9843	9.0037	18.679
0.05	10^{-1}	4.5518	6.6726	16.843	2.9687	5.8207	18.047
	10^0	4.5518	7.0598	18.688	2.9690	7.4238	18.048
	10^1	4.5518	7.1817	19.364	2.9691	8.5934	18.054
	10^2	4.5518	7.1886	19.444	2.9692	8.7857	18.055
	10^3	4.5518	7.1878	19.495	2.9692	8.8088	18.056
	10^4	4.5518	7.2143	19.457	2.9692	8.8104	18.054
0.10	10^{-1}	4.4076	6.3041	15.549	2.9233	5.5674	16.450
	10^0	4.4077	6.6868	17.012	2.9235	7.1204	16.453
	10^1	4.4078	6.7927	17.019	2.9238	8.1643	16.455
	10^2	4.4078	6.8034	17.018	2.9238	8.3348	16.456
	10^3	4.4078	6.8054	17.018	2.9239	8.3503	16.456
	10^4	4.4078	6.8064	17.018	2.9239	8.3667	16.456

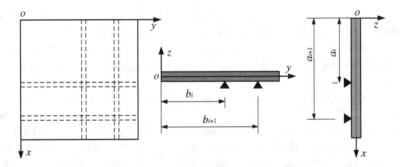

Fig. 4.12 Schematic diagram of laminated rectangular plates with arbitrary internal line supports

against the line support location parameter x_1/a are depicted. Six types of edge conditions used in the investigation are: F–S, S–F, S–S, C–C, F–C and C–F. It is obvious that the frequency parameters of the plate are significantly affected by the position of the line support, and this effect can vary with the boundary conditions. And for different modes, the effects of line support location are quite different. For the sake of completeness, the similar studies for the considered plate with lower

Table 4.13 Comparison of the first five frequency parameters $\Omega = \omega ab \sqrt{\rho h / D}$ for a square, $[0°/90°]$ laminated plate with a central line support in each direction ($h/a = 0.001$)

Boundary conditions	Method	Mode number				
		1	2	3	4	5
FFFF	Present	2.636	6.569	6.569	9.005	26.046
	Cheung and Zhou (2001a)	2.631	6.572	6.572	9.006	26.067
	Error (%)	0.19	0.05	0.05	0.01	0.08
SDSDSDSD	Present	24.438	31.348	31.383	37.078	68.539
SDSDSDSD	Cheung and Zhou (2001a)	24.440	31.333	31.333	36.989	68.549
	Error (%)	0.01	0.05	0.16	0.24	0.01
CCCC	Present	36.971	45.505	45.563	52.703	87.040
	Cheung and Zhou (2001a)	37.002	45.528	45.528	52.695	87.499
	Error (%)	0.08	0.05	0.08	0.02	0.52

length-width ratio ($a/b = 1$) are presented in Fig. 4.14. Comparing Figs. 4.13 with 4.14, we can see that the influence of the line support location on the frequency parameters vary with length-width ratios and boundary conditions.

4.4 Fundamental Equations of Laminated Sectorial, Annular and Circular Plates

The sectorial, annular and circular plates are plates of circular peripheries. They are used quite often in aerospace crafts, naval vessels, civil constructions and other fields of modern technology. When dealing with these plates, it is expedient to use polar coordinate system in the formulation. Sectorial, annular and circular plates made of isotropic materials have received considerable attention in the literature. When compared with the amount of information available for isotropic sectorial, annular and circular plates, studies reported on the vibration analysis of orthotropic and laminated sectorial, annular and circular plates are very limited. This can be due to the difficulty in constructing such plates. There commonly exist two types of orthotropic sectorial, annular and circular plates. The first type is to actually construct the plates with rectangular orthotropy and then cut the plates in sectorial, annular or circular shapes. On the other hand, one can also construct circular plates that the material principle directions take a circular shape around the center of the plates, thus results in polar orthotropy, and then cut the plates in sectorial, annular or circular shapes with desired geometry dimensions. The section deals with vibration of laminated sectorial, annular and circular plates made of layers having

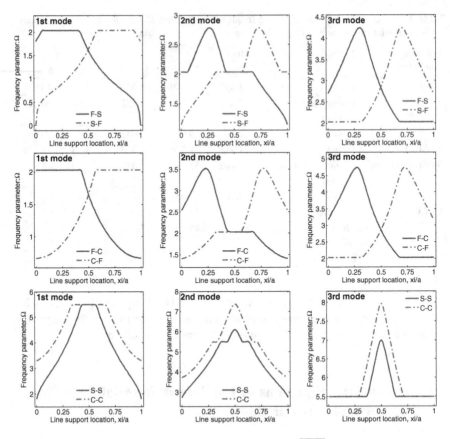

Fig. 4.13 Variation of the frequency parameters $\Omega = \omega ab \sqrt{\rho h/D}$ versus the line support locations for a long ($a/b = 5$), [0°/90°/0°] laminated plate with a y direction internal line support

polar orthotropy (i.e., lamination angle of 0° or 90°). The corresponding isotropic ones are considered as well.

Annular and circular plates can be treated as special cases of sectorial plates with circumferential direction of full circle (2π). As shown in Fig. 4.15, a general laminated sectorial plate of circumferential direction θ_0, constant thickness h, inner radius R_0 and outer radius R_1 is selected as the analysis model. Polar coordinate system (r, θ and z) located at the middle surface of the plate is used for plate coordinates, in which z is parallel to the thickness direction. The middle surface displacements of the plate in the r, θ and z directions are denoted by u, v and w, respectively. The laminated sectorial plate is assumed to be composed of N_L layers of polar orthotropic laminae.

Consider the sectorial plate in Fig. 4.15 and its polar coordinate system. The following geometry parameters can be applied to the shell equations derived in Chap. 1 to obtain those of sectorial, annular and circular plates.

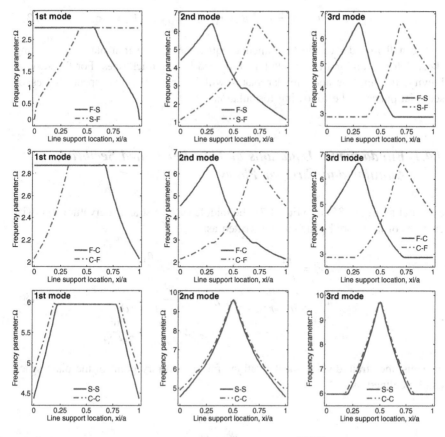

Fig. 4.14 Variation of the frequency parameters $\Omega = \omega ab \sqrt{\rho h/D}$ versus the line support locations for a square, [0°/90°/0°] laminated plate with a y direction internal line support

Fig. 4.15 Geometry and coordinate system of laminated sectorial plates with polar orthotropy

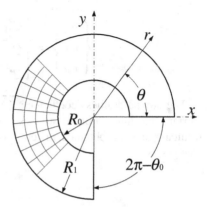

$$\alpha = r, \quad \beta = \theta, \quad A = 1, \quad B = r, \quad R_\alpha = \infty, \quad R_\beta = \infty \qquad (4.35)$$

We will first derive the fundamental equations for the thin plates (neglecting shear deformation and rotary inertia), followed by the thick ones. For the sake of brevity, the annular and circular plates will be treated as the special cases of sectorial plates in the following formulation.

4.4.1 Fundamental Equations of Thin Laminated Sectorial, Annular and Circular Plates

Substituting Eq. (4.35) into Eq. (1.7), the middle surface strains, curvature and twist changes of a sectorial plate can be written as:

$$
\begin{aligned}
\varepsilon_r^0 &= \frac{\partial u}{\partial r}, & \chi_r &= -\frac{\partial^2 w}{\partial r^2} \\
\varepsilon_\theta^0 &= \frac{\partial v}{r\partial \theta} + \frac{u}{r}, & \chi_\theta &= -\frac{\partial^2 w}{r^2 \partial \theta^2} - \frac{\partial w}{r\partial r} \\
\gamma_{r\theta}^0 &= \frac{\partial v}{\partial r} + \frac{\partial u}{r\partial \theta} - \frac{v}{r}, & \chi_{r\theta} &= -2\frac{\partial^2 w}{r\partial r\partial \theta} + \frac{\partial w}{2r^2 \partial \theta}
\end{aligned}
\qquad (4.36)
$$

Thus, the strain-displacement relations for an arbitrary point in the plate space can be defined as:

$$
\begin{aligned}
\varepsilon_r &= \varepsilon_r^0 + z\chi_r \\
\varepsilon_\theta &= \varepsilon_\theta^0 + z\chi_\theta \\
\gamma_{r\theta} &= \gamma_{r\theta}^0 + z\chi_{r\theta}
\end{aligned}
\qquad (4.37)
$$

Considering Hooke's law, the corresponding stresses in the plate space are:

$$
\left\{
\begin{array}{c}
\sigma_r \\
\sigma_\theta \\
\tau_{r\theta}
\end{array}
\right\}_k
=
\left[
\begin{array}{ccc}
\overline{Q}_{11}^k & \overline{Q}_{12}^k & \overline{Q}_{16}^k \\
\overline{Q}_{12}^k & \overline{Q}_{22}^k & \overline{Q}_{26}^k \\
\overline{Q}_{16}^k & \overline{Q}_{26}^k & \overline{Q}_{66}^k
\end{array}
\right]
\left\{
\begin{array}{c}
\varepsilon_r \\
\varepsilon_\theta \\
\gamma_{r\theta}
\end{array}
\right\}_k
\qquad (4.38)
$$

The lamina stiffness coefficients $\overline{Q}_{ij}^k$ ($i, j = 1, 2, 6$) are given in Eq. (1.12). By carrying the integration of stresses over the cross-section, the force and moment resultants can be obtained in terms of the middle surface strains and curvature changes as

$$
\begin{bmatrix} N_r \\ N_\theta \\ N_{r\theta} \\ M_r \\ M_\theta \\ M_{r\theta} \end{bmatrix} = \begin{bmatrix} A_{11} & A_{12} & A_{16} & B_{11} & B_{12} & B_{16} \\ A_{12} & A_{22} & A_{26} & B_{12} & B_{22} & B_{26} \\ A_{16} & A_{26} & A_{66} & B_{16} & B_{26} & B_{66} \\ B_{11} & B_{12} & B_{16} & D_{11} & D_{12} & D_{16} \\ B_{12} & B_{22} & B_{26} & D_{12} & D_{22} & D_{26} \\ B_{16} & B_{26} & B_{66} & D_{16} & D_{26} & D_{66} \end{bmatrix} \begin{bmatrix} \varepsilon_r^0 \\ \varepsilon_\theta^0 \\ \gamma_{r\theta}^0 \\ \chi_r \\ \chi_\theta \\ \chi_{r\theta} \end{bmatrix} \tag{4.39}
$$

where N_r, N_θ and $N_{r\theta}$ are the normal and shear force resultants and M_r, M_θ, $M_{r\theta}$ denote the bending and twisting moment resultants. A_{ij}, B_{ij}, and D_{ij} are the stiffness coefficients arising from the piecewise integration over the plate thickness direction, they can be written as in Eq. (1.15).

Substituting Eq. (4.35) into Eqs. (1.16), (1.17), (1.19) and (1.21), the energy functions of a sectorial plate with general boundary conditions can be obtained.

$$
U_s = \frac{1}{2} \int_r \int_\theta \left\{ \begin{matrix} N_r \varepsilon_r^0 + N_\theta \varepsilon_\theta^0 + N_{r\theta} \gamma_{r\theta}^0 \\ + M_r \chi_r + M_\theta \chi_\theta + M_{r\theta} \chi_{r\theta} \end{matrix} \right\} r dr d\theta \tag{4.40a}
$$

$$
T = \frac{1}{2} \int_r \int_\theta I_0 \left\{ (\partial u/\partial t)^2 + (\partial v/\partial t)^2 + (\partial w/\partial t)^2 \right\} r dr d\theta \tag{4.40b}
$$

$$
W_e = \int_r \int_\theta \{ q_r u + q_\theta v + q_z w \} r dr d\theta \tag{4.40c}
$$

$$
\begin{aligned}
U_{sp} = &\frac{1}{2} \int_\theta \left\{ \begin{matrix} \left[k_{r0}^u u^2 + k_{r0}^v v^2 + k_{r0}^w w^2 + K_{r0}^w (\partial w/\partial r)^2 \right]_{r=R_0} \\ + \left[k_{r1}^u u^2 + k_{r1}^v v^2 + k_{r1}^w w^2 + K_{r1}^w (\partial w/\partial r)^2 \right]_{r=R_1} \end{matrix} \right\} r d\theta \\
+ &\frac{1}{2} \int_r \left\{ \begin{matrix} \left[k_{\theta0}^u u^2 + k_{\theta0}^v v^2 + k_{\theta0}^w w^2 + K_{\theta0}^w (\partial w/r\partial\theta)^2 \right]_{\theta=0} \\ + \left[k_{\theta1}^u u^2 + k_{\theta1}^v v^2 + k_{\theta1}^w w^2 + K_{\theta1}^w (\partial w/r\partial\theta)^2 \right]_{\theta=\theta_0} \end{matrix} \right\} dr
\end{aligned} \tag{4.40d}
$$

where U_s and T are the stain and kinetic energy functions. W_e represents external work done by the external loads, in which q_r, q_θ and q_z denote the external loads in the r, θ and z directions, respectively. U_{sp} is boundary spring deformation energy function introduced by the artificial spring boundary technique (see Sect. 1.2.3 and Fig. 4.16).

Specializing Eqs. (1.33) and (1.34) to those of sectorial plates results in following governing equations

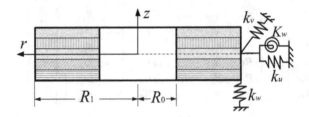

Fig. 4.16 Boundary conditions of a thin laminated sectorial plate

$$\frac{\partial(rN_r)}{\partial r} + \frac{\partial(N_{r\theta})}{\partial\theta} - N_\theta + rq_r = rI_0\frac{\partial^2 u}{\partial t^2}$$

$$\frac{\partial(rN_{r\theta})}{\partial r} + \frac{\partial(N_\theta)}{\partial\theta} + N_{r\theta} + rq_\theta = rI_0\frac{\partial^2 v}{\partial t^2} \qquad (4.41)$$

$$\frac{\partial(rQ_r)}{\partial r} + \frac{\partial(Q_\theta)}{\partial\theta} + rq_z = rI_0\frac{\partial^2 w}{\partial t^2}$$

where

$$Q_r = \frac{\partial(rM_r)}{r\partial r} + \frac{\partial(M_{r\theta})}{r\partial\theta} - \frac{M_\theta}{r}$$

$$Q_\theta = \frac{\partial(M_\theta)}{r\partial\theta} + \frac{\partial(rM_{r\theta})}{r\partial r} + \frac{M_{r\theta}}{r} \qquad (4.42)$$

For a sectorial plate symmetrically laminated with respect to the middle surface, the first two governing equations are decoupled from the third one, i.e., the in-plane vibration (u, v) will be decoupled from the bending vibration (w). The corresponding boundary conditions of thin laminated sectorial plates are:

$$r = R_0: \begin{cases} N_r - k_{r0}^u u = 0 \\ N_{r\theta} - k_{r0}^v v = 0 \\ Q_r + \dfrac{\partial M_{r\theta}}{r\partial\theta} - k_{r0}^w w = 0 \\ -M_r - K_{r0}^w\dfrac{\partial w}{\partial r} = 0 \end{cases} \quad r = R_1: \begin{cases} N_r + k_{r1}^u u = 0 \\ N_{r\theta} + k_{r1}^v v = 0 \\ Q_r + \dfrac{\partial M_{r\theta}}{r\partial\theta} + k_{r1}^w w = 0 \\ -M_r + K_{r1}^w\dfrac{\partial w}{\partial r} = 0 \end{cases}$$

$$\theta = 0: \begin{cases} N_{r\theta} - k_{\theta0}^u u = 0 \\ N_\theta - k_{\theta0}^v v = 0 \\ Q_\theta + \dfrac{\partial M_{r\theta}}{\partial r} - k_{\theta0}^w w = 0 \\ -M_\theta - K_{\theta0}^w\dfrac{\partial w}{r\partial\theta} = 0 \end{cases} \quad \theta = \theta_0: \begin{cases} N_{r\theta} + k_{\theta1}^u u = 0 \\ N_\theta + k_{\theta1}^v v = 0 \\ Q_\theta + \dfrac{\partial M_{r\theta}}{\partial r} + k_{\theta1}^w w = 0 \\ -M_\theta + K_{\theta1}^w\dfrac{\partial w}{r\partial\theta} = 0 \end{cases}$$

(4.43)

4.4.2 Fundamental Equations of Thick Laminated Sectorial, Annular and Circular Plates

Fundamental equations of thick sectorial plates are obtained by directly substituting Eq. (4.35) into those of thick laminated shells given in Sect. 1.3.

In the framework of first-order shear deformation plate theory, the displacement field in an arbitrary point of a thick laminated sectorial plate is expressed in terms of middle surface displacements and rotation components as:

$$
\begin{aligned}
U(r,\theta,z) &= u(r,\theta) + z\phi_r(r,\theta) \\
V(r,\theta,z) &= v(r,\theta) + z\phi_\theta(r,\theta) \\
W(r,\theta,z) &= w(r,\theta)
\end{aligned}
\tag{4.44}
$$

where ϕ_r and ϕ_θ represent the rotations of transverse normal respect to θ and r directions. Thus, the corresponding strains at this point are defined in terms of middle surface strains, curvature and twist changes as

$$
\begin{aligned}
\varepsilon_r &= \varepsilon_r^0 + z\chi_r, & \gamma_{rz} &= \gamma_{rz}^0 \\
\varepsilon_\theta &= \varepsilon_\theta^0 + z\chi_\theta, & \gamma_{\theta z} &= \gamma_{\theta z}^0 \\
\gamma_{r\theta} &= \gamma_{r\theta}^0 + z\chi_{r\theta}
\end{aligned}
\tag{4.45}
$$

where the middle surface strains and curvature and twist changes are written as:

$$
\begin{aligned}
\varepsilon_r^0 &= \frac{\partial u}{\partial r}, & \chi_r &= \frac{\partial \phi_r}{\partial r} \\
\varepsilon_\theta^0 &= \frac{\partial v}{r\partial \theta} + \frac{u}{r}, & \chi_\theta &= \frac{\partial \phi_\theta}{r\partial \theta} + \frac{\phi_r}{r} \\
\gamma_{r\theta}^0 &= \frac{\partial v}{\partial r} + \frac{\partial u}{r\partial \theta} - \frac{v}{r}, & \chi_{r\theta} &= \frac{\partial \phi_\theta}{\partial r} + \frac{\partial \phi_r}{r\partial \theta} - \frac{\phi_\theta}{r} \\
\gamma_{rz}^0 &= \frac{\partial w}{\partial r} + \phi_r \\
\gamma_{\theta z}^0 &= \frac{\partial w}{r\partial \theta} + \phi_\theta
\end{aligned}
\tag{4.46}
$$

According to Eqs. (1.46) and (1.47), the force and moment resultant equations of thick sectorial plates become:

$$
\begin{bmatrix}
N_r \\ N_\theta \\ N_{r\theta} \\ M_r \\ M_\theta \\ M_{r\theta}
\end{bmatrix}
=
\begin{bmatrix}
A_{11} & A_{12} & A_{16} & B_{11} & B_{12} & B_{16} \\
A_{12} & A_{22} & A_{26} & B_{12} & B_{22} & B_{26} \\
A_{16} & A_{26} & A_{66} & B_{16} & B_{26} & B_{66} \\
B_{11} & B_{12} & B_{16} & D_{11} & D_{12} & D_{16} \\
B_{12} & B_{22} & B_{26} & D_{12} & D_{22} & D_{26} \\
B_{16} & B_{26} & B_{66} & D_{16} & D_{26} & D_{66}
\end{bmatrix}
\begin{bmatrix}
\varepsilon_r^0 \\ \varepsilon_\theta^0 \\ \gamma_{r\theta}^0 \\ \chi_r \\ \chi_\theta \\ \chi_{r\theta}
\end{bmatrix}
\tag{4.47}
$$

$$\begin{bmatrix} Q_\theta \\ Q_r \end{bmatrix} = \begin{bmatrix} A_{44} & A_{45} \\ A_{45} & A_{55} \end{bmatrix} \begin{bmatrix} \gamma_{\theta z}^0 \\ \gamma_{rz}^0 \end{bmatrix} \tag{4.48}$$

Substituting Eq. (4.35) into Eqs. (1.49), (1.51), (1.53) and (1.54) yields the energy functions of a sectorial plate including shear deformation and rotary inertia:

$$U_s = \frac{1}{2} \int_r \int_\theta \left\{ \begin{array}{l} N_r \varepsilon_r^0 + N_\theta \varepsilon_\theta^0 + N_{r\theta} \gamma_{r\theta}^0 + M_r \chi_r \\ + M_\theta \chi_\theta + M_{r\theta} \chi_{r\theta} + Q_r \gamma_{rz}^0 + Q_\theta \gamma_{\theta z}^0 \end{array} \right\} r dr d\theta \tag{4.49a}$$

$$T = \frac{1}{2} \int_r \int_\theta \left\{ \begin{array}{l} I_0 \left(\frac{\partial u}{\partial t}\right)^2 + I_0 \left(\frac{\partial v}{\partial t}\right)^2 + I_0 \left(\frac{\partial w}{\partial t}\right)^2 + I_2 \left(\frac{\partial \phi_r}{\partial t}\right)^2 \\ + 2I_1 \frac{\partial u}{\partial t} \frac{\partial \phi_r}{\partial t} + 2I_1 \frac{\partial v}{\partial t} \frac{\partial \phi_\theta}{\partial t} + I_2 \left(\frac{\partial \phi_\theta}{\partial t}\right)^2 \end{array} \right\} r dr d\theta \tag{4.49b}$$

$$W_e = \int_r \int_\theta \{ q_r u + q_\theta v + q_z w + m_r \phi_r + m_\theta \phi_\theta \} r dr d\theta \tag{4.49c}$$

$$\begin{aligned} U_{sp} = &\frac{1}{2} \int_\theta \left\{ \begin{array}{l} \left[k_{r0}^u u^2 + k_{r0}^v v^2 + k_{r0}^w w^2 + K_{r0}^r \phi_r^2 + K_{r0}^\theta \phi_r^2 \right]_{r=R_0} \\ + \left[k_{r1}^u u^2 + k_{r1}^v v^2 + k_{r1}^w w^2 + K_{r1}^r \phi_r^2 + K_{r1}^\theta \phi_r^2 \right]_{r=R_1} \end{array} \right\} r d\theta \\ + &\frac{1}{2} \int_r \left\{ \begin{array}{l} \left[k_{\theta 0}^u u^2 + k_{\theta 0}^v v^2 + k_{\theta 0}^w w^2 + K_{\theta 0}^r \phi_r^2 + K_{\theta 0}^\theta \phi_r^2 \right]_{\theta=0} \\ + \left[k_{\theta 1}^u u^2 + k_{\theta 1}^v v^2 + k_{\theta 1}^w w^2 + K_{\theta 1}^r \phi_r^2 + K_{\theta 1}^\theta \phi_r^2 \right]_{\theta=\theta_0} \end{array} \right\} dr \end{aligned} \tag{4.49d}$$

where U_s, T and W_e are the stain energy, kinetic energy and external work functions. U_{sp} represents boundary spring deformation energy function introduced by the artificial spring boundary technique (see Sect. 1.3.3 and Fig. 4.17).

The governing equations and boundary conditions for thick laminated sectorial plates are obtained by substituting Eq. (4.35) into Eqs. (1.59), (1.60) and (1.61). The governing equations are

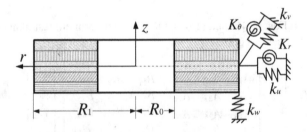

Fig. 4.17 Boundary conditions of a thick laminated sectorial plate

$$\frac{\partial(rN_r)}{\partial r} + \frac{\partial(N_{r\theta})}{\partial \theta} - N_\theta + rq_r = rI_0\frac{\partial^2 u}{\partial t^2} + rI_1\frac{\partial^2 \phi_r}{\partial t^2}$$

$$\frac{\partial(rN_{r\theta})}{\partial r} + \frac{\partial(N_\theta)}{\partial \theta} + N_{r\theta} + rq_\theta = rI_0\frac{\partial^2 v}{\partial t^2} + rI_1\frac{\partial^2 \phi_\theta}{\partial t^2}$$

$$\frac{\partial(rQ_r)}{\partial r} + \frac{\partial(Q_\theta)}{\partial \theta} + rq_z = rI_0\frac{\partial^2 w}{\partial t^2} \tag{4.50}$$

$$\frac{\partial(rM_r)}{\partial r} + \frac{\partial(M_{r\theta})}{\partial \theta} - M_\theta - rQ_r + rm_r = rI_1\frac{\partial^2 u}{\partial t^2} + rI_2\frac{\partial^2 \phi_r}{\partial t^2}$$

$$\frac{\partial(rM_{r\theta})}{\partial r} + \frac{\partial(M_\theta)}{\partial \theta} + M_{r\theta} - rQ_\theta + rm_\theta = rI_1\frac{\partial^2 v}{\partial t^2} + rI_2\frac{\partial^2 \phi_\theta}{\partial t^2}$$

And the corresponding boundary conditions are

$$r = R_0 : \begin{cases} N_r - k_{r0}^u u = 0 \\ N_{r\theta} - k_{r0}^v v = 0 \\ Q_r - k_{r0}^w w = 0 \\ M_r - K_{r0}^r \phi_r = 0 \\ M_{r\theta} - K_{r0}^\theta \phi_\theta = 0 \end{cases} \quad r = R_1 : \begin{cases} N_r + k_{r1}^u u = 0 \\ N_{r\theta} + k_{r1}^v v = 0 \\ Q_r + k_{r1}^w w = 0 \\ M_r + K_{r1}^r \phi_r = 0 \\ M_{r\theta} + K_{r1}^\theta \phi_\theta = 0 \end{cases}$$

$$\theta = 0 : \begin{cases} N_{r\theta} - k_{\theta 0}^u u = 0 \\ N_\theta - k_{\theta 0}^v v = 0 \\ Q_\theta - k_{\theta 0}^w w = 0 \\ M_{r\theta} - K_{\theta 0}^r \phi_r = 0 \\ M_\theta - K_{\theta 0}^\theta \phi_\theta = 0 \end{cases} \quad \theta = \theta_0 : \begin{cases} N_{r\theta} + k_{\theta 1}^u u = 0 \\ N_\theta + k_{\theta 1}^v v = 0 \\ Q_\theta + k_{\theta 1}^w w = 0 \\ M_{r\theta} + K_{\theta 1}^r \phi_r = 0 \\ M_\theta + K_{\theta 1}^\theta \phi_\theta = 0 \end{cases} \tag{4.51}$$

4.5 Vibration of Laminated Sectorial, Annular and Circular Plates

4.5.1 Vibration of Laminated Annular and Circular Plates

In this section, we consider free vibration of homogeneous and laminated annular and circular plates with general boundary conditions. The homogeneous circular plates are treated as special cases of the laminated ones in the studies. Only solutions considering effects of shear deformation are given in this section. The weak form solution procedure is adopted in the calculations.

There are two boundaries in an annular plate, in this work a two-letter character is employed to represent the boundary condition of an annular plate, such as FC identifies the plate with inner edge free and outer edge clamped. Unless otherwise stated, the natural frequencies of the considered plates are expressed in the non-dimensional parameters as $\Omega = \omega R_1\sqrt{\rho h/D_{11}}$. Unless otherwise stated, material properties of the layers of laminated annular and circular plates under consideration are given as: $E_2 = 10$ GPa, $E_1/E_2 =$ open, $\mu_{12} = 0.25$, $G_{12} = 0.6E_2$, $G_{13} = 0.6E_2$, $G_{23} = 0.5E_2$ and $\rho = 1{,}500$ kg/m^3 (subscript 1 and 2 represent the principle directions of the material and they are paralleled to r and θ directions, respectively).

Considering the circumferential symmetry of annular and circular plates, each displacement and rotation component of a laminated annular or circular plate is expanded as a modified Fourier series in the following form:

$$u(r, \theta) = \sum_{m=0}^{M} \sum_{n=0}^{N} A_{mn} \cos \lambda_m r \cos n\theta + \sum_{l=1}^{2} \sum_{n=0}^{N} a_{ln} P_l(r) \cos n\theta$$

$$v(r, \theta) = \sum_{m=0}^{M} \sum_{n=0}^{N} B_{mn} \cos \lambda_m r \, \sin n\theta + \sum_{l=1}^{2} \sum_{n=0}^{N} b_{ln} P_l(r) \sin n\theta$$

$$w(r, \theta) = \sum_{m=0}^{M} \sum_{n=0}^{N} C_{mn} \cos \lambda_m r \cos n\theta + \sum_{l=1}^{2} \sum_{n=0}^{N} c_{ln} P_l(r) \cos n\theta \qquad (4.52)$$

$$\phi_r(r, \theta) = \sum_{m=0}^{M} \sum_{n=0}^{N} D_{mn} \cos \lambda_m r \cos n\theta + \sum_{l=1}^{2} \sum_{n=0}^{N} d_{ln} P_l(r) \cos n\theta$$

$$\phi_\theta(r, \theta) = \sum_{m=0}^{M} \sum_{n=0}^{N} E_{mn} \cos \lambda_m r \, \sin n\theta + \sum_{l=1}^{2} \sum_{n=0}^{N} e_{ln} P_l(r) \sin n\theta$$

where $\lambda_m = m\pi/\Delta R$, ($\Delta R = R_1 - R_0$). n represents the circumferential wave number of the corresponding mode. It should be noted that n is non-negative integer. Interchanging of sin $n\theta$ and cos $n\theta$ in Eq. (4.52), another set of free vibration modes (anti-symmetric modes) can be obtained. $P_l(r)$ denote the auxiliary polynomial functions introduced to remove all the discontinuities potentially associated with the first-order derivatives at the boundaries. These auxiliary functions are in the same forms as those of Eq. (4.33). Note that the modified Fourier series presented in Eq. (4.52) are complete series defined over the domain [0, ΔR]. Therefore, linear transformations for coordinates from $r \in [R_0, R_1]$ to [0, ΔR] need to be introduced for the practical programming and computing.

In Table 4.14 (Jin et al. 2014a), the first six frequencies (Hz) of a single-layered, moderately thick composite annular plate with completely free boundary conditions and different truncated configurations are chosen to demonstrate the convergence of the current method. Considering the circumferential symmetry of the annular plate, the expression terms with respect to θ in displacements and rotation components automatically satisfy the governing equations and boundary conditions. Thus, the convergence only needs to be checked in the axial direction (r). The geometric and material constants of the plate are: $E_1/E_2 = 15$, $R_0 = 1$ m, $R_1 = 3$ m, $h = 0.1$ m. In all the following computations, the zero frequencies corresponding to the rigid body modes were omitted from the results. The table shows the present solutions has fast convergence behavior.

Table 4.14 Convergence of the first six frequencies (Hz) of a single-layered composite annular plate with FF boundary conditions

$M \times N$	Mode number					
	1	2	3	4	5	6
11×10	7.2421	11.244	18.650	28.423	33.407	51.069
12×10	7.2417	11.244	18.649	28.417	33.405	51.067
13×10	7.2415	11.244	18.649	28.415	33.405	51.067
14×10	7.2414	11.244	18.648	28.413	33.404	51.066
15×10	7.2414	11.244	18.648	28.413	33.404	51.066

Table 4.15 Comparison of the fundamental frequency parameters Ω for composite laminated annular plates with different boundary conditions and lamination schemes ($R_0/R_1 = 0.5$)

B.C.	Method	h/R_1	Lamination schemes			
			[I]	[II]	[I/II/I/II]	[I/II/II/I]
CC	Lin and Tseng (1998)	0.10	84.977	66.057	75.706	79.226
		0.05	102.62	82.416	92.216	98.656
		0.02	110.04	89.776	99.272	107.37
		0.01	111.24	91.002	100.43	108.83
	Present	0.10	87.134	68.693	78.711	81.861
		0.05	103.58	83.591	94.013	99.876
		0.02	110.19	89.928	100.22	107.58
		0.01	111.21	90.960	101.22	108.81
SC	Lin and Tseng (1998)	0.10	76.596	54.319	65.559	71.644
		0.05	88.679	63.002	75.585	84.914
		0.02	93.307	66.337	79.362	90.250
		0.01	94.033	66.860	79.987	91.102
	Present	0.10	77.603	55.899	67.959	73.015
		0.05	89.256	63.675	77.562	85.670
		0.02	93.571	66.549	80.997	90.541
		0.01	94.189	66.994	81.528	91.265
FC	Lin and Tseng (1998)	0.10	37.329	20.636	30.173	35.010
		0.05	41.246	21.851	32.847	39.320
		0.02	42.591	22.233	33.737	40.850
		0.01	42.794	22.290	33.870	41.084
	Present	0.10	38.150	20.936	30.979	35.896
		0.05	41.582	21.995	33.371	39.689
		0.02	42.728	22.323	34.151	40.993
		0.01	42.899	22.371	34.267	41.190

To validate the accuracy and reliability of current solutions, comparison of the fundamental frequency parameters Ω for a composite laminated annular plate ($R_0/R_1 = 0.5$) when the outer edge is clamped and the inner edge is either clamped, simply supported or free is presented in Table 4.15 (Jin et al. 2014a), in which four

Table 4.16 Frequency parameters Ω of a $[0°/90°]$ laminated circular plate with different boundary conditions ($R_1 = 1$)

h/R_1	Mode	$E_1/E_2 = 1$ (isotropic)			$E_1/E_2 = 15$		
		F	S	C	F	S	C
0.01	1	6.0186	4.8598	10.21379	3.4024	3.8685	8.8441
	2	8.8888	14.596	22.087	6.8133	8.3294	13.555
	3	13.747	27.196	36.53708	7.8292	15.542	21.015
	4	21.634	29.650	39.74355	11.828	24.076	30.300
0.05	1	5.9897	4.8524	10.16563	3.3735	3.8493	8.6463
	2	8.8615	14.525	21.86786	6.7621	8.2339	13.184
	3	13.631	26.945	35.95093	7.7342	15.230	20.229
	4	21.431	29.366	39.10696	11.630	23.360	28.832
0.1	1	5.9332	4.8295	10.01998	3.3116	3.7898	8.0874
	2	8.7785	14.311	21.22982	6.6109	7.9643	12.215
	3	13.387	26.221	34.32761	7.4922	14.397	18.329
	4	20.906	28.539	37.32406	11.107	21.526	25.448

different thickness-radius ratios are included, i.e., $h/R_1 = 0.1, 0.05, 0.02$ and 0.01, corresponding to thick to thin laminated annular plates. Four types of lamination schemes included in the comparison are: [I], [II], [I/II/I/II] and [I/II /II/I], where I and II represent two kinds of composite laminate, their material properties are given as: for material I: $E_2/E_1 = 50$, $G_{12} = 0.6613E_1$, $G_{13} = G_{23} = 0.5511E_1$, $\mu_{12} = 0.006$; for material II: $E_2/E_1 = 5$, $G_{12} = 0.35E_1$, $G_{13} = G_{23} = 0.292E_1$, $\mu_{12} = 0.06$. By comparing, we can find that the discrepancies between present results and solutions reported by Lin and Tseng (1998) based on an eight node isoparametric finite element method are acceptable. The discrepancy in the results may be attributed to different solution approaches are used in the literature. The table also shows that natural frequencies are influenced by stacking sequence, the order of the magnitude of the fundamental frequencies for the four different lamination schemes being [I] > [I/II/II/I] > [I/II/I/II] > [II].

For a circular plate, there is only one boundary. Table 4.16 shows the lowest three frequency parameters Ω for a $[0°/90°]$ laminated circular plate with different boundary conditions and thickness-to-outer radius ratios (h/R_1). The F, S and C boundary conditions, two orthotropy ratio $E_1/E_2 = 1$ and 25, and thickness-to-outer radius ratios of 0.01, 0.05 and 0.1 are used. As seen in the table, the frequency parameters for the circular plate with $E_1/E_2 = 25$ are smaller than those of $E_1/E_2 = 1$. And the frequency parameters decrease with the thickness-to-outer radius ratio increases. The corresponding contour mode shapes for the plate with orthotropy ratio $E_1/E_2 = 25$ and thickness-to-outer radius ratios $h/R_1 = 0.05$ are given in Fig. 4.18. As seen in the figure, the mode shapes for S boundary conditions are similar to those of C boundary conditions.

Table 4.17 presents the first three frequency parameters Ω of a $[0°/90°]$ laminated annular plate with different boundary conditions and various inner-to-outer

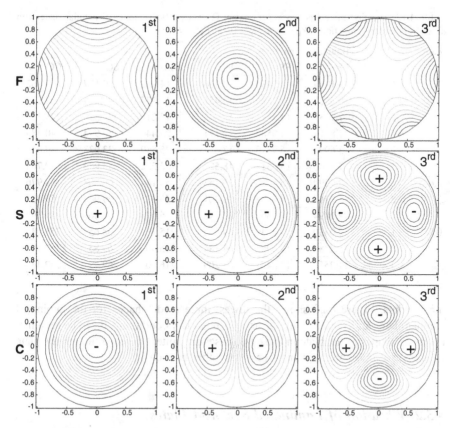

Fig. 4.18 Mode shapes for a [0°/90°] laminated circular plate with different boundary conditions ($h/R_1 = 0.05$, $E_1/E_2 = 15$)

radius ratios (R_0/R_1). The considered plate is assumed to be made of composite layers with following parameters: $R_1 = 1$, $h/\Delta R = 0.05$, $E_1/E_2 = 15$. Four inner-to-outer radius ratios, i.e., $R_0/R_1 = 0.2, 0.4, 0.6$ and 0.8 and six sets of boundary conditions, i.e., FF, FS, FC, SS, SC and CC are studied. From the table, we can see that frequency parameters for the plate with FC, SS, SC and CC boundary conditions increase with inner-to-outer radius ratio increases. However, for FF boundary conditions, the maximum fundamental, second and third frequency parameters occur at inner-to-outer radius ratios of 0.2, 0.6 and 0.8, respectively and for FS boundary conditions, the minimum fundamental, second and third frequency parameters separately occur at inner-to-outer radius ratios of 0.2, 0.4 and 0.8, respectively. In order to enhance our understanding of the effects of the inner-to-outer radius ratios on vibrations of annular plates, the lowest three mode shapes for the laminated annular plate with FC boundary conditions are given in Fig. 4.19. Due to the circumferential symmetry of the annular plate, the mode shapes are symmetrical as well. These figures also show that mode shapes are influenced by

Table 4.17 The first three frequency parameters Ω of a $[0°/90°]$ laminated annular plate with different boundary conditions ($R_1 = 1$, $h/\Delta R = 0.05$, $E_1/E_2 = 15$)

R_0/R_1	Mode	Boundary conditions					
		FF	FS	FC	SS	SC	CC
0.2	1	3.1899	3.5483	9.3364	14.912	20.177	23.162
	2	5.4501	7.7562	12.451	15.777	20.824	24.238
	3	7.7137	14.481	19.806	17.211	22.320	24.421
0.4	1	2.8314	4.0587	12.171	25.444	32.994	40.389
	2	5.6991	6.9513	12.853	25.519	33.033	40.637
	3	7.4366	13.351	18.527	26.532	33.932	40.768
0.6	1	2.4299	5.4710	20.153	55.144	69.808	88.976
	2	6.7767	7.5738	20.301	55.323	69.963	89.035
	3	7.1894	12.736	22.748	56.113	70.660	89.392
0.8	1	2.0330	9.8049	64.064	215.18	268.65	351.93
	2	5.8389	12.098	64.184	215.40	268.85	352.06
	3	11.255	17.545	65.209	216.10	269.49	352.47

inner radius. Figure 4.20 shows the lowest three mode shapes for the laminated annular plate ($R_0/R_1 = 0.4$) with CF and FC boundary conditions. It can be seen that the mode shapes of the plate in these two cases are quite different although the plate is camped at one boundary and free at the other boundary.

4.5.2 Vibration of Laminated Sectorial Plates

In this section, we consider free vibration of laminated sectorial plates with general boundary conditions. Similar to the studies performed earlier for laminated annular and circular plates, the results given in this section are obtained by using the shear deformation plate theory (SDPT) and weak form solution procedure. Only plates having polar orthotropy will be studied.

For a general sectorial plate, there exist four boundaries, i.e., $r = R_0$, $r = R_1$, $\theta = 0$ and $\theta = \theta_0$. For the sake of brevity, a four-letter character is employed to represent the boundary condition of a sectorial plate, such as FCSC identifies the plate with F, C, S and C boundary conditions at boundaries $r = R_0$, $\theta = 0$, $r = R_1$ and $\theta = \theta_0$, respectively. In addition, unless otherwise stated, the non-dimensional frequency parameter $\Omega = \omega R_1 \sqrt{\rho h/D_{11}}$ is used and material properties of the layers of laminated sectorial plates under consideration are given as: $E_2 = 10$ GPa, $E_1/E_2 = $ open, $\mu_{12} = 0.25$, $G_{12} = 0.6E_2$, $G_{13} = 0.6E_2$, $G_{23} = 0.5E_2$ and $\rho = 1{,}500$ kg/m^3.

For a circular or annular plate, the assumed 2D displacement field can be reduced to a quasi 1D problem through Fourier decomposition of the circumferential wave motion. However, for a general sectorial plate, the assumption of whole periodic wave numbers in the circumferential direction is inappropriate, and thus, a set of complete two-dimensional analysis is required and resort must be made to a full two-dimensional solution scheme. Therefore, each displacement and rotation component of the laminated sectorial plates is expanded as a two-dimensional modified Fourier series as:

$$u(r,\theta) = \sum_{m=0}^{M} \sum_{n=0}^{N} A_{mn} \cos \lambda_m r \cos \lambda_n \theta + \sum_{l=1}^{2} \sum_{n=0}^{N} a_{ln} P_l(r) \cos \lambda_n \theta$$
$$+ \sum_{l=1}^{2} \sum_{m=0}^{M} b_{lm} P_l(\theta) \cos \lambda_m r$$

$$v(r,\theta) = \sum_{m=0}^{M} \sum_{n=0}^{N} B_{mn} \cos \lambda_m r \cos \lambda_n \theta + \sum_{l=1}^{2} \sum_{n=0}^{N} c_{ln} P_l(r) \cos \lambda_n \theta$$
$$+ \sum_{l=1}^{2} \sum_{m=0}^{M} d_{lm} P_l(\theta) \cos \lambda_m r$$

$$w(r,\theta) = \sum_{m=0}^{M} \sum_{n=0}^{N} C_{mn} \cos \lambda_m r \cos \lambda_n \theta + \sum_{l=1}^{2} \sum_{n=0}^{N} e_{ln} P_l(r) \cos \lambda_n \theta$$
$$+ \sum_{l=1}^{2} \sum_{m=0}^{M} f_{lm} P_l(\theta) \cos \lambda_m r$$

$$\phi_r(r,\theta) = \sum_{m=0}^{M} \sum_{n=0}^{N} D_{mn} \cos \lambda_m r \cos \lambda_n \theta + \sum_{l=1}^{2} \sum_{n=0}^{N} g_{ln} P_l(r) \cos \lambda_n \theta$$
$$+ \sum_{l=1}^{2} \sum_{m=0}^{M} h_{lm} P_l(\theta) \cos \lambda_m r$$

$$\phi_\theta(r,\theta) = \sum_{m=0}^{M} \sum_{n=0}^{N} E_{mn} \cos \lambda_m r \cos \lambda_n \theta + \sum_{l=1}^{2} \sum_{n=0}^{N} i_{ln} P_l(r) \cos \lambda_n \theta$$
$$+ \sum_{l=1}^{2} \sum_{m=0}^{M} j_{lm} P_l(\theta) \cos \lambda_m r$$

(4.53)

where $\lambda_m = m\pi/\Delta R$ and $\lambda_n = n\pi/\theta_0$. $P_l(r)$ and $P_l(\theta)$ denote the auxiliary polynomial functions introduced to remove all the discontinuities potentially associated with the first-order derivatives at the boundaries. These auxiliary functions are in the same form as those of Eq. (4.33). Similarly, linear transformations for coordinates from $r \in [R_0, R_1]$ to $[0, \Delta R]$ need to be introduced for the practical programming and computing.

Table 4.18 shows a convergence and comparison study of the lowest six natural frequencies (Hz) for an isotropic ($E = 210$ GPa, $\mu = 0.3$ and $\rho = 7,800$ kg/m^3) sectorial plate with FFFF and CCCC boundary conditions. The sectorial plate having inner radius $R_0 = 0.5$ m, outer radius $R_1 = 1$ m, thickness $h = 0.01$ m and circumferential dimension $\theta_0 = \pi$. The present results are compared with those of

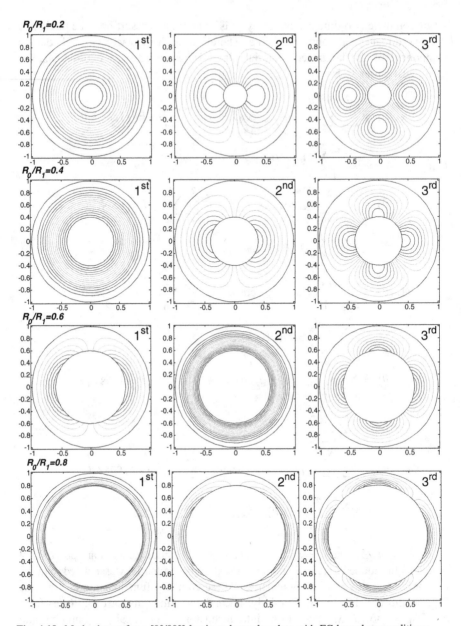

Fig. 4.19 Mode shapes for a [0°/90°] laminated annular plate with FC boundary conditions

FEM analysis (ANSYS with element type of SHELL63 and element size of 0.01 m). As can be seen from Table 4.18, the frequencies have converged mono-tonically up to four significant figures as the truncation numbers increase. The table also shows good agreements in the comparison.

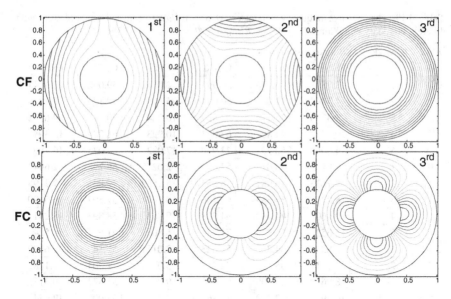

Fig. 4.20 Mode shapes for a [0°/90°] laminated annular plate ($R_0/R_1 = 0.4$) with CF and FC boundary conditions

Table 4.18 Convergence and comparison of frequencies (Hz) for an isotropic sectorial plate with FFFF and CCCC boundary conditions ($R_0 = 0.5$ m, $R_1/R_0 = 2$, $h/R_0 = 0.02$, $\theta_0 = \pi$, $E = 210$ GPa, $\mu = 0.3$ and $\rho = 7,800$ kg/m^3)

Boundary conditions	$M \times N$	Mode number					
		1	2	3	4	5	6
FFFF	15 × 15	17.775	18.739	38.540	42.003	69.740	71.841
	15 × 16	17.760	18.738	38.537	41.999	69.706	71.838
	15 × 17	17.757	18.736	38.537	41.990	69.699	71.833
	15 × 18	17.747	18.735	38.535	41.988	69.683	71.831
	16 × 18	17.742	18.733	38.524	41.983	69.677	71.810
	17 × 18	17.739	18.730	38.520	41.982	69.675	71.807
	18 × 18	17.735	18.728	38.510	41.979	69.670	71.790
	ANSYS	17.744	18.729	38.677	42.006	69.741	72.094
CCCC	15 × 15	225.33	234.63	251.37	276.82	311.91	357.42
	15 × 16	225.33	234.62	251.32	276.67	311.73	356.71
	15 × 17	225.33	234.60	251.27	276.59	311.47	356.47
	15 × 18	225.32	234.59	251.24	276.51	311.36	356.08
	16 × 18	225.31	234.57	251.23	276.49	311.34	356.06
	17 × 18	225.31	234.57	251.22	276.48	311.33	356.06
	18 × 18	225.30	234.56	251.21	276.47	311.32	356.05
	ANSYS	225.97	235.24	251.88	277.09	311.76	356.24

Table 4.19 The first five frequency parameters Ω of a $[0°/90°]$ laminated sectorial plate with different boundary conditions ($R_0 = 0.5$ m, $R_1/R_0 = 2$, $\theta_0 = \pi$, $E_1/E_2 = 15$)

$h/\Delta R$	Mode	Boundary conditions					
		FFFF	SSSS	CCCC	FCFC	CFCF	FSFS
0.01	1	3.9802	36.096	59.994	2.0332	59.062	2.6844
	2	4.7963	37.086	60.694	5.1950	59.126	7.2936
	3	8.3330	39.772	62.835	10.284	60.041	9.6338
	4	10.548	44.802	67.063	10.993	61.283	13.394
	5	14.906	52.593	73.970	17.032	64.075	16.789
0.05	1	3.9119	36.015	57.657	2.0275	56.796	2.6443
	2	4.7635	36.981	58.354	5.1583	56.856	7.2195
	3	8.2257	39.587	60.436	10.177	57.697	9.5120
	4	10.460	44.442	64.528	10.863	58.893	13.241
	5	14.652	51.929	71.121	16.809	61.577	16.524
0.10	1	3.8324	34.360	51.823	2.0151	51.149	2.6109
	2	4.7117	35.297	52.512	5.0908	51.204	7.1182
	3	8.0639	37.800	54.508	9.9678	51.872	9.3185
	4	10.292	42.399	58.366	10.579	53.000	12.974
	5	14.162	49.357	64.439	16.332	55.494	15.953

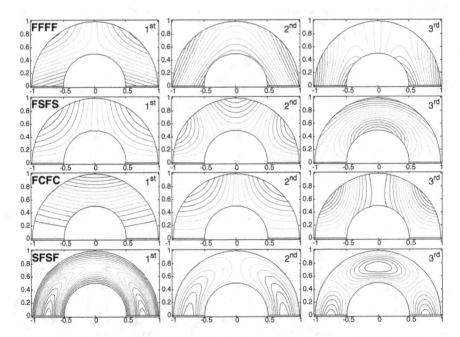

Fig. 4.21 Mode shapes for a $[0°/90°]$ laminated sectorial plate ($R_0/R_1 = 0.5$, $h/\Delta R = 0.05$) with various boundary conditions

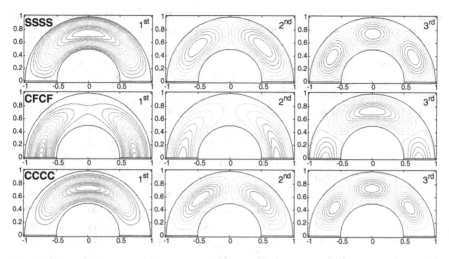

Fig. 4.21 (continued)

The effects of thickness ratio ($h/\Delta R$) on the frequency parameters of sectorial plates are studied by the following example. Table 4.19 shows the lowest five frequency parameters Ω for a [0°/90°] laminated sectorial plate with different boundary conditions and various thickness ratios $h/\Delta R$. The sectorial plate is assume to be made of polar orthotropic layer of orthotropy ratio $E_1/E_2 = 15$. Three types of thickness ratios, i.e., $h/\Delta R = 0.01$, 0.05 and 0.1, corresponding to thin to moderately thick sectorial plates are considered in the investigation. From the table, we can see that the frequency parameters of the plate in all cases decrease with the thickness ratio $h/\Delta R$ increases. Despite this, it should be noted that, the natural frequencies of the plate with the thickness ratio $h/\Delta R$ increases due to the fact that the stiffness of the plate get larger. The corresponding contour mode shapes for the plate with thickness ratio $h/\Delta R = 0.05$ are given in Fig. 4.21 as well. As seen in the figure, the mode shapes for the SSSS boundary conditions are similar to those for boundary conditions of CCCC. Due to the symmetry in the boundary conditions and geometry, the corresponding mode shapes of the plate are symmetrical as well.

The effects of circumferential dimension (i.e., circumferential included angle θ_0) on the frequency parameters of sectorial plates are also investigated. Table 4.20 shows the lowest five frequency parameters Ω for a [0°/90°] laminated sectorial plate with different boundary conditions and circumferential dimensions. The geometry and material parameters of the considered sectorial plate are similar to those used in Table 4.20 except that the current plate having a thickness-to-inner radius ratio $h/\Delta R = 0.1$. From the table, we can see that the fundamental frequency parameter of the plate having CFFF boundary conditions increases with circumferential included angle θ_0 increases while for the plate with CCCF boundary

Table 4.20 The first five frequency parameters Ω of a [0°/90°] laminated sectorial plate with different boundary conditions and circumferential dimensions ($R_0 = 0.5$ m, $R_1/R_0 = 2$, $h/R_0 = 0.1$, $E_1/E_2 = 15$)

θ_0	Mode	Boundary conditions					
		FFFC	FFCC	FCCC	CFFF	CCFF	CCCF
$\pi/4$	1	6.3800	18.779	56.096	7.4494	10.329	52.191
	2	15.500	57.237	86.414	10.749	29.959	69.542
	3	30.226	61.331	108.72	32.499	51.239	115.45
	4	43.421	94.202	149.20	33.550	69.149	123.79
	5	58.385	114.59	151.39	48.868	75.023	137.53
$3\pi/4$	1	0.8700	12.794	17.871	7.9348	8.0795	51.175
	2	2.5360	18.789	27.405	8.0873	9.2724	52.190
	3	6.1158	28.739	41.720	9.3391	12.500	54.578
	4	9.4626	42.736	58.147	13.085	19.445	60.033
	5	10.027	56.919	61.123	20.115	29.359	69.307
$5\pi/4$	1	0.3527	12.685	15.313	7.9946	8.0151	51.179
	2	0.8757	15.430	17.926	8.0296	8.7689	51.749
	3	2.5561	18.667	23.342	8.7377	9.2431	52.120
	4	4.6200	24.160	30.564	8.9771	10.502	53.205
	5	4.9623	31.311	39.131	10.708	13.310	55.254

conditions, the minimum fundamental frequency parameter occurs at $\theta_0 = 3\pi/4$. In other cases, the frequency parameters decrease with circumferential included angle θ_0 increases. This may be attributed to the stiffness of the sectorial plate reduce with circumferential included angle θ_0 increases.

For a solid sectorial plate, there exist only three boundaries, i.e., $r = R_1$, $\theta = 0$ and $\theta = \theta_0$. The results on this subject are very limited in the open literature. The lowest three frequency parameters and contour plots of the corresponding mode shapes for a [0°/90°] laminated solid sectorial plate are given in Fig. 4.22. The orthotropy ratio is chosen to be $E_1/E_2 = 15$. The thickness ratio $h/R_1 = 0.05$ is used in the calculation. The circumferential included angle θ_0 is varied from $\pi/2$ (90°) to $3\pi/2$ (270°) by a step of $\pi/4$ (45°). The solid laminated sectorial is completely clamped (C) at boundary $r = R_1$ and completely free (F) at the other two boundaries. It is noticed that vary θ_0 from $\pi/2$ to $3\pi/2$ increases the fundamental frequency parameter of the plate. For the second mode, it can be found that the minimum frequency parameter occurs at $\theta_0 = 5\pi/4$. Considering the third mode, it is observed that increasing θ_0 from $\pi/2$ to $3\pi/2$ decreases the value of the frequency parameter. In addition, the mode shapes in all the subfigures are symmetrical about geometric center line. It is attributed to the symmetry in the boundary conditions and geometry of the plate. The similar observations can be seen in Figs. 4.18, 4.19, 4.20 and 4.21 as well.

In conclusion, vibration of laminated plates is studied in this chapter, including the rectangular, circular, annular and sectorial plates. A variety of vibration results including frequencies and mode shapes for laminated plates with classical and

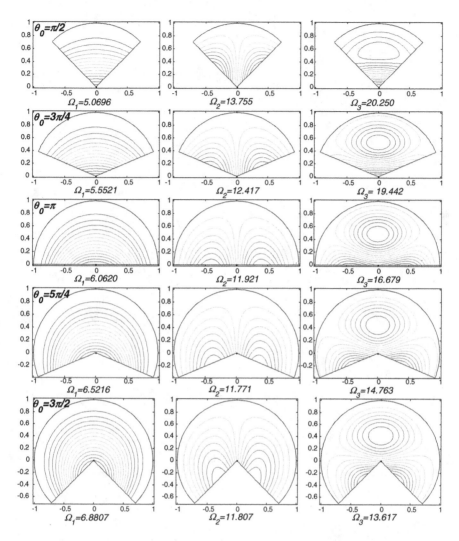

Fig. 4.22 Mode shapes and frequency parameters for a [0°/90°] laminated solid sectorial plate with various circumferential dimensions ($R_1 = 1$, $h/R_1 = 0.05$, $E_1/E_2 = 15$)

elastic boundary conditions are presented for different geometric and material parameters and lamination schemes, which may serve as benchmark solution for future researches. It is found that due to the symmetry in the boundary conditions and geometry, mode shapes of the annular and circular plates are symmetrical as well.

Chapter 5
Cylindrical Shells

Cylindrical shells are the simplest case of shells of revolution. A cylindrical shell is formed by revolving a straight line (generator) around an axis that is paralleled to the line itself. The surface obtained from the revolution of the generator defined the cylindrical shell's middle surface. Cylindrical shells may have different geometrical shapes determined by the revolution routes and circumferential included angles. By appropriately selecting the revolution routes, cylindrical shells with desired cross-sections can be produced, such as circular, elliptic, rectangular, polygon, etc. In the literature and engineering applications, cylindrical shells with circular cross-sections are most frequently encountered. In this type of cylindrical shells, each point on the middle surface maintains a similar distance from the axis. The current chapter is devoted to dealing with closed and open circular cylindrical shells.

In recent decades, composite materials have found increasing application with the rapid development of industries because they offer advantages over conventional materials. As one of the important structural components, composite laminated circular cylindrical shells are widely used in various engineering applications, such as naval vehicles aircrafts, and civil industries. The vibration analysis of them is often required and has always been one important research subject in dynamic behaviors and optimal design of complex composite shells. The literature on the vibration analysis of cylindrical shells is vast. A large variety of classical and modern theories and different computational methods have been proposed by researchers, and extensive studies have been carried out based on these theories and methods. There are mainly three major theories which are usually known as: the classical shell theories (CSTs), the first-order shear deformation theories (FSDTs) and the higher-order shear deformation theories (HSDTs). The CSTs are based on the Kirchhoff–Love assumptions, in which transverse normal and shear deformations are neglected. Unlike thin plates, where only one theory (i.e., classical plate theory) is agreed upon by most researchers, thin shells have a variety of sub-category thin shell theories developed through different assumptions and simplifications, such as Reissner-Naghdi's linear shell theory, Donner-Mushtari's theory, Flügge's theory, Sanders' theory, etc., about which detailed descriptions are

© Science Press, Beijing and Springer-Verlag Berlin Heidelberg 2015
G. Jin et al., *Structural Vibration*, DOI 10.1007/978-3-662-46364-2_5

available in the monograph by Leissa (1973). Many of the previous studies regarding laminated cylindrical shells are based on the CSTs (for example, Jin et al. 2013a; Lam and Loy 1995a, b, 1998; Liu et al. 2012; Qu et al. 2013c; Zhang 2001, 2002).

Although sufficiently accurate vibration results for thin shells can be achieved using the CSTs with appropriate solution procedures, they are inadequate for the vibration analysis of composite laminated shells which are rather thick or when they are made from materials with a high degree of anisotropic. In such cases, the effects of transverse shear deformations must be considered. Thus, FSDTs based on Reissner-Mindlin's displacement assumptions were developed to take into account the effects of transverse shear deformations. There exist considerable research efforts devoted to the laminated cylindrical shells based on the FSDTs (such as, Ferreira et al. 2007; Jin et al. 2013b, 2014a; Khdeir 1995; Khdeir and Reddy 1989; Qatu 1999; Qu et al. 2013a, b; Reddy 1984; Soldatos and Messina 2001; Zenkour 1998; Zenkour and Fares 2001; Ye et al. 2014b).

Since the transverse shear strains in the conventional FSDTs are assumed to be constant through the thickness, shear correction factors have to be incorporated to adjust the transverse shear stiffness. To overcome the deficiency of the FSDTs and further improve the dynamic analysis of shell structures, a number of HSDTs with varying degree of refinements of the kinematics of deformation were developed. Noticeably, significant contributions to the higher order shear deformation theories of composite shells have been made recently by many researchers (Ferreira et al. 2006; Mantari et al. 2011; Pinto Correia et al. 2003; Reddy and Liu 1985; Schmidt and Reddy 1988; Viola et al. 2012, 2013). As pointed out by Reddy (2003), although the HSDTs are capable of solving the global dynamic problem of shells more accurately, they introduce rather sophisticated formulations and boundary terms that are not easily applicable or yet understood. And these theories require more computational demanding compared to those FSDTs. Therefore, such theories should be used only when necessary. When the main emphasis of the analysis is to determine the vibration frequencies and mode shapes, the FSDTs may be a recommendable compromise between the solution accuracy and effort.

The development of researches on this subject has been well documented in several monographs respectively by Carrera et al. (2011), Qatu (2004), Reddy (2003), Ye (2003), Soedel (2004) and reviews (Carrera 2002, 2003; Liew et al. 2011; Qatu 2002a, b; Qatu et al. 2010; Toorani and Lakis 2000).

This chapter considers vibrations of laminated circular cylindrical shells with general boundary conditions. Equations of thin (classical shell theory, CST) and thick (shear deformation shell theory, SDST) laminated cylindrical shells are given in the first and second sections, respectively, by specializing the corresponding equations of the general shells (Chap. 1) to those of cylindrical shells. On the basis of SDST, several vibration results are presented for closed laminated cylindrical shells with different boundary conditions, lamination schemes and geometry

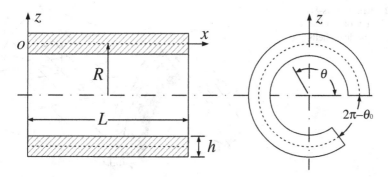

Fig. 5.1 Geometry notations and coordinate system of open laminated circular cylindrical shells

parameters by using the modified Fourier series and weak form solution procedure developed in Chap. 2 (Table 3.3 shows that convergence of solutions with weak form solution procedure is faster than the strong form one). Effects of boundary conditions, geometry parameters and material properties are studied as well. Vibration of shallow and deep open laminated cylindrical shells will then be treated in the latest section of this chapter.

Closed cylindrical shells are special cases of the open ones. Consider an open laminated circular cylindrical shell as shown in Fig. 5.1. The length, mean radius, total thickness and circumferential included angle of the shell are represented by L, R, h and θ_0, respectively. The middle surface of the shell where an orthogonal coordinate system (x, θ and z) is fixed is taken as the reference surface of the shell. The x, θ and z axes are taken in the axial, circumferential and radial directions accordingly. The middle surface displacements of the shell in the axial, circumferential and radial directions are denoted by u, v and w, respectively. The shell is assumed to be composed of arbitrary number of liner orthotropic laminas which are perfectly bonded together. The principal coordinates of the composite material in each layer are denoted by 1, 2 and 3, respectively, and the angle between the material axis and the x-axis of the shell is devoted by ϑ. The distances from the top surface and the bottom surface of the kth layer to the referenced middle surface are represented by Z_{k+1} and Z_k accordingly. The partial cross-sectional view of the cylindrical shell is given in Fig. 5.2.

Consider the cylindrical shell in Fig. 5.1 and its cylindrical coordinate system. The coordinates, characteristics of the Lamé parameters and radii of curvatures are:

$$\alpha = x, \quad \beta = \theta, \quad A = 1, \quad B = R, \quad R_\alpha = \infty, \quad R_\beta = R \quad (5.1)$$

The above geometry parameters can be directly applied to the general shell equations derived in Chap. 1 to obtain those of cylindrical shells.

Fig. 5.2 Partial cross-sectional view of laminated circular cylindrical shells

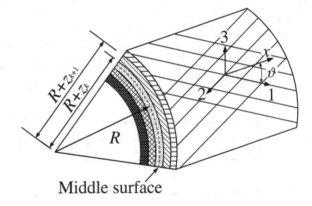

Middle surface

5.1 Fundamental Equations of Thin Laminated Cylindrical Shells

We first derive the fundamental equations for thin laminated cylindrical shells (neglecting shear deformation and rotary inertia), followed by the thick ones. We will only consider laminated composite layers with cylindrical orthotropy. The equations are formulated for the general dynamic analysis of laminated cylindrical open shells. It can be readily applied to static (i.e., letting frequency ω equal to zero) and free vibration (i.e., neglecting the external load) analysis.

5.1.1 Kinematic Relations

Substituting Eq. (5.1) into Eq. (1.7), the middle surface strains and curvature changes of thin cylindrical shells can be specialized from those of general shells. They are formed in terms of middle surface displacements as:

$$
\begin{aligned}
\varepsilon_x^0 &= \frac{\partial u}{\partial x} & \chi_x &= -\frac{\partial^2 w}{\partial x^2} \\
\varepsilon_\theta^0 &= \frac{\partial v}{R\partial\theta} + \frac{w}{R} & \chi_\theta &= \frac{\partial v}{R^2\partial\theta} - \frac{\partial^2 w}{R^2\partial\theta^2} \\
\gamma_{x\theta}^0 &= \frac{\partial v}{\partial x} + \frac{\partial u}{R\partial\theta} & \chi_{x\theta} &= \frac{\partial v}{R\partial x} - 2\frac{\partial^2 w}{R\partial x\partial\theta}
\end{aligned}
\tag{5.2}
$$

where ε_x^0, ε_θ^0 and $\gamma_{x\theta}^0$ denote the middle surface normal and shear strains. χ_x, χ_θ and $\chi_{x\theta}$ are the middle surface curvature and twist changes. Thus, the strain-displacement relations for an arbitrary point in the kth layer of a thin laminated cylindrical shell can be defined as:

$$\varepsilon_x = \varepsilon_x^0 + z\chi_x$$
$$\varepsilon_\theta = \varepsilon_\theta^0 + z\chi_\theta \qquad (5.3)$$
$$\gamma_{x\theta} = \gamma_{x\theta}^0 + z\chi_{x\theta}$$

where $Z_{k+1} < z < Z_k$.

5.1.2 Stress-Strain Relations and Stress Resultants

The corresponding stress-strain relations in the kth layer can be obtained according to generalized Hooke's law as:

$$\left\{ \begin{array}{c} \sigma_x \\ \sigma_\theta \\ \tau_{x\theta} \end{array} \right\}_k = \left[\begin{array}{ccc} \overline{Q}_{11}^k & \overline{Q}_{12}^k & \overline{Q}_{16}^k \\ \overline{Q}_{12}^k & \overline{Q}_{22}^k & \overline{Q}_{26}^k \\ \overline{Q}_{16}^k & \overline{Q}_{26}^k & \overline{Q}_{66}^k \end{array} \right] \left\{ \begin{array}{c} \varepsilon_x \\ \varepsilon_\theta \\ \gamma_{x\theta} \end{array} \right\}_k \qquad (5.4)$$

where σ_x and σ_θ are the normal stresses in the x and θ directions, respectively. $\tau_{x\theta}$ is the corresponding shear stress. The lamina stiffness coefficients $\overline{Q}_{ij}^k$ ($i,j = 1, 2, 6$) represent the elastic properties of the material of the layer. They are written as in Eq. (1.12). Integrating the stresses over the thickness or substituting Eq. (5.1) into Eq. (1.14), the force and moment resultants of thin laminated cylindrical shells can be obtained in a matrix form as

$$\left[\begin{array}{c} N_x \\ N_\theta \\ N_{x\theta} \\ M_x \\ M_\theta \\ M_{x\theta} \end{array} \right] = \left[\begin{array}{cccccc} A_{11} & A_{12} & A_{16} & B_{11} & B_{12} & B_{16} \\ A_{12} & A_{22} & A_{26} & B_{12} & B_{22} & B_{26} \\ A_{16} & A_{26} & A_{66} & B_{16} & B_{26} & B_{66} \\ B_{11} & B_{12} & B_{16} & D_{11} & D_{12} & D_{16} \\ B_{12} & B_{22} & B_{26} & D_{12} & D_{22} & D_{26} \\ B_{16} & B_{26} & B_{66} & D_{16} & D_{26} & D_{66} \end{array} \right] \left[\begin{array}{c} \varepsilon_x^0 \\ \varepsilon_\theta^0 \\ \gamma_{x\theta}^0 \\ \chi_x \\ \chi_\theta \\ \chi_{x\theta} \end{array} \right] \qquad (5.5)$$

where N_x, N_θ and $N_{x\theta}$ are the normal and shear force resultants and M_x, M_θ and $M_{x\theta}$ denote the bending and twisting moment resultants. The stiffness coefficient A_{ij}, B_{ij}, and D_{ij} are given in Eq. (1.15). It should be noted that for a laminated cylindrical shell which is symmetrically laminated with respect to the middle surface, the constants B_{ij} equal to zero.

5.1.3 Energy Functions

The strain energy function for thin laminated cylindrical shells during vibration can be written as:

$$U_s = \frac{1}{2} \int_x \int_\theta \left\{ \begin{array}{l} N_x \varepsilon_x^0 + N_\theta \varepsilon_\theta^0 + N_{x\theta} \gamma_{x\theta}^0 \\ + M_x \chi_x + M_\theta \chi_\theta + M_{x\theta} \chi_{x\theta} \end{array} \right\} R d\theta dx \qquad (5.6)$$

Substituting Eqs. (5.2) and (5.5) into Eq. (5.6), the strain energy function can be written in terms of middle surface displacements and divided to three components $U_s = U_{ss} + U_{sb} + U_{sbs}$, where U_{ss}, U_{sb} and U_{sbs} represent the stretching strain energy, bending strain energy and bending–stretching coupling energy, respectively.

$$U_{ss} = \frac{1}{2} \int_x \int_\theta \left\{ \begin{array}{l} A_{11}\left(\frac{\partial u}{\partial x}\right)^2 + \frac{A_{22}}{R^2}\left(\frac{\partial v}{\partial \theta} + w\right)^2 + A_{66}\left(\frac{\partial v}{\partial x} + \frac{\partial u}{R\partial \theta}\right)^2 \\ + \frac{2A_{12}}{R}\left(\frac{\partial v}{\partial \theta} + w\right)\left(\frac{\partial u}{\partial x}\right) + 2A_{16}\left(\frac{\partial v}{\partial x} + \frac{\partial u}{R\partial \theta}\right)\left(\frac{\partial u}{\partial x}\right) \\ + \frac{2A_{26}}{R}\left(\frac{\partial v}{\partial x} + \frac{\partial u}{R\partial \theta}\right)\left(\frac{\partial v}{\partial \theta} + w\right) \end{array} \right\} R d\theta dx \qquad (5.7a)$$

$$U_{sb} = \frac{1}{2} \int_x \int_\theta \left\{ \begin{array}{l} D_{11}\left(\frac{\partial^2 w}{\partial x^2}\right)^2 - \frac{2D_{12}}{R^2}\left(\frac{\partial v}{\partial \theta} - \frac{\partial^2 w}{\partial \theta^2}\right)\left(\frac{\partial^2 w}{\partial x^2}\right) - \frac{2D_{16}}{R} \\ \times \left(\frac{\partial^2 w}{\partial x^2}\right)\left(\frac{\partial v}{\partial x} - 2\frac{\partial^2 w}{\partial x\partial \theta}\right) + \frac{D_{22}}{R^4}\left(\frac{\partial v}{\partial \theta} - \frac{\partial^2 w}{\partial \theta^2}\right)^2 + \frac{D_{66}}{R^2} \\ \times \left(\frac{\partial v}{\partial x} - 2\frac{\partial^2 w}{\partial x\partial \theta}\right)^2 + \frac{2D_{26}}{R^3}\left(\frac{\partial v}{\partial x} - 2\frac{\partial^2 w}{\partial x\partial \theta}\right)\left(\frac{\partial v}{\partial \theta} - \frac{\partial^2 w}{\partial \theta^2}\right) \end{array} \right\} R d\theta dx \qquad (5.7b)$$

$$U_{sbs} = \int_x \int_\theta \left\{ \begin{array}{l} -B_{11}\left(\frac{\partial^2 w}{\partial x^2}\right)\left(\frac{\partial u}{\partial x}\right) + \frac{B_{12}}{R^2}\left(\frac{\partial v}{\partial \theta} - \frac{\partial^2 w}{\partial \theta^2}\right)\left(\frac{\partial u}{\partial x}\right) + \frac{B_{16}}{R} \\ \times \left(\frac{\partial v}{\partial x} - 2\frac{\partial^2 w}{\partial x\partial \theta}\right)\left(\frac{\partial u}{\partial x}\right) - \frac{B_{12}}{R}\left(\frac{\partial^2 w}{\partial x^2}\right)\left(\frac{\partial v}{\partial \theta} + w\right) + \frac{B_{22}}{R^3} \\ \times \left(\frac{\partial v}{\partial \theta} - \frac{\partial^2 w}{\partial \theta^2}\right)\left(\frac{\partial v}{\partial \theta} + w\right) + \frac{B_{26}}{R^2}\left(\frac{\partial v}{\partial x} - 2\frac{\partial^2 w}{\partial x\partial \theta}\right) \\ \times \left(\frac{\partial v}{\partial \theta} + w\right) - B_{16}\left(\frac{\partial^2 w}{\partial x^2}\right)\left(\frac{\partial v}{\partial x} + \frac{\partial u}{R\partial \theta}\right) + \frac{B_{26}}{R^2} \\ \times \left(\frac{\partial v}{\partial \theta} - \frac{\partial^2 w}{\partial \theta^2}\right)\left(\frac{\partial v}{\partial x} + \frac{\partial u}{R\partial x}\right) \\ + \frac{B_{66}}{R}\left(\frac{\partial v}{\partial x} + \frac{\partial u}{R\partial x}\right)\left(\frac{\partial v}{\partial x} - 2\frac{\partial^2 w}{\partial x\partial \theta}\right) \end{array} \right\} R d\theta dx \qquad (5.7c)$$

The corresponding kinetic energy (*T*) of the shells can be written as:

$$T = \frac{1}{2} \int_x \int_\theta I_0 \left\{ \left(\frac{\partial u}{\partial t}\right)^2 + \left(\frac{\partial v}{\partial t}\right)^2 + \left(\frac{\partial w}{\partial t}\right)^2 \right\} R d\theta dx \qquad (5.8)$$

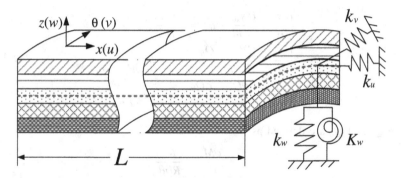

Fig. 5.3 Boundary conditions of thin laminated cylindrical shells

The inertia term I_0 is the same as in Eq. (1.19). Suppose q_x, q_θ and q_z are the external loads in the x, θ and z directions, respectively. Thus, the external work can be expressed as:

$$W_e = \int\limits_x \int\limits_\theta \{q_x u + q_\theta v + q_z w\} R d\theta dx \tag{5.9}$$

The same as usual, the general boundary conditions of a cylindrical shell are implemented by using the artificial spring boundary technique (see Sect. 1.2.3). Letting symbols $k_\psi^u, k_\psi^v, k_\psi^w$ and k_ψ^w ($\psi = x0$, $\theta0$, $x1$ and $\theta1$) to indicate the stiffness of the boundary springs at the boundaries $x = 0$, $\theta = 0$, $x = L$ and $\theta = \theta_0$, respectively (see Fig. 5.3). Therefore, the deformation strain energy stored in the boundary springs (U_{sp}) during vibration can be defined as:

$$U_{sp} = \frac{1}{2} \int\limits_\theta \left\{ \begin{array}{c} \left[k_{x0}^u u^2 + k_{x0}^v v^2 + k_{x0}^w w^2 + K_{x0}^w (\partial w/\partial x)\right]|_{x=0} \\ + \left[k_{x1}^u u^2 + k_{x1}^v v^2 + k_{x1}^w w^2 + K_{x1}^w (\partial w/\partial x)\right]|_{x=L} \end{array} \right\} R d\theta$$
$$+ \frac{1}{2} \int\limits_x \left\{ \begin{array}{c} \left[k_{\theta0}^u u^2 + k_{\theta0}^v v^2 + k_{\theta0}^w w^2 + K_{\theta0}^w (\partial w/R\partial\theta)\right]|_{\theta=0} \\ + \left[k_{\theta1}^u u^2 + k_{\theta1}^v v^2 + k_{\theta1}^w w^2 + K_{\theta1}^w (\partial w/R\partial\theta)\right]|_{\theta=\theta_0} \end{array} \right\} dx \tag{5.10}$$

5.1.4 Governing Equations and Boundary Conditions

The governing equations and boundary conditions of thin laminated cylindrical shells can be obtained by specializing form the governing equations of thin general laminated shells or applying the Hamilton's principle in the same manner as described in Sect. 1.2.4. Substituting Eq. (5.1) into Eq. (1.28) and then simplifying the expressions. Thus, the governing equations of thin laminated cylindrical shells become:

$$\frac{\partial N_x}{\partial x} + \frac{\partial N_{x\theta}}{R\partial\theta} + q_x = I_0\frac{\partial^2 u}{\partial t^2}$$

$$\frac{\partial N_{x\theta}}{\partial x} + \frac{\partial N_\theta}{R\partial\theta} + \frac{Q_\theta}{R} + q_\theta = I_0\frac{\partial^2 v}{\partial t^2} \qquad (5.11)$$

$$-\frac{N_\theta}{R} + \frac{\partial Q_x}{\partial x} + \frac{\partial Q_\theta}{R\partial\theta} + q_z = I_0\frac{\partial^2 w}{\partial t^2}$$

where Q_x and Q_θ are defined as

$$Q_x = \frac{\partial M_x}{\partial x} + \frac{\partial M_{x\theta}}{R\partial\theta}$$

$$Q_\theta = \frac{\partial M_{x\theta}}{\partial x} + \frac{\partial M_\theta}{R\partial\theta} \qquad (5.12)$$

Substituting Eqs. (5.2), (5.5) and (5.13) into Eq. (5.12), the governing equations can be written in terms of middle surface displacements. These equations are proved useful when exact solutions are desired. These equations can be written as:

$$\left(\begin{bmatrix} L_{11} & L_{12} & L_{13} \\ L_{21} & L_{22} & L_{23} \\ L_{31} & L_{32} & L_{33} \end{bmatrix} - \omega^2 \begin{bmatrix} -I_0 & 0 & 0 \\ 0 & -I_0 & 0 \\ 0 & 0 & -I_0 \end{bmatrix}\right) \begin{bmatrix} u \\ v \\ w \end{bmatrix} = \begin{bmatrix} -p_x \\ -p_\theta \\ -p_z \end{bmatrix} \qquad (5.13)$$

The coefficients of the linear operator L_{ij} are written as

$$L_{11} = A_{11}\frac{\partial^2}{\partial x^2} + \frac{2A_{16}}{R}\frac{\partial^2}{\partial x\partial\theta} + \frac{A_{66}}{R^2}\frac{\partial^2}{\partial\theta^2}$$

$$L_{12} = \left(A_{16} + \frac{B_{16}}{R}\right)\frac{\partial^2}{\partial x^2} + \left(\frac{A_{12}}{R} + \frac{A_{66}}{R} + \frac{B_{12}+B_{66}}{R^2}\right)\frac{\partial^2}{\partial x\partial\theta}$$

$$+ \left(\frac{A_{26}}{R^2} + \frac{B_{26}}{R^3}\right)\frac{\partial^2}{\partial\theta^2} = L_{21}$$

$$L_{13} = -B_{11}\frac{\partial^3}{\partial x^3} - \frac{B_{26}}{R^3}\frac{\partial^3}{\partial\theta^3} - \frac{3B_{16}}{R}\frac{\partial^3}{\partial x^2\partial\theta} - \frac{B_{12}+2B_{66}}{R^2}\frac{\partial^3}{\partial x\partial\theta^2}$$

$$+ \frac{A_{12}}{R}\frac{\partial}{\partial x} + \frac{A_{26}}{R^2}\frac{\partial}{\partial\theta} = -L_{31} \qquad (5.14)$$

$$L_{22} = \left(A_{66} + \frac{2B_{66}}{R}\right)\frac{\partial^2}{\partial x^2} + \left(\frac{2A_{26}}{R} + \frac{4B_{26}}{R^2}\right)\frac{\partial^2}{\partial x\partial\theta} + \left(\frac{A_{22}}{R^2} + \frac{2B_{22}}{R^3}\right)\frac{\partial^2}{\partial\theta^2}$$

$$L_{23} = -\left(B_{16} + \frac{D_{16}}{R}\right)\frac{\partial^3}{\partial x^3} - \left(\frac{B_{22}}{R^3} + \frac{D_{22}}{R^4}\right)\frac{\partial^3}{\partial\theta^3} - 3\left(\frac{B_{26}}{R^2} + \frac{D_{26}}{R^3}\right)\frac{\partial^3}{\partial x\partial\theta^2}$$

$$- \left(\frac{B_{12}+2B_{66}}{R} + \frac{D_{12}+2D_{66}}{R^2}\right)\frac{\partial^3}{\partial x^2\partial\theta} + \left(\frac{A_{26}}{R} + \frac{B_{26}}{R^2}\right)\frac{\partial}{\partial x}$$

$$+ \left(\frac{A_{22}}{R^2} + \frac{B_{22}}{R^3}\right)\frac{\partial}{\partial\theta} = -L_{32}$$

$$L_{33} = 2\left(\frac{B_{12}}{R}\frac{\partial^2}{\partial x^2} + \frac{2B_{26}}{R^2}\frac{\partial^2}{\partial x\partial\theta} + \frac{B_{22}}{R^3}\frac{\partial^2}{\partial\theta^2}\right) - \frac{A_{22}}{R^2} - \frac{4D_{16}}{R}\frac{\partial^4}{\partial x^3\partial\theta}$$

$$- D_{11}\frac{\partial^4}{\partial x^4} - 2\left(\frac{D_{12} + 2D_{66}}{R^2}\right)\frac{\partial^4}{\partial x^2\partial\theta^2} - \frac{4D_{26}}{R^3}\frac{\partial^4}{\partial x\partial\theta^3} - \frac{D_{22}}{R^4}\frac{\partial^4}{\partial\theta^4}$$

And the general boundary conditions of thin laminated cylindrical shells are:

$$x = 0:\begin{cases} N_x - k_{x0}^u u = 0 \\ N_{x\theta} + \frac{M_{x\theta}}{R} - k_{x0}^v v = 0 \\ Q_x + \frac{\partial M_{x\theta}}{R\partial\theta} - k_{x0}^w w = 0 \\ -M_x - K_{x0}^w \frac{\partial w}{\partial x} = 0 \end{cases} \quad x = L:\begin{cases} N_x + k_{x1}^u u = 0 \\ N_{x\theta} + \frac{M_{x\theta}}{R} + k_{x1}^v v = 0 \\ Q_x + \frac{\partial M_{x\theta}}{R\partial\theta} + k_{x1}^w w = 0 \\ -M_x + K_{x1}^w \frac{\partial w}{\partial x} = 0 \end{cases}$$

$$\theta = 0:\begin{cases} N_{x\theta} - k_{\theta0}^u u = 0 \\ N_\theta + \frac{M_\theta}{R} - k_{\theta0}^v v = 0 \\ Q_\theta + \frac{\partial M_{x\theta}}{\partial x} - k_{\theta0}^w w = 0 \\ -M_\theta - K_{\theta0}^w \frac{\partial w}{R\partial\theta} = 0 \end{cases} \quad \theta = \theta_0:\begin{cases} N_{x\theta} + k_{\theta1}^u u = 0 \\ N_\theta + \frac{M_\theta}{R} + k_{\theta1}^v v = 0 \\ Q_\theta + \frac{\partial M_{x\theta}}{\partial x} + k_{\theta1}^w w = 0 \\ -M_\theta + K_{\theta1}^w \frac{\partial w}{R\partial\theta} = 0 \end{cases}$$

(5.15)

The classification of the classical boundary conditions shown in Table 1.3 for thin laminated general shells is applicable for thin laminated cylindrical shells. Taking boundaries $x =$ constant for example, the possible combinations for each classical boundary conditions are given in Table 5.1. Similar boundary conditions can be obtained for boundaries $\theta =$ constant.

It is obvious that there exists a huge number of possible combinations of boundary conditions for a thin laminated cylindrical shell. In contrast to most existing solution procedures, the artificial spring boundary technique offers a unified operation for laminated cylindrical shells with general boundary conditions. The stiffness of the boundary springs can take any value from zero to infinity to better model many real-world restraint conditions. Taking edge $x = 0$ for example, the widely encountered classical boundary conditions, such as F (completely free), S (simply-supported), SD (shear-diaphragm) and C (completely clamped) boundaries can be readily realized by assigning the stiffness of the boundary springs as:

$$\text{F: } k_{x0}^u = k_{x0}^v = k_{x0}^w = K_{x0}^w = 0$$
$$\text{SD: } k_{x0}^v = k_{x0}^w = 10^7 D, k_{x0}^u = K_{x0}^w = 0$$
$$\text{S: } k_{x0}^u = k_{x0}^v = k_{x0}^w = 10^7 D, K_{x0}^w = 0$$
$$\text{C: } k_{x0}^u = k_{x0}^v = k_{x0}^w = K_{x0}^w = 10^7 D$$

(5.16)

where $D = E_1 h^3/12(1 - \mu_{12}\mu_{21})$ is the flexural stiffness of the cylindrical shell.

Table 5.1 Possible classical boundary conditions for thin laminated cylindrical shells at boundaries $x = $ constant

Boundary type	Conditions
Free boundary conditions	
F	$N_x = N_{x\theta} + \frac{M_{x\theta}}{R} = Q_x + \frac{\partial M_{x\theta}}{R\partial\theta} = M_x = 0$
F2	$u = N_{x\theta} + \frac{M_{x\theta}}{R} = Q_x + \frac{\partial M_{x\theta}}{R\partial\theta} = M_x = 0$
F3	$N_x = v = Q_x + \frac{\partial M_{x\theta}}{R\partial\theta} = M_x = 0$
F4	$u = v = Q_x + \frac{\partial M_{x\theta}}{R\partial\theta} = M_x = 0$
Simply supported boundary conditions	
S	$u = v = w = M_x = 0$
SD	$N_x = v = w = M_x = 0$
S3	$u = N_{x\theta} + \frac{M_{x\theta}}{R} = w = M_x = 0$
S4	$N_x = N_{x\theta} + \frac{M_{x\theta}}{R} = w = M_x = 0$
Clamped boundary conditions	
C	$u = v = w = \frac{\partial w}{\partial x} = 0$
C2	$N_x = v = w = \frac{\partial w}{\partial x} = 0$
C3	$u = N_{x\theta} + \frac{M_{x\theta}}{R} = w = \frac{\partial w}{\partial x} = 0$
C4	$N_x = N_{x\theta} + \frac{M_{x\theta}}{R} = w = \frac{\partial w}{\partial x} = 0$

5.2 Fundamental Equations of Thick Laminated Cylindrical Shells

The fundamental equations of thin laminated cylindrical shells presented in previous section are applicable only when the total thickness of the shell is smaller than 1/20 of the smallest of the wave lengths and/or radii of curvature (Qatu 2004) due to the fact that both shear deformation and rotary inertia are neglected in the formulation. This section presents fundamental equations of thick laminated cylindrical shells considering the effects of shear deformation and rotary inertia. The treatment that follows is a specialization of the general first-order shear deformation shell theory (SDST) given in Sect. 1.3 to those of the thick laminated cylindrical shells.

5.2.1 Kinematic Relations

On the basis of the shell model given in Fig. 5.1 and the assumptions of SDST, the displacement field of thick laminated cylindrical shells is defined in terms of the displacements and rotation components of the middle surface as

$$\begin{aligned}
U(x,\theta,z) &= u(x,\theta) + z\phi_x \\
V(x,\theta,z) &= v(x,\theta) + z\phi_\theta \\
W(x,\theta,z) &= w(x,\theta)
\end{aligned} \tag{5.17}$$

where u, v and w are the middle surface displacements of the shell in the axial, circumferential and radial directions, respectively, and ϕ_x and ϕ_θ represent the rotations of transverse normal respect to θ- and x-axes. Specializing Eqs. (1.33) and (1.34) to those of cylindrical shells, the normal and shear strains at any point of the shell space can be defined in terms of the middle surface strains and curvature changes as:

$$\varepsilon_x = \varepsilon_x^0 + z\chi_x$$

$$\varepsilon_\theta = \frac{1}{(1+z/R)}\left(\varepsilon_\theta^0 + z\chi_\theta\right)$$

$$\gamma_{x\theta} = \left(\gamma_{x\theta}^0 + z\chi_{x\theta}\right) + \frac{1}{(1+z/R)}\left(\gamma_{\theta x}^0 + z\chi_{\theta x}\right) \qquad (5.18)$$

$$\gamma_{xz} = \gamma_{xz}^0$$

$$\gamma_{\theta z} = \frac{\gamma_{\theta z}^0}{(1+z/R)}$$

where γ_{xz}^0 and $\gamma_{\theta z}^0$ represent the transverse shear strains. The middle surface strains and curvature changes are defined as:

$$\begin{aligned}
\varepsilon_x^0 &= \frac{\partial u}{\partial x}, & \chi_x &= \frac{\partial \phi_x}{\partial x} \\
\varepsilon_\theta^0 &= \frac{\partial v}{R\partial \theta} + \frac{w}{R}, & \chi_\theta &= \frac{\partial \phi_\theta}{R\partial \theta} \\
\gamma_{x\theta}^0 &= \frac{\partial v}{\partial x}, & \chi_{x\theta} &= \frac{\partial \phi_\theta}{\partial x} \\
\gamma_{\theta x}^0 &= \frac{\partial u}{R\partial \theta}, & \chi_{\theta x} &= \frac{\partial \phi_x}{R\partial \theta} \\
\gamma_{xz}^0 &= \frac{\partial w}{\partial x} + \phi_x & & \\
\gamma_{\theta z}^0 &= \frac{\partial w}{R\partial \theta} - \frac{v}{R} + \phi_\theta & &
\end{aligned} \qquad (5.19)$$

5.2.2 Stress-Strain Relations and Stress Resultants

According to Hooke's law, the corresponding stresses in the kth layer of a thick laminated cylindrical shell can be obtained as:

$$\left\{\begin{array}{c} \sigma_x \\ \sigma_\theta \\ \tau_{\theta z} \\ \tau_{xz} \\ \tau_{x\theta} \end{array}\right\}_k = \begin{bmatrix} \overline{Q}_{11}^k & \overline{Q}_{12}^k & 0 & 0 & \overline{Q}_{16}^k \\ \overline{Q}_{12}^k & \overline{Q}_{22}^k & 0 & 0 & \overline{Q}_{26}^k \\ 0 & 0 & \overline{Q}_{44}^k & \overline{Q}_{45}^k & 0 \\ 0 & 0 & \overline{Q}_{45}^k & \overline{Q}_{55}^k & 0 \\ \overline{Q}_{16}^k & \overline{Q}_{26}^k & 0 & 0 & \overline{Q}_{66}^k \end{bmatrix} \left\{\begin{array}{c} \varepsilon_x \\ \varepsilon_\theta \\ \gamma_{\theta z} \\ \gamma_{xz} \\ \gamma_{x\theta} \end{array}\right\}_k \qquad (5.20)$$

where τ_{xz} and $\tau_{\theta z}$ are the corresponding shear stress components. $\overline{Q}_{ij}^k$ $(i, j = 1, 2, 4, 5, 6)$ are known as the transformation stiffness coefficients, which represent the elastic properties of the material of the layer. They are given in Eq. (1.39). By carrying the integration of stresses over the cross-section, the force and moment resultants can be obtained:

$$
\begin{bmatrix} N_x \\ N_{x\theta} \\ Q_x \end{bmatrix} = \int\limits_{-h/2}^{h/2} \begin{bmatrix} \sigma_x \\ \tau_{x\theta} \\ \tau_{xz} \end{bmatrix} \left(1 + \tfrac{z}{R}\right) dz, \qquad
\begin{bmatrix} N_\theta \\ N_{\theta x} \\ Q_\theta \end{bmatrix} = \int\limits_{-h/2}^{h/2} \begin{bmatrix} \sigma_\theta \\ \tau_{\theta x} \\ \tau_{\theta z} \end{bmatrix} dz
$$

$$
\begin{bmatrix} M_x \\ M_{x\theta} \end{bmatrix} = \int\limits_{-h/2}^{h/2} \begin{bmatrix} \sigma_x \\ \tau_{x\theta} \end{bmatrix} \left(1 + \tfrac{z}{R}\right) z\, dz, \qquad
\begin{bmatrix} M_\theta \\ M_{\theta x} \end{bmatrix} = \int\limits_{-h/2}^{h/2} \begin{bmatrix} \sigma_\theta \\ \tau_{\theta x} \end{bmatrix} z\, dz
\tag{5.21}
$$

It should be stressed that although $\tau_{x\theta}$ equals to $\tau_{\theta x}$ from the symmetry of the stress tensor, it is obvious from Eq. (5.21) that the shear force resultants $N_{x\theta}$ and $N_{\theta x}$ are not equal. Similarly, the twisting moment resultants $M_{x\theta}$ and $M_{\theta x}$ are not equal too. Carrying out the integration over the thickness, from layer to layer, yields

$$
\begin{bmatrix} N_x \\ N_\theta \\ N_{x\theta} \\ N_{\theta x} \\ M_x \\ M_\theta \\ M_{x\theta} \\ M_{\theta x} \end{bmatrix} =
\begin{bmatrix}
A_{11} & \overline{A_{12}} & \overline{A_{16}} & A_{16} & B_{11} & \overline{B_{12}} & \overline{B_{16}} & B_{16} \\
A_{12} & A_{22} & A_{26} & \overline{A_{26}} & B_{12} & \overline{B_{22}} & \overline{B_{26}} & B_{26} \\
A_{16} & A_{26} & A_{66} & \overline{A_{66}} & B_{16} & \overline{B_{26}} & \overline{B_{66}} & B_{66} \\
A_{16} & \overline{A_{26}} & A_{66} & \overline{A_{66}} & B_{16} & \overline{B_{26}} & B_{66} & \overline{B_{66}} \\
\overline{B_{11}} & B_{12} & \overline{B_{16}} & B_{16} & \overline{D_{11}} & D_{12} & \overline{D_{16}} & D_{16} \\
B_{12} & \overline{B_{22}} & B_{26} & \overline{B_{26}} & D_{12} & \overline{D_{22}} & D_{26} & \overline{D_{26}} \\
\overline{B_{16}} & B_{26} & \overline{B_{66}} & \overline{B_{66}} & \overline{D_{16}} & D_{26} & \overline{D_{66}} & D_{66} \\
B_{16} & \overline{B_{26}} & B_{66} & \overline{B_{66}} & D_{16} & \overline{D_{26}} & D_{66} & \overline{D_{66}}
\end{bmatrix}
\begin{bmatrix} \varepsilon_x^0 \\ \varepsilon_\theta^0 \\ \gamma_{x\theta}^0 \\ \gamma_{\theta x}^0 \\ \chi_x \\ \chi_\theta \\ \chi_{x\theta} \\ \chi_{\theta x} \end{bmatrix}
\tag{5.22}
$$

$$
\begin{bmatrix} Q_\theta \\ Q_x \end{bmatrix} =
\begin{bmatrix} \overline{\overline{A_{44}}} & A_{45} \\ A_{45} & A_{55} \end{bmatrix}
\begin{bmatrix} \gamma_{\theta z}^0 \\ \gamma_{xz}^0 \end{bmatrix}
\tag{5.23}
$$

The stiffness coefficients A_{ij}, B_{ij} and D_{ij} are given as in Eq. (1.43). The stiffness coefficients $\overline{A_{ij}}, \overline{B_{ij}}, \overline{D_{ij}}, \overline{\overline{A_{ij}}}, \overline{\overline{B_{ij}}}$ and $\overline{\overline{D_{ij}}}$ are defined as:

$$
\overline{A_{ij}} = A_{ij} + B_{ij}/R
$$
$$
\overline{B_{ij}} = B_{ij} + D_{ij}/R
$$
$$
\overline{D_{ij}} = D_{ij} + E_{ij}/R
$$
$$
\overline{\overline{A_{ij}}} = R \sum_{k=1}^{N_L} \overline{Q_{ij}^k} \ln\left(\frac{R + z_{k+1}}{R + z_k}\right) \qquad (i,j = 1,2,6)
$$
$$
\overline{\overline{A_{ij}}} = K_s R \sum_{k=1}^{N_L} \overline{Q_{ij}^k} \ln\left(\frac{R + z_{k+1}}{R + z_k}\right) \qquad (i,j = 4,5)
\tag{5.24}
$$
$$
\overline{\overline{B_{ij}}} = R \sum_{k=1}^{N_L} \overline{Q_{ij}^k} \left[(z_{k+1} - z_k) - R \ln\left(\frac{R + z_{k+1}}{R + z_k}\right)\right]
$$
$$
\overline{\overline{D_{ij}}} = \sum_{k=1}^{N_L} \overline{Q_{ij}^k} \left[-R^2(z_{k+1} - z_k) + \frac{R}{2}(z_{k+1}^2 - z_k^2) + R^3 \ln\left(\frac{R + z_{k+1}}{R + z_k}\right)\right]
$$

where K_s is the shear correction factor. N_L represents the mount of the layers. E_{ij} are defined as

$$E_{ij} = \frac{1}{4} \sum_{k=1}^{N_L} \overline{Q}_{ij}^k \left[z_{k+1}^4 - z_k^4 \right], \quad (i,j = 1,2,6) \tag{5.25}$$

The above equations include the effects of the deepness term z/R, and thus create complexities in carrying out the integration and the follow-up programming. Figures 3.7 and 3.8 showed that the effects of the deepness term z/R on the frequency parameters of an extremely deep, unsymmetrically laminated curved beam ($\theta_0 = 286.48°$) with thickness-to-radius ratio $h/R = 0.1$ are very small and the maximum effect is less than 0.41 % for the worse case. These results can be used in establishing the limits of shell theories and shell equations as well. When the FSDT is applied to thin and moderately thick cylindrical shells, the deepness term z/R can be neglected (Qu et al. 2013a, b; Jin et al. 2013b, 2014a; Ye et al. 2014b), in such case, the force and moment resultants of cylindrical shells (Eq. (5.23)) can be rewritten as:

$$
\begin{bmatrix} N_x \\ N_\theta \\ N_{x\theta} \\ N_{\theta x} \\ M_x \\ M_\theta \\ M_{x\theta} \\ M_{\theta x} \end{bmatrix} =
\begin{bmatrix}
A_{11} & A_{12} & A_{16} & A_{16} & B_{11} & B_{12} & B_{16} & B_{16} \\
A_{12} & A_{22} & A_{26} & A_{26} & B_{12} & B_{22} & B_{26} & B_{26} \\
A_{16} & A_{26} & A_{66} & A_{66} & B_{16} & B_{26} & B_{66} & B_{66} \\
A_{16} & A_{26} & A_{66} & A_{66} & B_{16} & B_{26} & B_{66} & B_{66} \\
B_{11} & B_{12} & B_{16} & B_{16} & D_{11} & D_{12} & D_{16} & D_{16} \\
B_{12} & B_{22} & B_{26} & B_{26} & D_{12} & D_{22} & D_{26} & D_{26} \\
B_{16} & B_{26} & B_{66} & B_{66} & D_{16} & D_{26} & D_{66} & D_{66} \\
B_{16} & B_{26} & B_{66} & B_{66} & D_{16} & D_{26} & D_{66} & D_{66}
\end{bmatrix}
\begin{bmatrix} \varepsilon_x^0 \\ \varepsilon_\theta^0 \\ \gamma_{x\theta}^0 \\ \gamma_{\theta x}^0 \\ \chi_x \\ \chi_\theta \\ \chi_{x\theta} \\ \chi_{\theta x} \end{bmatrix}
\tag{5.26}
$$

$$
\begin{bmatrix} Q_\theta \\ Q_x \end{bmatrix} = \begin{bmatrix} A_{44} & A_{45} \\ A_{45} & A_{55} \end{bmatrix} \begin{bmatrix} \gamma_{\theta z}^0 \\ \gamma_{xz}^0 \end{bmatrix}
\tag{5.27}
$$

5.2.3 Energy Functions

The strain energy (U_s) of laminated cylindrical shells including shear deformation effects during vibration can be defined in terms of the middle surface strains and curvature changes and stress resultants as

$$U_s = \frac{1}{2} \int\limits_x \int\limits_\theta \left\{ \begin{array}{l} N_x \varepsilon_x^0 + N_\theta \varepsilon_\theta^0 + N_{x\theta} \gamma_{x\theta}^0 + N_{\theta x} \gamma_{\theta x}^0 + M_x \chi_x \\ + M_\theta \chi_\theta + M_{x\theta} \chi_{x\theta} + M_{\theta x} \chi_{\theta x} + Q_\theta \gamma_{\theta z} + Q_x \gamma_{xz} \end{array} \right\} R d\theta dx \tag{5.28}$$

Substituting Eqs. (5.19), (5.22) and (5.23) into Eq. (5.28), the strain energy of the shell can be expressed in terms of middle surface displacements (u, v, w) and rotation components (ϕ_x, ϕ_θ).

The kinetic energy (T) of the cylindrical shells including rotary inertia is written as:

$$T = \int_x \int_\theta \int_{-h/2}^{h/2} \frac{\rho^k}{2} \left\{ \left(\frac{\partial u}{\partial t} + z \frac{\partial \phi_x}{\partial t} \right)^2 + \left(\frac{\partial w}{\partial t} \right)^2 + \left(\frac{\partial v}{\partial t} + z \frac{\partial \phi_\theta}{\partial t} \right)^2 \right\} (R + z) dz d\theta dx$$

$$= \frac{1}{2} \int_x \int_\theta \left\{ \begin{array}{l} \overline{I}_0 \left(\frac{\partial u}{\partial t}\right)^2 + 2\overline{I}_1 \frac{\partial u}{\partial t}\frac{\partial \phi_x}{\partial t} + \overline{I}_2 \left(\frac{\partial \phi_x}{\partial t}\right)^2 + \overline{I}_0 \left(\frac{\partial v}{\partial t}\right)^2 \\ + 2\overline{I}_1 \frac{\partial v}{\partial t}\frac{\partial \phi_\theta}{\partial t} + \overline{I}_2 \left(\frac{\partial \phi_\theta}{\partial t}\right)^2 + \overline{I}_0 \left(\frac{\partial w}{\partial t}\right)^2 \end{array} \right\} R d\theta dx \qquad (5.29)$$

where the inertia terms are:

$$\overline{I}_0 = I_0 + I_1/R$$
$$\overline{I}_1 = I_1 + I_2/R$$
$$\overline{I}_2 = I_2 + I_3/R \qquad (5.30)$$

$$[I_0, I_1, I_2, I_3] = \sum_{k=1}^{N_L} \int_{z_k}^{z_{k+1}} \rho^k [1, z, z^2, z^3] dz$$

in which ρ^k is the mass of the kth layer per unit middle surface area. Assuming the distributed external forces q_x, q_θ and q_z are in the x, θ and z directions, respectively, and m_x and m_θ represent the external couples in the middle surface, thus, the work done by the external forces and moments is

$$W_e = \int_x \int_\theta \{q_x u + q_\theta v + q_z w + m_x \phi_x + m_\theta \phi_\theta\} R d\theta dx \qquad (5.31)$$

Using the artificial spring boundary technique as described earlier, let $k_\psi^u, k_\psi^u, k_\psi^w$, K_ψ^x, and K_ψ^θ ($\psi = x0, \theta0, x1$ and $\theta1$) to represent the rigidities (per unit length) of the boundary springs at the boundaries $x = 0$, $\theta = 0$, $x = L$ and $\theta = \theta_0$, respectively. Therefore, the deformation strain energy (U_{sp}) stored in the boundary springs during vibration is defined as (Fig. 5.4):

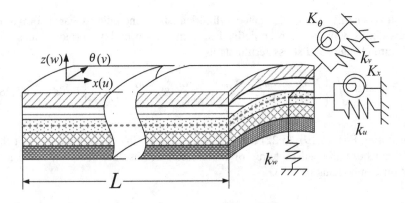

Fig. 5.4 Boundary conditions of thick laminated cylindrical shells

$$U_{sp} = \frac{1}{2} \int_\theta \left\{ \begin{array}{l} \left[k_{x0}^u u^2 + k_{x0}^v v^2 + k_{x0}^w w^2 + K_{x0}^x \phi_x^2 + K_{x0}^\theta \phi_\theta^2 \right]|_{x=0} \\ + \left[k_{x1}^u u^2 + k_{x1}^v v^2 + k_{x1}^w w^2 + K_{x1}^x \phi_x^2 + K_{x1}^\theta \phi_\theta^2 \right]|_{x=L} \end{array} \right\} R d\theta$$

$$+ \frac{1}{2} \int_x \left\{ \begin{array}{l} \left[k_{\theta 0}^u u^2 + k_{\theta 0}^v v^2 + k_{\theta 0}^w w^2 + K_{\theta 0}^x \phi_x^2 + K_{\theta 0}^\theta \phi_\theta^2 \right]|_{\theta=0} \\ + \left[k_{\theta 1}^u u^2 + k_{\theta 1}^v v^2 + k_{\theta 1}^w w^2 + K_{\theta 1}^x \phi_x^2 + K_{\theta 1}^\theta \phi_\theta^2 \right]|_{\theta=\theta_0} \end{array} \right\} dx \quad (5.32)$$

5.2.4 Governing Equations and Boundary Conditions

The governing equations and boundary conditions of thick laminated cylindrical shells can be obtained by substituting Eq. (5.1) into those of laminated general thick shells (Eq. (1.59)). According to Eq. (1.59), we have

$$\frac{\partial N_x}{\partial x} + \frac{\partial N_{\theta x}}{R \partial \theta} + q_x = \overline{I}_0 \frac{\partial^2 u}{\partial t^2} + \overline{I}_1 \frac{\partial^2 \phi_x}{\partial t^2}$$

$$\frac{\partial N_{x\theta}}{\partial x} + \frac{\partial N_\theta}{R \partial \theta} + \frac{Q_\theta}{R} + q_\theta = \overline{I}_0 \frac{\partial^2 v}{\partial t^2} + \overline{I}_1 \frac{\partial^2 \phi_\theta}{\partial t^2}$$

$$-\frac{N_\theta}{R} + \frac{\partial Q_x}{\partial x} + \frac{\partial Q_\theta}{R \partial \theta} + q_z = \overline{I}_0 \frac{\partial^2 w}{\partial t^2} \qquad (5.33)$$

$$\frac{\partial M_x}{\partial x} + \frac{\partial M_{\theta x}}{R \partial \theta} - Q_x + m_x = \overline{I}_1 \frac{\partial^2 u}{\partial t^2} + \overline{I}_2 \frac{\partial^2 \phi_x}{\partial t^2}$$

$$\frac{\partial M_{x\theta}}{\partial x} + \frac{\partial M_\beta}{R \partial \theta} - Q_\theta + m_\theta = \overline{I}_1 \frac{\partial^2 v}{\partial t^2} + \overline{I}_2 \frac{\partial^2 \phi_\theta}{\partial t^2}$$

Substituting Eqs. (5.19), (5.22) and (5.23) into above equations, the governing equations of thick cylindrical shells can be written in terms of displacements as

$$\left(\begin{bmatrix} L_{11} & L_{12} & L_{13} & L_{14} & L_{15} \\ L_{21} & L_{22} & L_{23} & L_{24} & L_{25} \\ L_{31} & L_{32} & L_{33} & L_{34} & L_{35} \\ L_{41} & L_{42} & L_{43} & L_{44} & L_{45} \\ L_{51} & L_{52} & L_{53} & L_{54} & L_{55} \end{bmatrix} - \omega^2 \begin{bmatrix} M_{11} & 0 & 0 & M_{14} & 0 \\ 0 & M_{22} & 0 & 0 & M_{25} \\ 0 & 0 & M_{33} & 0 & 0 \\ M_{41} & 0 & 0 & M_{44} & 0 \\ 0 & M_{52} & 0 & 0 & M_{55} \end{bmatrix} \right) \begin{bmatrix} u \\ v \\ w \\ \phi_x \\ \phi_\theta \end{bmatrix}$$

$$= \begin{bmatrix} -p_x \\ -p_y \\ -p_z \\ -m_x \\ -m_\theta \end{bmatrix} \qquad (5.34)$$

The coefficients of the linear operator L_{ij} and M_{ij} are given as

$$L_{11} = \overline{A_{11}}\frac{\partial^2}{\partial x^2} + \frac{2A_{16}}{R}\frac{\partial^2}{\partial x\partial\theta} + \frac{\overline{\overline{A_{66}}}}{R^2}\frac{\partial^2}{\partial\theta^2}$$

$$L_{12} = L_{21} = \overline{A_{16}}\frac{\partial^2}{\partial x^2} + \left(\frac{A_{12}}{R} + \frac{A_{66}}{R}\right)\frac{\partial^2}{\partial x\partial\theta} + \frac{\overline{\overline{A_{26}}}}{R^2}\frac{\partial^2}{\partial\theta^2}$$

$$L_{13} = -L_{31} = \frac{A_{12}}{R}\frac{\partial}{\partial x} + \frac{\overline{\overline{A_{26}}}}{R^2}\frac{\partial}{\partial\theta}$$

$$L_{14} = L_{41} = \overline{B_{11}}\frac{\partial^2}{\partial x^2} + \frac{2B_{16}}{R}\frac{\partial^2}{\partial x\partial\theta} + \frac{\overline{\overline{B_{66}}}}{R^2}\frac{\partial^2}{\partial\theta^2}$$

$$L_{15} = L_{51} = \overline{B_{16}}\frac{\partial^2}{\partial x^2} + \left(\frac{B_{12}}{R} + \frac{B_{66}}{R}\right)\frac{\partial^2}{\partial x\partial\theta} + \frac{\overline{\overline{B_{26}}}}{R^2}\frac{\partial^2}{\partial\theta^2}$$

$$L_{22} = \overline{A_{66}}\frac{\partial^2}{\partial x^2} + \frac{2A_{26}}{R}\frac{\partial^2}{\partial x\partial\theta} + \frac{\overline{\overline{A_{22}}}}{R^2}\frac{\partial^2}{\partial\theta^2} - \frac{\overline{A_{44}}}{R^2}$$

$$L_{23} = -L_{32} = \left(\frac{A_{26} + A_{45}}{R}\right)\frac{\partial}{\partial x} + \left(\frac{\overline{\overline{A_{22}}} + \overline{\overline{A_{44}}}}{R^2}\right)\frac{\partial}{\partial\theta}$$

$$L_{24} = L_{42} = \overline{B_{16}}\frac{\partial^2}{\partial x^2} + (\frac{B_{12} + B_{66}}{R})\frac{\partial^2}{\partial x\partial\theta} + \frac{\overline{\overline{B_{26}}}}{R^2}\frac{\partial^2}{\partial\theta^2} + \frac{A_{45}}{R}$$

$$L_{25} = L_{52} = \overline{B_{66}}\frac{\partial^2}{\partial x^2} + \frac{2B_{26}}{R}\frac{\partial^2}{\partial x\partial\theta} + \frac{\overline{\overline{B_{22}}}}{R^2}\frac{\partial^2}{\partial\theta^2} + \frac{\overline{A_{44}}}{R^2}$$

$$L_{33} = \overline{A_{55}}\frac{\partial^2}{\partial x^2} + \frac{2A_{45}}{R}\frac{\partial^2}{\partial x\partial\theta} + \frac{\overline{\overline{A_{44}}}}{R^2}\frac{\partial^2}{\partial\theta^2} - \frac{\overline{\overline{A_{22}}}}{R^2}$$

$$L_{34} = -L_{43} = (\overline{A_{55}} - \frac{B_{12}}{R})\frac{\partial}{\partial x} + (\frac{A_{45}}{R} - \frac{\overline{\overline{B_{26}}}}{R^2})\frac{\partial}{\partial\theta}$$

$$L_{35} = -L_{53} = (A_{45} - \frac{B_{26}}{R})\frac{\partial}{\partial x} + (\frac{\overline{A_{44}}}{R} - \frac{\overline{\overline{B_{22}}}}{R^2})\frac{\partial}{\partial\theta}$$

$$L_{44} = \overline{D_{11}}\frac{\partial^2}{\partial x^2} + \frac{2D_{16}}{R}\frac{\partial^2}{\partial x\partial\theta} + \frac{\overline{\overline{D_{66}}}}{R^2}\frac{\partial^2}{\partial\theta^2} - \overline{A_{55}}$$

$$L_{45} = L_{54} = \overline{D_{16}}\frac{\partial^2}{\partial x^2} + \left(\frac{D_{12}}{R} + \frac{D_{66}}{R}\right)\frac{\partial^2}{\partial x\partial\theta} + \frac{\overline{\overline{D_{26}}}}{R^2}\frac{\partial^2}{\partial\theta^2} - A_{45}$$

$$L_{55} = \overline{D_{66}}\frac{\partial^2}{\partial x^2} + \frac{2D_{26}}{R}\frac{\partial^2}{\partial x\partial\theta} + \frac{\overline{\overline{D_{22}}}}{R^2}\frac{\partial^2}{\partial\theta^2} - \overline{A_{44}}$$

$$M_{11} = M_{22} = M_{33} = -\overline{I_0}$$

$$M_{14} = M_{41} = M_{15} = M_{51} = -\overline{I_1}$$

$$M_{44} = M_{55} = -\overline{I_2}$$

$$(5.35)$$

According to Eqs. (1.60) and (1.61), the general boundary conditions of thick laminated cylindrical shells are

$$
\begin{aligned}
x = 0:&
\begin{cases}
N_x - k_{x0}^u u = 0 \\
N_{x\theta} - k_{x0}^v v = 0 \\
Q_x - k_{x0}^w w = 0 \\
M_x - K_{x0}^x \phi_x = 0 \\
M_{x\theta} - K_{x0}^\theta \phi_\theta = 0
\end{cases}
&
x = L:&
\begin{cases}
N_x + k_{x1}^u u = 0 \\
N_{x\theta} + k_{x1}^v v = 0 \\
Q_x + k_{x1}^w w = 0 \\
M_x + K_{x1}^x \phi_x = 0 \\
M_{x\theta} + K_{x1}^\theta \phi_\theta = 0
\end{cases}
\\[2mm]
\theta = 0:&
\begin{cases}
N_{\theta x} - k_{\theta 0}^u u = 0 \\
N_\theta - k_{\theta 0}^v v = 0 \\
Q_\theta - k_{\theta 0}^w w = 0 \\
M_{\theta x} - K_{\theta 0}^x \phi_x = 0 \\
M_\theta - K_{\theta 0}^\theta \phi_\theta = 0
\end{cases}
&
\theta = \theta_0:&
\begin{cases}
N_{\theta x} + k_{\theta 1}^u u = 0 \\
N_\theta + k_{\theta 1}^v v = 0 \\
Q_\theta + k_{\theta 1}^w w = 0 \\
M_{\theta x} + K_{\theta 1}^x \phi_x = 0 \\
M_\theta + K_{\theta 1}^\theta \phi_\theta = 0
\end{cases}
\end{aligned}
\tag{5.36}
$$

Thick cylindrical shells can have 24 possible classical boundary conditions at each edge. This yields a numerous combinations of boundary conditions, particular for open cylindrical shells. Table 5.2 shows the 24 possible classical boundary conditions for boundaries $x =$ constant, similar boundary conditions can be obtained for boundaries $\theta =$ constant (only existing in open cylindrical shells).

Classical boundary conditions can be seen as special cases of the elastic ones. By using the artificial spring boundary technique, all the aforementioned classical boundary conditions can be readily generated by assigning the boundary springs at proper stiffness. Although we can obtain accurate solutions for laminated cylindrical shells with arbitrary classical conditions and their combinations, in this chapter we will consider four typical boundary conditions that are frequently encountered in practices, namely, F, SD, S and C boundary conditions. Taking edge $x = 0$ for example, the corresponding spring stiffness for the four classical boundary conditions are given below (N/m and N/rad are utilized as the units of the stiffness about the translational springs and rotational springs, respectively):

$$
\begin{aligned}
&\text{F: } k_{x0}^u = k_{x0}^v = k_{x0}^w = K_{x0}^x = K_{x0}^\theta = 0 \\
&\text{SD: } k_{x0}^v = k_{x0}^w = K_{x0}^\theta = 10^7 D, \quad k_{x0}^u = K_{x0}^x = 0 \\
&\text{S: } k_{x0}^u = k_{x0}^v = k_{x0}^w = K_{x0}^\theta = 10^7 D, \quad K_{x0}^x = 0 \\
&\text{C: } k_{x0}^u = k_{x0}^v = k_{x0}^w = K_{x0}^x = K_{x0}^\theta = 10^7 D
\end{aligned}
\tag{5.37}
$$

5.3 Vibration of Laminated Closed Cylindrical Shells

Thin and thick laminated closed cylindrical shells with different boundary conditions, lamination schemes and geometry parameters are treated in the subsequent analysis. Both solutions in the framework of the CST and SDST (neglecting the

Table 5.2 Possible classical boundary conditions for thick cylindrical shells at each boundary of x = constant

Boundary type	Conditions
Free boundary conditions	
F	$N_x = N_{x\theta} = Q_x = M_x = M_{x\theta} = 0$
F2	$u = N_{x\theta} = Q_x = M_x = M_{x\theta} = 0$
F3	$N_x = v = Q_x = M_x = M_{x\theta} = 0$
F4	$u = v = Q_x = M_x = M_{x\theta} = 0$
F5	$N_x = N_{x\theta} = Q_x = M_x = \phi_\theta = 0$
F6	$u = N_{x\theta} = Q_x = M_x = \phi_\theta = 0$
F7	$N_x = v = Q_x = M_x = \phi_\theta = 0$
F8	$u = v = Q_x = M_x = \phi_\theta = 0$
Simply supported boundary conditions	
S	$u = v = w = M_x = \phi_\theta = 0$
SD	$N_x = v = w = M_x = \phi_\theta = 0$
S3	$u = N_{x\theta} = w = M_x = \phi_\theta = 0$
S4	$N_x = N_{x\theta} = w = M_x = \phi_\theta = 0$
S5	$u = v = w = M_x = M_{x\theta} = 0$
S6	$N_x = v = w = M_x = M_{x\theta} = 0$
S7	$u = N_{x\theta} = w = M_x = M_{x\theta} = 0$
S8	$N_x = N_{x\theta} = w = M_x = M_{x\theta} = 0$
Clamped boundary conditions	
C	$u = v = w = \phi_x = \phi_\theta = 0$
C2	$N_x = v = w = \phi_x = \phi_\theta = 0$
C3	$u = N_{x\theta} = w = \phi_x = \phi_\theta = 0$
C4	$N_x = N_{x\theta} = w = \phi_x = \phi_\theta = 0$
C5	$u = v = w = \phi_x = M_{x\theta} = 0$
C6	$N_x = v = w = \phi_x = M_{x\theta} = 0$
C7	$u = N_{x\theta} = w = \phi_x = M_{x\theta} = 0$
C8	$N_x = N_{x\theta} = w = \phi_x = M_{x\theta} = 0$

deepness term z/R, i.e., Eq. (5.26)) are presented. Natural frequencies and mode shapes of the cylindrical shells are obtained by applying the previous developed modified Fourier series and weak form solution procedure. For a laminated closed cylindrical shell, there exist two boundaries, i.e., $x = 0$ and $x = L$. For the sake of simplicity, a two-letter string is employed to represent the boundary condition of a cylindrical shell, such as F–C indicates the shell with edges $x = 0$ and $x = L$ having F and C boundary conditions, respectively. Unless otherwise stated, the natural frequencies of the considered shells are expressed in the non-dimensional parameters as $\Omega = (\omega L^2/h)\sqrt{\rho/E_2}$ and the material properties of the layers of laminated cylindrical shells under consideration are given as: $E_2 = 10$ GPa, $E_1/E_2 =$ open, $\mu_{12} = 0.25$, $G_{12} = 0.6\,E_2$, $G_{13} = G_{23} = 0.5\,E_2$, $\rho = 1{,}450$ kg/m^3.

Considering the circumferential symmetry of closed circular cylindrical shells, the assumed 2D displacement field can be reduced to a quasi 1D problem through Fourier decomposition of the circumferential wave motion. Thus, taking the FSDT displacement field for example, each displacement and rotation component of a closed circular cylindrical shell is expanded as the following form:

$$u(x, \theta) = \sum_{m=0}^{M} \sum_{n=0}^{N} A_{mn} \cos \lambda_m x \cos n\theta + \sum_{l=1}^{2} \sum_{n=0}^{N} a_{ln} P_l(x) \cos n\theta$$

$$v(x, \theta) = \sum_{m=0}^{M} \sum_{n=0}^{N} B_{mn} \cos \lambda_m x \sin n\theta + \sum_{l=1}^{2} \sum_{n=0}^{N} b_{ln} P_l(x) \sin n\theta$$

$$w(x, \theta) = \sum_{m=0}^{M} \sum_{n=0}^{N} C_{mn} \cos \lambda_m x \cos n\theta + \sum_{l=1}^{2} \sum_{n=0}^{N} c_{ln} P_l(x) \cos n\theta \qquad (5.38)$$

$$\phi_x(x, \theta) = \sum_{m=0}^{M} \sum_{n=0}^{N} D_{mn} \cos \lambda_m x \cos n\theta + \sum_{l=1}^{2} \sum_{n=0}^{N} d_{ln} P_l(x) \cos n\theta$$

$$\phi_\theta(x, \theta) = \sum_{m=0}^{M} \sum_{n=0}^{N} E_{mn} \cos \lambda_m x \sin n\theta + \sum_{l=1}^{2} \sum_{n=0}^{N} e_{ln} P_l(x) \sin n\theta$$

where $\lambda_m = m\pi/L$. n represents the circumferential wave number of the corresponding mode. It should be note that n is a non-negative integer. Interchanging of $\sin n\theta$ and $\cos n\theta$ in Eq. (5.38), another set of free vibration modes (anti-symmetric modes) can be obtained. A_{mn}, B_{mn}, C_{mn}, D_{mn} and E_{mn} are expansion coefficients of standard cosine Fourier series. a_{ln}, b_{ln}, c_{ln}, d_{ln} and e_{ln} are the corresponding supplement coefficients. M and N denote the truncation numbers with respect to variables x and θ, respectively. According to Eq. (5.35), it is obvious that each displacement and rotation component in the FSDT displacement field is required to have up to the second derivatives. Therefore, two auxiliary polynomial functions $P_l(x)$ are introduced in each displacement expression to remove all the discontinuities potentially associated with the first-order derivatives at the boundaries. These auxiliary functions are defined as (Ye et al. 2014a, d, e)

$$P_1(x) = x\left(\frac{x}{L} - 1\right)^2 \quad P_2(x) = \frac{x^2}{L}\left(\frac{x}{L} - 1\right) \qquad (5.39)$$

According to Eq. (5.14), it can be seen that the displacements in the CST displacement field are required to have up to the third (u and v) or fourth (w) derivatives. In such case, the axial, circumferential and radial displacements of a closed circular cylindrical shell are expanded as a 1-D modified Fourier series in which four auxiliary polynomial functions should be introduced to remove all the discontinuities potentially associated with the first-order and third-order derivatives at the boundaries:

$$u(x, \theta) = \sum_{m=0}^{M} \sum_{n=0}^{N} A_{mn} \cos \lambda_m x \cos n\theta + \sum_{l=1}^{4} \sum_{n=0}^{N} a_{ln} P_l(x) \cos n\theta$$

$$v(x, \theta) = \sum_{m=0}^{M} \sum_{n=0}^{N} B_{mn} \cos \lambda_m x \sin n\theta + \sum_{l=1}^{4} \sum_{n=0}^{N} b_{ln} P_l(x) \sin n\theta \qquad (5.40)$$

$$w(x, \theta) = \sum_{m=0}^{M} \sum_{n=0}^{N} C_{mn} \cos \lambda_m x \cos n\theta + \sum_{l=1}^{4} \sum_{n=0}^{N} c_{ln} P_l(x) \cos n\theta$$

where the four auxiliary polynomial functions $P_l(x)$ are given as (Zhang and Li 2009)

$$P_1(x) = \frac{9L}{4\pi} \sin(\frac{\pi x}{2L}) - \frac{L}{12\pi} \sin(\frac{3\pi x}{2L})$$

$$P_2(x) = -\frac{9L}{4\pi} \cos(\frac{\pi x}{2L}) - \frac{L}{12\pi} \cos(\frac{3\pi x}{2L})$$

$$P_3(x) = \frac{L^3}{\pi^3} \sin(\frac{\pi x}{2L}) - \frac{L^3}{3\pi^3} \sin(\frac{3\pi x}{2L}) \qquad (5.41)$$

$$P_4(x) = -\frac{L^3}{\pi^3} \cos(\frac{\pi x}{2L}) - \frac{L^3}{3\pi^3} \cos(\frac{3\pi x}{2L})$$

5.3.1 Convergence Studies and Result Verification

Considering the circumferential symmetry of the circular shells, the convergence only needs to be checked in the axial direction (i.e., x). In Table 5.3, the first six natural frequencies (Hz) for a two-layered, [0°/90°] laminated cylindrical shell with six truncation schemes (i.e., $M = 6, 7, 8, 9, 10, 11, N = 10$) are presented. The geometric and material constants of the shell are: $R = 1$ m, $L/R = 5$, $h/R = 0.1$, $E_2 = 10$ GPa, $E_1/E_2 = 15$, $\mu_{12} = 0.27$, $G_{12} = 0.5E_2$, $G_{13} = 0.5E_2$, $G_{23} = 0.2E_2$, $\rho = 1,700$ kg/m^3. The natural frequencies of the shell are calculated by MATAB on a notebook. The configuration of the computer is: Inter Core2 Duo CPU and 2 GB RAM. It is obviously that the modified Fourier series solution has an excellent convergence, and is sufficient accuracy even when only a small number of terms are included in the series expression. The maximum difference between the '$M = 6$' and '$M = 11$' form results for the F–F and C–C boundary conditions are less than 0.15 and 0.08 %, respectively. Furthermore, from the table, we can see that although the series are truncated as much as 11 × 10, the computing time is less 1.53 s. In addition, convergence of the C–C solutions is faster than the case of F–F boundary conditions. Unless otherwise stated, the truncated number of the displacement expressions will be uniformly selected as $M = 11$ in the following discussions.

To validate the accuracy and reliability of current method, the current solutions are compared with those reported by other researchers. First, let us consider a

Table 5.3 Convergence of the first six frequencies (Hz) for a [0°/90°] laminated cylindrical shell with F–F and C–C boundary conditions

B.C.	M	Mode number						Time (s)
		1	2	3	4	5	6	
F–F	6	3.4476	52.387	54.131	135.37	144.13	146.02	1.11
	7	3.4476	52.387	54.131	135.35	144.13	146.02	1.17
	8	3.4446	52.387	54.127	135.35	144.13	146.02	1.25
	9	3.4446	52.387	54.127	135.34	144.13	146.02	1.35
	10	3.4423	52.387	54.123	135.34	144.13	146.02	1.42
	11	3.4423	52.387	54.123	135.33	144.13	146.02	1.53
C–C	6	91.565	118.86	154.66	160.87	185.97	235.66	1.10
	7	91.514	118.77	154.65	160.87	185.97	235.65	1.19
	8	91.514	118.77	154.65	160.80	185.94	235.54	1.25
	9	91.500	118.74	154.65	160.79	185.94	235.54	1.34
	10	91.501	118.74	154.65	160.77	185.93	235.52	1.42
	11	91.494	118.73	154.65	160.77	185.94	235.51	1.52

Table 5.4 Comparison of the frequency parameters Ω for a [0°/90°/0°] laminated cylindrical shell with different boundary conditions

n	F–F			C–C		
	Messina and Soldatos (1999c)	Present	Error (%)	Messina and Soldatos (1999c)	Present	Error (%)
1	304.13	304.16	0.01	159.31	159.44	0.08
2	26.58	26.56	−0.09	107.71	107.89	0.17
3	74.91	74.78	−0.17	108.05	108.11	0.06
4	142.93	142.51	−0.29	157.23	156.94	−0.18
5	229.74	228.70	−0.45	237.70	236.76	−0.40
7	456.60	452.69	−0.86	460.98	457.16	−0.83
10	917.18	902.24	−1.63	920.32	905.47	−1.61

three-layered, cross-ply [0°/90°/0°] cylindrical shell. The elementally material parameters and geometric properties of the layers of the shell are given as: $L/R = 5$, $h/R = 0.05$, $E_1/E_2 = 25$, $\mu_{12} = 0.25$, $G_{12} = 0.5E_2$, $G_{13} = 0.5E_2$, $G_{23} = 0.2E_2$. Comparisons of the shell with two sets of classical boundary conditions (i.e., F–F and C–C) for the first longitudinal mode (i.e., $m = 1$) and six different circumference wave numbers (i.e., $n = 1$–5 and 10, $n = 6$, 8 and 9 were not considered in referential paper) are presented in Table 5.4. From the table, we can see that the present solutions agree very well with results obtained by Messina and Soldatos (1999c). The differences between these two results are very small, and do not exceed 1.63 % for the worst case. The small differences may be caused by a different solution procedure were used by Messina and Soldatos (1999c).

Table 5.5 Comparison of frequency parameters $\Omega = \omega h / \pi \sqrt{\rho / G_{12}}$ for two S–S supported, cross-ply laminated cylindrical shells with different thick-radius ratios ($R = 1$ m, $L/R = 1$, $E_1/E_2 = 40$)

h/R	[0°/90°/90°/0°]			[90°/0°/0°/90°]		
	Present	Ye (2003)	Ich Thinh and Nguyen (2013)	Present	Ye (2003)	Ich Thinh and Nguyen (2013)
0.1	0.0639	0.0646	0.0640	0.0533	0.0527	0.0531
	0.0657	0.0663	0.0657	0.0592	0.0591	0.0591
	0.0789	0.0793	0.0789	0.0710	0.0707	0.0709
0.2	0.1588	0.1638	0.1589	0.1335	0.1302	0.1333
	0.1678	0.1709	0.1683	0.1528	0.1507	0.1527
	0.1727	0.1752	0.1726	0.1593	0.1589	0.1592
0.3	0.2542	0.2630	0.2546	0.2275	0.2188	0.2273
	0.2670	0.2729	0.2669	0.2430	0.2364	0.2428
	0.2788	0.2838	0.2797	0.2701	0.2683	0.2699

The previous example is presented as thin laminated composite cylindrical shell with classical boundary conditions. The validity of the proposed method for vibration analysis of thick laminated composite cylindrical shells will be proved in following examples. In Table 5.5, the detail comparisons between results obtained by present method and those provided by Ye (2003) with 3D elasticity theory as well as Ich Thinh and Nguyen (2013) by the FSDT for certain cross-ply cylindrical shells with S–S boundary conditions are presented, in which two types of lamination schemes ([0°/90°/90°/0°] and [90°/0°/0°/90°]) are included. The material parameters and geometric properties of the layers of the considered shells are assumed to be: $R = 1$ m, $L/R = 1$, $E_1/E_2 = 40$. The comparisons are conducted for thickness-to-radius ratios $h/R = 0.1, 0.2, 0.3$, respectively. The lowest three frequencies of the shells, which are expressed in terms of non-dimension parameters, $\Omega = \omega h / \pi \sqrt{\rho / G_{12}}$ are computed. It is obvious that the current results match very well with the referential data. The proposed solution is efficient and accurate in predicting nature frequencies of thick laminated cylindrical shells. The differences between the present results and those reported by Ye (2003) are bigger than those relative to Ich Thinh and Nguyen (2013). It is attributed to different shell theories were used in the literature.

Then, a further comparison is performed for laminated composite shells with different boundary conditions and length-to-radius ratios. The used material constants are the same as previous example. Two types of lamination schemes ([0°/90°] and [0°/90°/0°]) and two kinds of length-to-radius ratios (i.e., $L/R = 1$ and 2) are considered. The fundamental frequency parameters $\Omega = (\omega L^2 / 100h) \sqrt{\rho / E_2}$ for the shells with thickness-to-radius ratio $h/R = 0.2$ are calculated. In Table 5.6, comparisons between the present solutions and results reported by Khdeir et al. (1989), Ich Thinh and Nguyen (2013) are presented, in which two sets of classical boundary conditions i.e., S–C and C–C are considered. A good agreement can be

Table 5.6 Comparison of the fundamental frequency parameters $\Omega = (\omega L^2/100h)\sqrt{\rho/E_2}$ of two certain cross-ply laminated cylindrical shells with different length-radius ratios and boundary conditions ($R = 1$ m, $h/R = 0.2$)

Lamination schemes	Theories	S–C		C–C	
		$L/R = 1$	$L/R = 2$	$L/R = 1$	$L/R = 2$
$[0°/90°]$	HSDT (Khdeir et al. 1989)	0.0938	0.1726	0.1085	0.1928
	FSDT (Khdeir et al. 1989)	0.0893	0.1697	0.1002	0.1876
	CST (Khdeir et al. 1989)	0.1152	0.1841	0.1048	0.2120
	FSDT (Ich Thinh and Nguyen 2013)	0.0823	0.1661	0.0982	0.1737
	Present	0.0921	0.1639	0.0982	0.1738
$[0°/90°/0°]$	HSDT (Khdeir et al. 1989)	0.1087	0.1972	0.1192	0.2191
	FSDT (Khdeir et al. 1989)	0.1036	0.1945	0.1093	0.2129
	CST (Khdeir et al. 1989)	0.1850	0.2662	0.2049	0.3338
	FSDT (Ich Thinh and Nguyen 2013)	0.1025	0.1950	0.1083	0.2083
	Present	0.1028	0.1991	0.1086	0.2084

seen from the table. The small deviations in the results are caused by different computation methods and shell theories were used in the literature.

5.3.2 Effects of Shear Deformation and Rotary Inertia

In this section, effects of the shear deformation and rotary inertia which are neglected in the CST will be investigated. Figures 5.5 and 5.6 show the differences between the lowest three frequency parameters Ω obtained by CST and SDST of a short ($R/L = 1$), two-layered, $[0°/90°]$ laminated cylindrical shell with different thickness-to-radius ratios (h/R) and orthotropy ratios (E_1/E_2) for F–F and C–C boundary conditions, respectively. Two orthotropy ratios, i.e., $E_1/E_2 = 1$ and 40, corresponding to isotropic and composite shells are shown in each figure. The thickness-to-radius ratio h/R is varied from 0.001 to 0.2, corresponding to very thin to thick cylindrical shells. As expected, the figures show that the difference between the CST and SDST solutions increases as h/R increases. In addition, we can see that the effects of the shear deformation and rotary inertia increase as the orthotropy ratio (E_1/E_2) increases. From Fig. 5.5, we can see that when h/R is less than 0.01, the maximum difference between the frequency parameters Ω obtained by CST and SDST is less than 0.21 % for the worst case. However, when h/R is equal to 0.1, the maximum difference can be as many as 3.75 and 4.3 % for the cylindrical shell with orthotropy ratios of $E_1/E_2 = 1$ and $E_1/E_2 = 40$, respectively. Figure 5.5 also shows that the maximum differences between these two results can be as many as 8.12 and 10.22 % for a thickness-to-radius ratio of 0.2. Figure 3.6 shows that the maximum

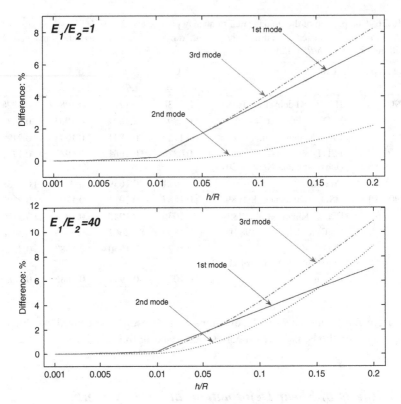

Fig. 5.5 Differences between the lowest three frequency parameters Ω obtained by CST and SDST for a short [0°/90°] laminated cylindrical shell with F–F boundary conditions ($R/L = 1$)

difference between these CST and SDST results can be as many as 16.3 and 59.9 % for the C–C cylindrical shell with orthotropy ratios of $E_1/E_2 = 1$ and $E_1/E_2 = 40$, respectively. In such case, the CST results are utterly inaccurate.

For the sake of completeness, Figs. 5.7 and 5.8 show the similar studies for a long ($R/L = 5$), [0°/90°] laminated cylindrical shell. Constantly, the effects of shear deformation and rotary inertia increase with thickness-to-span length ratio increases. Comparing Figs. 5.7 and 5.8 with Figs. 5.5 and 5.6, we can find that the effects of shear deformation and rotary inertia decrease as length-to-radius ratio increases. The maximum difference between the CST and SDST results is less than 7.5 % for the worst case, which is only one eighth of those of cylindrical shell with length-to-radius ratio $L/R = 1$. These investigations show that the CST is only applicable for thin cylindrical shells as well as the long moderately thick ones. For cylindrical shells with higher thickness ratios or smaller length-to-radius ratios, both shear deformation and rotary inertia effects should be included in the calculation.

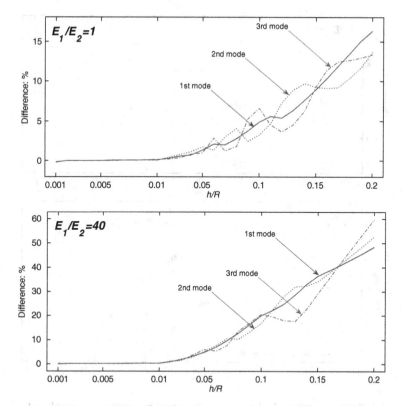

Fig. 5.6 Differences between the lowest three frequency parameters Ω obtained by CST and SDST for a short [0°/90°] laminated cylindrical shell with C–C boundary conditions ($R/L = 1$)

5.3.3 Laminated Closed Cylindrical Shells with General End Conditions

First, let us consider a three-layered, [0°/90°/0°] cylindrical shell with various thickness-to-radius ratios. The material properties and geometric constants of the layers of the shell are: $E_1/E_2 = 15$, $R = 1$ m, $L/R = 2$. In Table 5.7, the lowest four frequency parameters Ω for the shell are presented, in which six sets of classical boundary combinations (i.e., F–F, F–S, F–C, S–S, S–C and C–C) and five kinds of thickness-to-radius ratios ($h/R = 0.01$, 0.02, 0.05, 0.1 and 0.15) are included in the table. It can be seen from the table that the frequency parameters increase in general as the thickness-radius ratio increases. The second observation that needs to be made here is that the effects of thickness-to-radius ratio are much higher for the thicker cylindrical shells than it is for the thinner ones. In addition, it is obviously that boundary conditions have a conspicuous effect on the vibration frequencies of the shell. Increasing the restraint stiffness always results in increment of the frequency parameters.

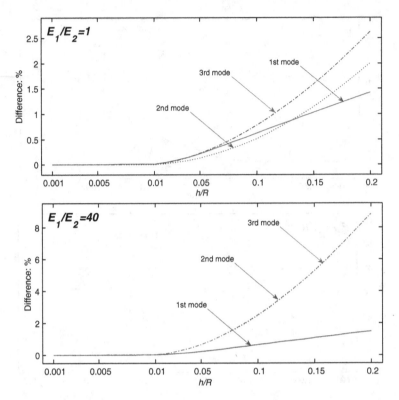

Fig. 5.7 Differences between the lowest three frequency parameters Ω obtained by CST and SDST for a long [0°/90°] laminated cylindrical shell with F–F boundary conditions ($R/L = 5$)

Table 5.8 shows the lowest four frequency parameters Ω of a moderately thick cylindrical shell with different lamination schemes for various boundary conditions. Five lamination schemes, i.e., [0°], [90°], [0°/90°], [0°/90°/0°] and [0°/90°/0°/90°] are studied for all aforementioned boundary conditions (see Table 5.7). The geometric and material constants used in the study are: $R = 1$ m, $L/R = 1$, $h/R = 0.1$, $E_1/E_2 = 15$. When the shell with F–F, F–S and F–C boundary conditions, the [90°] lamination scheme yields higher frequency parameter than those of [0°], [0°/90°], [0°/90°/0°] and [0°/90°/0°/90°] (except the fundamental mode). For other boundary conditions, the frequency parameters of [90°] are the smallest among the five lamination schemes. Furthermore, the [0°] lamination scheme yields significantly lower frequency parameters than those of [0°/90°/0°]. The similar comparison results can be seen for [0°/90°] and [0°/90°/0°/90°] lamination schemes.

The effects of length-to-radius ratio on frequency parameters are also investigated. Table 5.9 shows the lowest four frequency parameters Ω for a two-layered, unsymmetrically laminated cross-ply cylindrical shell ([0°/90°]) with various length-to-radius ratios and different sets of boundary conditions. The shell is assumed to be made of composite layers with following geometric and material constants:

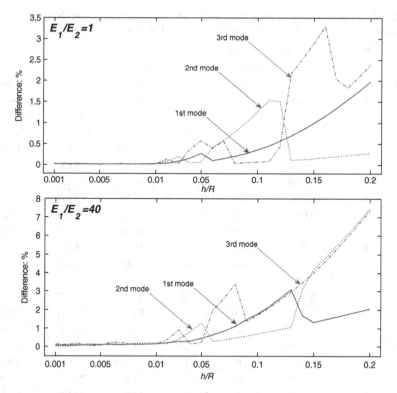

Fig. 5.8 Differences between the lowest three frequency parameters Ω obtained by CST and SDST for a long [0°/90°] laminated cylindrical shell with C–C boundary conditions ($R/L = 5$)

$R = 1$ m, $h/R = 0.05$, $E_1/E_2 = 15$. It is obviously that the length-to-radius ratio has a great influence on the vibration of the shell. The frequency parameters decrease when the length-radius ratio is increased.

The following numerical analysis is conducted to investigate the influence of fiber orientations on the frequency parameters Ω of laminated cylindrical shells. Let us consider a three-layered [0°/ϑ/0°] cylindrical shell with varying boundary conditions. The shell is modeled as both the top and bottom layers are of uniform principal direction which is paralleled to x-axis whilst the included angle between the fiber orientation of the middle layer and x-axis is changed from 0° to 180° step-by-step. In Fig. 5.9, the first three frequency parameters Ω for the shell with three types of boundary conditions (i.e., SD–SD, S–S and C–C) are plotted, respectively. The following geometric and material properties are used in the analysis: $R = 1$ m, $L/R = 2$, $h/R = 0.05$, $E_1/E_2 = 15$. Many interesting characteristics can be observed from the figure. The first observation is that all the figures are symmetrical about central line (i.e., $\vartheta = 90°$). The second observation is that the frequency parameter traces in all cases climb up and then decline, and reach their crests around $\vartheta = 50°$ when ϑ is increased from 0 to 90° except the second mode of SD-SD boundary

Table 5.7 Frequency parameters Ω for a [0°/90°/0°] laminated cylindrical shell with various thickness-to-radius ratios and boundary conditions ($R = 1$ m, $L/R = 2$, $E_1/E_2 = 15$)

h/R	Mode	Boundary conditions					
		F–F	F–S	F–C	S–S	S–C	C–C
0.01	1	0.6924	47.444	47.887	91.228	93.711	96.460
	2	3.8252	49.853	50.482	94.879	97.010	99.436
	3	5.7572	55.831	56.111	97.776	100.42	103.28
	4	10.818	65.872	66.601	107.00	108.73	110.75
0.02	1	0.6913	31.185	31.833	59.048	61.749	64.919
	2	3.8242	34.505	35.361	62.322	65.184	68.399
	3	5.7498	38.063	38.471	64.959	67.227	70.009
	4	10.812	50.185	51.009	76.562	79.194	82.038
0.05	1	0.6868	17.397	18.321	34.303	37.189	40.825
	2	3.8178	20.540	21.570	34.342	37.371	40.972
	3	5.7149	23.205	23.788	41.782	44.120	47.219
	4	10.772	33.448	34.046	43.830	46.327	49.130
0.10	1	0.6794	11.010	12.122	22.277	25.089	28.433
	2	3.7968	13.105	13.921	23.925	26.512	29.471
	3	5.6384	16.769	17.456	27.431	29.741	32.624
	4	10.638	21.360	21.857	34.142	35.582	37.150
0.15	1	0.6720	8.053	9.1254	17.725	19.981	22.435
	2	3.7633	11.216	11.919	18.755	20.955	23.424
	3	5.5395	11.945	12.638	23.400	24.722	26.135
	4	10.431	20.377	20.789	25.035	26.676	28.591

conditions, in which the maximum frequency parameters occur around $\vartheta = 60°$. In addition, it is obvious that, the second frequency parameters almost equal to the third ones when $\vartheta = 20°$ and the differences between the first frequency parameters and the second ones are very small for a fiber orientation of 90°.

Figure 5.10 performs the similar study for a two-layered [0°/ϑ] cylindrical shell. The similar characteristics observed in Fig. 5.9 can be found in this figure as well. The different observation is that the variation tendencies of the frequency parameters are more complicated than those of Fig. 5.9. It may be due to the shell is unsymmetrically laminated.

Laminated composite cylindrical shells with elastically restrained edges are widely encountered in engineering practices. The vibration analyses of these shells are necessary and of great significance. There are infinite types of possible combinations of elastic boundary conditions at the two edges of the cylindrical shells, and it is impossible to undertake an all-encompassing survey of them. Therefore, in this paper we choose three typical uniform elastic restraint conditions that are defined as follows (at $x = 0$):

Table 5.8 Frequency parameters Ω for a moderately thick laminated cylindrical shell with various lamination schemes and boundary conditions ($R = 1$ m, $L/R = 1$, $h/R = 0.1$, $E_1/E_2 = 15$)

Lamination schemes	Mode	Boundary conditions					
		F–F	F–S	F–C	S–S	S–C	C–C
[0°]	1	0.1973	4.6489	6.0143	12.092	14.494	16.976
	2	0.7703	5.0847	6.3089	12.366	14.746	17.220
	3	1.8917	5.6701	6.8472	12.575	14.863	17.242
	4	2.1636	7.0632	7.8891	13.256	15.555	17.986
[90°]	1	0.1993	5.3648	5.5490	11.244	11.865	12.571
	2	2.8163	7.9426	8.0825	11.354	11.956	12.691
	3	2.8341	8.1412	8.2330	15.357	15.793	16.357
	4	7.5175	13.087	13.667	16.394	16.750	17.132
[0°/90°]	1	0.1993	5.7888	5.9648	11.441	12.140	13.279
	2	1.3637	5.8632	5.9869	12.411	13.062	14.113
	3	2.0319	8.1790	8.2594	12.558	13.185	14.199
	4	3.7740	8.3929	8.5320	15.247	15.756	16.563
[0°/90°/0°]	1	0.1990	4.9791	6.3003	12.512	14.780	17.130
	2	0.9487	5.5288	6.7570	12.936	15.140	17.446
	3	1.9630	6.3494	7.4064	13.549	15.578	17.699
	4	2.6585	8.3430	9.0301	14.484	16.444	18.538
[0°/90°/0°/90°]	1	0.1992	5.8999	6.4928	12.256	13.833	15.742
	2	1.9420	6.8706	7.3300	12.748	14.224	15.985
	3	2.3103	8.4126	8.7915	14.384	15.745	17.451
	4	5.3078	10.520	10.806	16.692	17.654	18.795

E^1: The normal direction is elastically restrained ($u \neq 0$, $v = w = \phi_x = \phi_\theta = 0$), i.e.,
$\quad k_u = \Gamma$;

E^2: The transverse direction is elastically restrained ($w \neq 0$, $u = v = \phi_x = \phi_\theta = 0$), i.e.,
$\quad k_w = \Gamma$;

E^3: The rotation is elastically restrained ($\phi_x \neq 0$, $u = v = w = \phi_\theta = 0$), i.e., $K_x = \Gamma$.

Table 5.10 shows the lowest two frequency parameters Ω of a two-layered [0°/ϑ] cylindrical shell with different restrain parameters Γ and fiber orientations. Four different lamination schemes, i.e., $\vartheta = 0°$, 30°, 60° and 90° are performed in the calculation. The shell parameters used are $R = 1$ m, $L/R = 3$, $h/R = 0.05$, $E_1/E_2 = 15$. The shell is clamped at the edge of $x = L$ and with elastic boundary conditions at the other edge. The table shows that increasing restraint rigidities in the normal and transverse directions have very limited effects on the frequency parameters of the shell. When the normal restrained rigidity is varied from $10^{-1} * D$ to $10^3 * D$, the maximum increment in the table is less than 2.49 % for all cases. The further observation from the table is that for the shell with lamination scheme of [0°],

Table 5.9 The lowest four frequency parameters Ω for a [0°/90°] laminated cylindrical shell with various length-to-radius ratios and boundary conditions ($R = 1$ m, $h/R = 0.05$, $E_1/E_2 = 15$)

L/R	Mode	Boundary conditions					
		F–F	F–S	F–C	S–S	S–C	C–C
1	1	0.2032	8.8400	8.9230	16.894	17.614	18.718
	2	1.4017	9.9271	9.9750	17.726	18.459	19.555
	3	2.1356	10.866	10.982	18.653	19.283	20.253
	4	3.9373	13.394	13.428	21.924	22.558	23.462
2	1	0.6868	20.615	20.618	36.090	36.479	36.953
	2	5.6072	20.664	20.676	40.438	40.710	41.043
	3	6.8179	31.790	31.795	43.913	44.325	44.813
	4	15.751	33.172	33.173	54.095	54.271	54.483
3	1	1.2663	29.385	29.487	58.491	58.637	58.800
	2	12.617	39.266	39.324	65.558	65.780	66.026
	3	14.005	48.125	48.161	77.299	77.366	77.439
	4	35.441	66.797	67.810	97.770	99.680	101.60
4	1	1.8621	38.500	38.718	86.300	86.330	86.363
	2	22.430	61.348	61.495	86.869	86.961	87.057
	3	23.895	65.943	66.010	128.66	128.67	128.68
	4	63.007	94.226	94.610	133.41	134.70	134.98
5	1	2.4535	49.307	49.581	107.98	107.99	107.99
	2	35.047	72.760	73.053	120.62	120.62	120.62
	3	36.550	100.78	100.84	167.63	167.65	167.70
	4	98.451	127.71	127.81	172.83	173.61	174.42

the effect of the rotation direction restraint rigidity is more significant. When the rotation restraint rigidity is varied from $10^{-1} * D$ to $10^3 * D$, the increments of the first and second modes can be 5.57 and 4.99 %, respectively).

Table 5.11 shows similar studies for the shell with F boundary conditions at the edge of $x = L$. The table reveals that the restraint rigidity at normal direction has large effects on the frequency parameters of all the lamination schemes. When the normal direction restraint rigidity is varied from $10^{-1} * D$ to $10^3 * D$, the increments of these four lamination schemes for the first and second modes can be (1,037, 1,087, 1,118, 1,103 %) and (359, 393, 654, 879 %), respectively. This table also shows that the increments of the restraint rigidities in the transverse and rotation directions have very limited effects on the frequency parameters of the shell, with maximum increment less than 2.73 % in all cases when the rotation restrained rigidity is varied from $10^{-1} * D$ to $10^4 * D$. From the two tables, it is obvious that the effects of elastic restraint rigidity on the frequency parameters of laminated cylindrical shells is varied with mode sequences, lamination schemes and spring components.

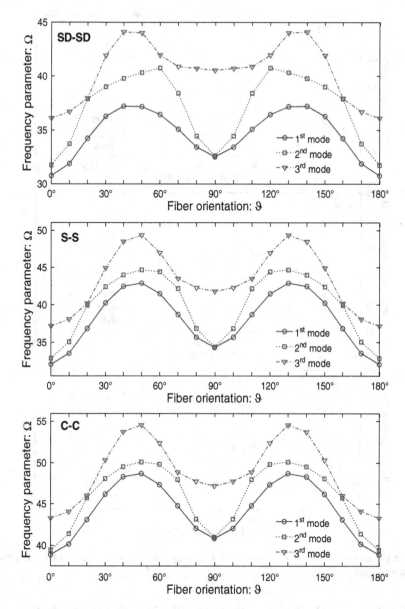

Fig. 5.9 Influence of fiber orientations on frequency parameters Ω of a three-layered $[0°/\vartheta/0°]$ cylindrical shell

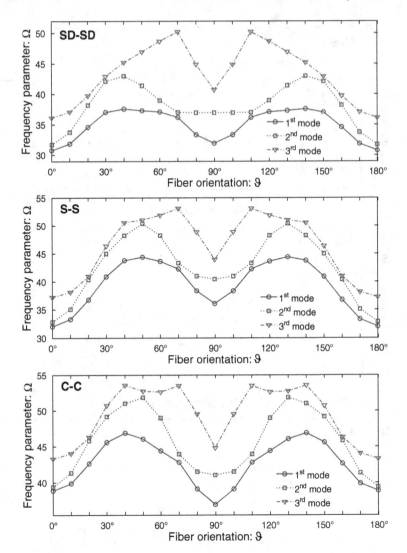

Fig. 5.10 Influence of fiber orientations on frequency parameters Ω of a two-layered $[0°/\vartheta]$ cylindrical shell

5.3.4 Laminated Closed Cylindrical Shells with Intermediate Ring Supports

Cylindrical shells with intermediate ring supports are widely used in engineering practices, such as pipelines for undersea transmission cables, irrigation pipes, and household water pipes. Without these intermediate supports, these structures may undergo large deformation, and violent vibration due to their low stiffness, and will

Table 5.10 The first two frequency parameters Ω for a two-layered $[0°/\vartheta]$ cylindrical shell with different restrain parameters Γ and fiber orientations ($R = 1$ m, $L/R = 3$, $h/R = 0.05$, $E_1/E_2 = 15$)

ϑ	Γ	$E^1 - C$		$E^2 - C$		$E^3 - C$	
		1	2	1	2	1	2
0°	$10^{-1} \times D$	54.351	57.998	55.146	58.474	53.390	56.641
	$10^0 \times D$	54.351	57.998	55.178	58.460	53.905	57.149
	$10^1 \times D$	54.358	58.002	55.265	58.548	55.473	58.649
	$10^2 \times D$	54.418	58.037	55.808	58.988	56.256	59.370
	$10^3 \times D$	54.871	58.311	56.282	59.393	56.366	59.470
30°	$10^{-1} \times D$	68.056	68.804	69.239	72.454	68.847	72.258
	$10^0 \times D$	68.057	68.805	69.257	72.582	69.142	72.550
	$10^1 \times D$	68.061	68.818	69.321	72.689	70.063	73.498
	$10^2 \times D$	68.108	68.946	69.804	73.229	70.536	74.008
	$10^3 \times D$	68.498	69.954	70.441	73.914	70.604	74.082
60°	$10^{-1} \times D$	64.869	73.191	70.924	75.311	71.434	75.639
	$10^0 \times D$	64.869	73.192	70.923	75.308	71.457	75.663
	$10^1 \times D$	64.891	73.198	70.944	75.316	71.518	75.726
	$10^2 \times D$	65.077	73.251	71.095	75.431	71.544	75.752
	$10^3 \times D$	66.487	73.688	71.420	75.663	71.548	75.755
90°	$10^{-1} \times D$	55.862	62.395	58.534	65.692	58.644	65.790
	$10^0 \times D$	55.863	62.396	58.535	65.693	58.687	65.850
	$10^1 \times D$	55.875	62.422	58.546	65.703	58.770	65.976
	$10^2 \times D$	56.000	62.632	58.611	65.783	58.796	66.019
	$10^3 \times D$	56.833	63.889	58.747	65.954	58.800	66.026

eventually lead to failure. The vibration analyses of these shells are necessary and particularly important. However, cylindrical shells with intermediate supports have received limited attention. Vibrations of laminated cylindrical shells with arbitrary intermediate ring supports will be considered in the subsequent analysis.

As shown in Fig. 5.11, a laminated cylindrical shell restrained by arbitrary intermediate ring supports is chosen as the analysis model. a_i represents the position of the i'th ring support along the axial direction of the shell. The displacement fields in the position of the ring support satisfy $w_0(a_i, \theta) = 0$ (Swaddiwudhipong et al. 1995; Zhang and Xiang 2006). This condition can be readily obtained by introducing a set of continuously distributed linear springs at the position of each ring support to restrict the displacement in radial direction and setting the stiffness of these springs equal to infinite (which is represented by a very large number, $10^7 D$). Thus, the potential energy (P_{irs}) stored in the ring supported springs can be depicted as:

$$P_{irs} = \frac{1}{2} \int_\theta \left\{ \sum_{i=1}^{N_i} k_w^i w(a_i, \theta)^2 \right\} R d\theta \tag{5.42}$$

Table 5.11 The first two frequency parameters Ω for a two-layered $[0°/\vartheta]$ cylindrical shell with different restrain parameters Γ and fiber orientations ($R = 1$ m, $L/R = 3$, $h/R = 0.05$, $E_1/E_2 = 15$)

ϑ	Γ	$E^1 - F$		$E^2 - F$		$E^3 - F$	
		1	2	1	2	1	2
0°	$10^{-1} \times D$	1.8409	5.1377	27.916	30.197	27.470	29.729
	$10^0 \times D$	5.5480	5.8210	27.918	30.206	27.585	29.860
	$10^1 \times D$	8.5948	10.415	27.936	30.234	27.925	30.281
	$10^2 \times D$	12.470	20.970	28.021	30.381	28.090	30.506
	$10^3 \times D$	20.946	23.605	28.099	30.515	28.114	30.539
30°	$10^{-1} \times D$	1.8409	5.2805	29.420	35.936	29.589	36.424
	$10^0 \times D$	5.6731	5.8208	29.422	35.937	29.622	36.470
	$10^1 \times D$	8.6521	11.695	29.439	35.986	29.730	36.625
	$10^2 \times D$	13.358	21.687	29.579	36.274	29.788	36.715
	$10^3 \times D$	21.870	26.044	29.755	36.636	29.796	36.728
60°	$10^{-1} \times D$	1.8409	4.4993	34.774	36.036	34.893	36.107
	$10^0 \times D$	4.9467	5.8205	34.776	36.037	34.920	36.112
	$10^1 \times D$	8.1479	13.429	34.784	36.040	34.995	36.126
	$10^2 \times D$	14.868	21.309	34.846	36.064	35.029	36.133
	$10^3 \times D$	22.433	33.538	34.980	36.111	35.034	36.134
90°	$10^{-1} \times D$	1.8409	3.8049	29.401	39.298	29.389	39.269
	$10^0 \times D$	4.3293	5.8205	29.401	39.298	29.414	39.284
	$10^1 \times D$	7.7880	13.965	29.404	39.299	29.466	39.313
	$10^2 \times D$	15.478	20.351	29.424	39.306	29.484	39.323
	$10^3 \times D$	22.150	37.252	29.468	39.319	29.486	39.324

where N_i is the amount of ring supports. k_w^i denotes the set of ring supported springs distributed at $x = a_i$. By adding the potential energy P_{irs} stored in the ring supported springs in the Lagrangian energy functional and applying the weak form solution procedure, the characteristic equation for the shell with arbitrary end conditions and ring supports can be readily obtained.

To validate the present analysis, results for C–F isotropic cylindrical shells with one ring support are compared with others in the literature. In Table 5.12, comparisons of the first three longitudinal modal (i.e., $m = 1, 2, 3$) non-dimensional frequency parameters $\Omega = \omega R \sqrt{\rho(1 - \mu^2)/E}$ of an isotropic cylindrical shell with one ($a_1/L = 1/2$) ring supports are presented. The lowest four circumference wave numbers (i.e., $n = 1, 2, 3, 4$) are considered in the analysis. The material and geometrical properties of the shell are given as: $L/R = 50$, $\mu = 0.3$. The comparisons are conducted for thickness ratios $h/R = 0.005, 0.05$, respectively. The results reported by Swaddiwudhipong et al. (1995) are considered in the comparison. A good agreement can be seen from the table. The small deviations in the results are caused by different computation methods and shell theories are used in the literature.

Fig. 5.11 Schematic diagram of a laminated cylindrical shell with intermediate ring supports

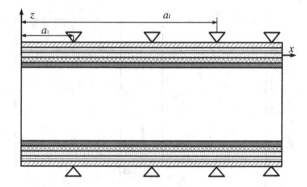

Table 5.12 Comparison of frequency parameters $\Omega = \omega R\sqrt{\rho(1-\mu^2)/E}$ of a C–F cylindrical shell with one ($a_1/L = 1/2$) ring support ($L/R = 50$, $\mu = 0.3$)

h/R	n	Swaddiwudhipong et al. (1995)			Present		
		$m = 1$	$m = 2$	$m = 3$	$m = 1$	$m = 2$	$m = 3$
0.005	1	0.00267	0.01599	0.02268	0.00265	0.01598	0.02243
	2	0.00394	0.00651	0.00847	0.00397	0.00652	0.00845
	3	0.01097	0.01125	0.01155	0.01097	0.01126	0.01156
	4	0.02101	0.02108	0.02114	0.02101	0.02108	0.02115
0.05	1	0.00267	0.01600	0.02269	0.00267	0.01601	0.02245
	2	0.03876	0.03930	0.03975	0.03871	0.03928	0.03973
	3	0.10950	0.10980	0.10990	0.10912	0.10935	0.10957
	4	0.21000	0.21020	0.21040	0.20850	0.20873	0.20894

The influence of ring support location on the frequencies of a four-layered, [45°/ −45°/45°/−45°] cylindrical shell with one intermediate ring support is also investigated. The material and geometric constants of the layers of the shell are: $L/R = 10$, $h/R = 0.1$, $E_1/E_2 = 25$, $\mu_{12} = 0.27$, $G_{12} = 0.5E_2$, $G_{13} = 0.5E_2$, $G_{23} = 0.2E_2$. In Fig. 5.12, variations of the lowest three longitudinal model ($n = 1$, $m = 1, 2, 3$) frequency parameters Ω of the considered cylindrical shell against the ring support location parameter a/L are depicted. Four types of end conditions included in the presentation are: F–C, C–F, S–S, and C–C. It is obvious that the frequency parameters of the shell significantly affected by the location of the ring support, and this effect varies with the end conditions. For a cylindrical shell with symmetrical boundary conditions imposed on both ends, such as S–S and C–C boundary conditions, the frequency parameter curves are symmetrical about the center line of the figure (i.e., $a/L = 0.5$). For a cylindrical shell with unsymmetrical end conditions, such as F–C and C–F boundary conditions, the frequency parameter curves are not symmetrical about the center line of the figure. Figures 5.13 and 5.14 show the similar studies for the second and third circumference numbers ($n = 2, 3$), respectively. Comparing with Fig. 5.12, it can be seen that the influence of the ring

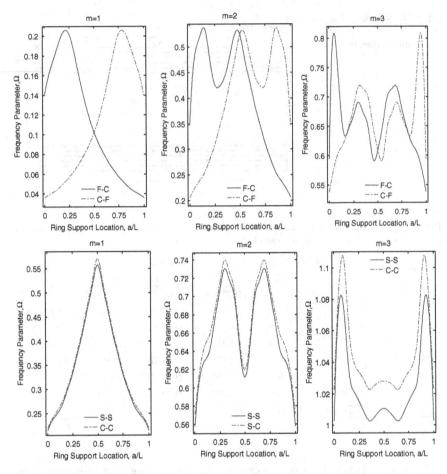

Fig. 5.12 Variations of frequency parameters Ω versus ring location (a/L) for a $[45°/-45°]_2$ laminated cylindrical shell with one intermediate ring support ($n = 1$)

support location varies with circumference numbers as well. In addition, the amount of the peak values of a frequency parameter curve is in direct proportion to the longitudinal mode number.

5.4 Vibration of Laminated Open Cylindrical Shells

Composite laminated open cylindrical shells have a wide range of engineering applications, particularly in aerospace crafts, military hardware and civil constructions. Table 5.2 shows that thick cylindrical shells can have 24 possible classical boundary conditions at each edge. This leads to 576 combinations of

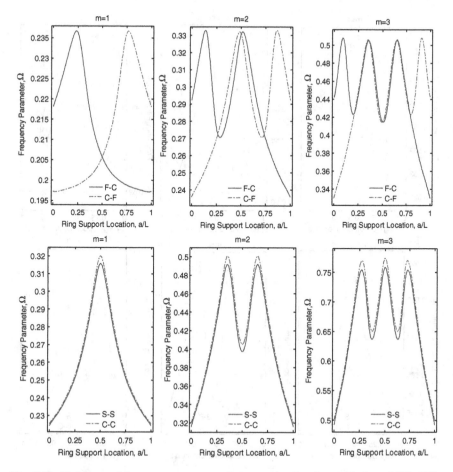

Fig. 5.13 Variations of frequency parameters Ω versus ring location (a/L) for a $[45°/-45°]_2$ laminated cylindrical shell with one intermediate ring support $(n = 2)$

boundary conditions when a cylindrical shell is closed. When the shells are open, this yields a higher number of combinations of boundary conditions for such shells.

In comparing to laminated closed cylindrical shells, information available for the vibrations of the shallow and deep open ones is very limited despite their practical importance. The most likely reason for this lacuna lies in the analytical difficulties involved: for a completely closed shell such as circular cylindrical, circular conical and spherical shells, the assumed 2D displacement field can be reduced to a quasi 1D problem through Fourier decomposition of the circumferential wave motion. However, for an open shell, the assumption of whole periodic wave numbers in the circumferential direction is inappropriate, and thus, a set of complete two-dimensional analysis is required and resort must be made to a full two-dimensional solution scheme. Such a scheme will inevitably be complicated further by the

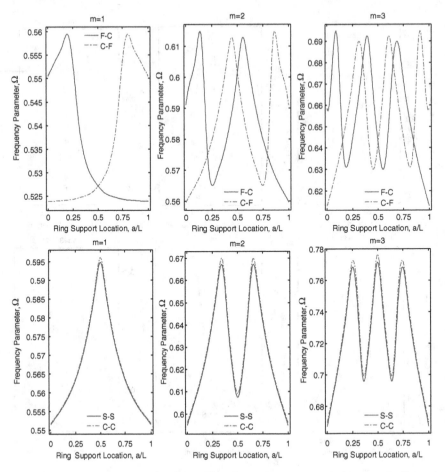

Fig. 5.14 Variations of frequency parameters Ω versus ring location (a/L) for a $[45°/-45°]_2$ laminated cylindrical shell with one intermediate ring support ($n = 3$)

dependence of the circumferential arc length on its meridional location (Bardell et al. 1998). This forms a major deterrent so that the analyses of open shells have not been widely available.

In this section, we consider free vibration of laminated deep open cylindrical shells. As was done previously for the closed cylindrical shells, only solutions in the framework of SDST which neglecting the effects of the deepness term z/R (i.e., Eq. (5.26)) are considered in this section. Under the modified Fourier series framework, regardless of boundary conditions, each displacement and rotation component of a laminated open cylindrical shell is expanded as a two-dimensional modified Fourier series as:

$$u(x, \theta) = \sum_{m=0}^{M} \sum_{n=0}^{N} A_{mn} \cos \lambda_m x \cos \lambda_n \theta + \sum_{l=1}^{2} \sum_{n=0}^{N} a_{ln} P_l(x) \cos \lambda_n \theta$$

$$+ \sum_{l=1}^{2} \sum_{m=0}^{M} b_{lm} P_l(\theta) \cos \lambda_m x$$

$$v(x, \theta) = \sum_{m=0}^{M} \sum_{n=0}^{N} B_{mn} \cos \lambda_m x \cos \lambda_n \theta + \sum_{l=1}^{2} \sum_{n=0}^{N} c_{ln} P_l(x) \cos \lambda_n \theta$$

$$+ \sum_{l=1}^{2} \sum_{m=0}^{M} d_{lm} P_l(\theta) \cos \lambda_m x$$

$$w(x, \theta) = \sum_{m=0}^{M} \sum_{n=0}^{N} C_{mn} \cos \lambda_m x \cos \lambda_n \theta + \sum_{l=1}^{2} \sum_{n=0}^{N} e_{ln} P_l(x) \cos \lambda_n \theta$$

$$+ \sum_{l=1}^{2} \sum_{m=0}^{M} f_{lm} P_l(\theta) \cos \lambda_m x$$

$$\phi_x(x, \theta) = \sum_{m=0}^{M} \sum_{n=0}^{N} D_{mn} \cos \lambda_m x \cos \lambda_n \theta + \sum_{l=1}^{2} \sum_{n=0}^{N} g_{ln} P_l(x) \cos \lambda_n \theta$$

$$+ \sum_{l=1}^{2} \sum_{m=0}^{M} h_{lm} P_l(\theta) \cos \lambda_m x$$

$$\phi_\theta(x, \theta) = \sum_{m=0}^{M} \sum_{n=0}^{N} E_{mn} \cos \lambda_m x \cos \lambda_n \theta + \sum_{l=1}^{2} \sum_{n=0}^{N} i_{ln} P_l(x) \cos \lambda_n \theta$$

$$+ \sum_{l=1}^{2} \sum_{m=0}^{M} j_{lm} P_l(\theta) \cos \lambda_m x$$

(5.43)

where $\lambda_m = m\pi/L$ and $\lambda_n = n\pi/\theta_0$. A_{mn}, B_{mn}, C_{mn}, D_{mn} and E_{mn} are expansion coefficients of standard cosine Fourier series. a_{ln}, b_{lm}, c_{ln}, d_{lm}, e_{ln}, f_{lm} g_{ln}, h_{lm}, i_{ln} and j_{lm} are the corresponding supplement coefficients. $P_l(x)$ and $P_l(\theta)$ denote two sets auxiliary polynomial functions introduced to remove all the discontinuities potentially associated with the first-order derivatives at the boundaries of $x = 0$, $x = L$ and $\theta = 0$, $\theta = \theta_0$, respectively. These auxiliary functions are in the similar forms as Eq. (5.41).

For a general open cylindrical shell, there exist four boundaries, i.e., $x = 0$, $x = L$, $\theta = 0$ and $\theta = \theta_0$. For the sake of brevity, a four-letter character is employed to represent the boundary condition of an open cylindrical shell, such as FCSC identifies the shell with F, C, S and C boundary conditions at boundaries $x = 0$, $\theta = 0$, $x = L$ and $\theta = \theta_0$, respectively. Unless otherwise stated, the natural frequencies of the considered shells are expressed as frequency parameter as $\Omega =$

$(\omega L^2/h)\sqrt{\rho/E_2}$ and the material properties of the layers of open cylindrical shells under consideration are taken as: $E_2 = 10$ GPa, $E_1/E_2 = $ open, $\mu_{12} = 0.25$, $G_{12} = G_{13} = 0.5E_2$, $G_{23} = 0.2E_2$, $\rho = 1{,}450$ kg/m^3.

5.4.1 Convergence Studies and Result Verification

Table 5.13 shows the convergence and comparison studies of frequency parameters $\Omega = \omega L^2/h\sqrt{\rho/E_1}$ for a four-layered, $[-60°/60°/60°/-60°]$ open cylindrical shell with FFFF and CCCC boundary conditions, respectively. The material and geometry constants of the layers of the shell are: $R = 2$ m, $L/R = 0.5$, $h/L = 0.01$, $\theta_0 = 2$ arcsin 0.25, $E_1 = 60.7$ GPa, $E_2 = 24.8$ GPa, $\mu_{12} = 0.23$, $G_{12} = G_{13} = G_{23} = 12$ GPa, $\rho = 1{,}700$ kg/m^3. It should be noted that the zero frequencies corresponding to the rigid body modes were omitted from the results. Excellent convergence of frequencies can be observed in the table. The convergence of the FFFF solutions is faster than those of CCCC boundary conditions. Numerical results reported by Zhao et al. (2003) by using mesh-free method and CST, Messina and Soldatos (1999a) based on HSDT as well as Qatu and Leissa (1991) based on shallow shell

Table 5.13 Convergence and comparison of frequency parameters $\Omega = \omega L^2/h\sqrt{\rho/E_1}$ of a $[-60°/60°/60°/-60°]$ e-glass/epoxy shallow open cylindrical shell

B.C.	$M \times N$	Mode number					
		1	2	3	4	5	6
FFFF	14 × 14	3.2450	5.5874	8.3715	11.120	12.505	15.272
	15 × 15	3.2448	5.5873	8.3700	11.118	12.502	15.271
	16 × 16	3.2439	5.5872	8.3694	11.117	12.500	15.271
	17 × 17	3.2438	5.5871	8.3681	11.115	12.497	15.270
	18 × 18	3.2430	5.5870	8.3676	11.114	12.496	15.270
	Zhao et al. (2003)	3.3016	5.7328	8.5087	11.133	12.626	15.724
	Messina and Soldatos (1999a)	3.2498	5.5910	8.3873	11.137	12.533	15.328
	Qatu and Leissa (1991)	3.2920	5.7416	8.5412	11.114	12.591	15.696
CCCC	14 × 14	24.510	29.276	36.634	37.998	43.387	46.288
	15 × 15	24.509	29.272	36.623	37.986	43.385	46.278
	16 × 16	24.505	29.255	36.619	37.975	43.384	46.270
	17 × 17	24.504	29.253	36.613	37.968	43.383	46.265
	18 × 18	24.502	29.243	36.610	37.962	43.382	46.261

Table 5.14 Comparison of frequency parameters $\Omega = \omega L^2 \sqrt{\rho/E_1 Rh}$ of a [90°/0°] laminated open cylindrical shell with SDSDSDSD boundary conditions ($L/R = 5$, $\theta_0 = 60°$)

h/R	Method	Mode number					
		1	2	3	4	5	6
0.10	Present	7.025	7.636	8.850	10.84	13.44	14.05
	Messina and Soldatos (1999b)	7.025	7.702	8.932	10.94	13.57	14.05
	Difference (%)	0.01	0.87	0.93	0.97	0.95	0.01
0.05	Present	5.919	8.140	9.934	11.11	14.43	18.01
	Messina and Soldatos (1999b)	5.932	8.153	9.935	11.12	14.45	18.03
	Difference (%)	0.21	0.15	0.01	0.10	0.13	0.13
0.02	Present	5.050	9.777	14.84	15.38	15.71	15.97
	Messina and Soldatos (1999b)	5.049	9.774	14.84	15.40	15.71	16.01
	Difference (%)	0.02	0.03	0.02	0.14	0.02	0.27
0.01	Present	5.724	11.04	12.29	13.03	14.64	17.63
	Messina and Soldatos (1999b)	5.724	11.03	12.27	13.03	14.62	17.61
	Difference (%)	0.00	0.05	0.13	0.04	0.14	0.13

theory are included in Table 5.13. These comparisons show the present solutions are in good agreement with the reference results, although different theories and methods were employed in the literature.

To further validate the accuracy and reliability of current solution, Table 5.14 shows the comparison of frequency parameters $\Omega = \omega L^2 \sqrt{\rho/E_1 Rh}$ of a two-layered, cross-ply [90°/0°] open cylindrical shell subjected to SDSDSDSD boundary conditions, with results provided by Messina and Soldatos (1999b) based on the conjunction of Ritz method and the Love-type version of a unified shear-deformable shell theory. The shell parameters used in the comparison are: $E_2 = 25E_2$, $R = 1$ m, $L/R = 5$, $\theta_0 = 60°$. The first six frequencies and four sets of thickness-radius ratios, i.e., $h/R = 0.1, 0.05, 0.02$ and 0.01, corresponding to thick to thin open cylindrical shells are performed in the comparison. It is clearly evident that the present solutions are generally in good agreement with the reference results, although a different shell theory is employed by Messina and Soldatos (1999b). The differences between these two results are very small, and do not exceed 0.97 % for the worst case.

5.4.2 Laminated Open Cylindrical Shells with General End Conditions

Some further numerical results for laminated open cylindrical shells with different boundary conditions and shell parameters, such as geometric properties, lamination schemes are given in the subsequent discussions.

Table 5.15 Frequency parameters Ω of a three-layered, cross-ply [0°/90°/0°] open cylindrical shell with various boundary conditions and circumferential included angles ($R = 1$ m, $L/R = 2$, $h/R = 0.1$)

θ_0	Mode	Boundary conditions						
		FFFF	FSFS	FCFC	SFSF	CFCF	SSSS	CCCC
$\pi/4$	1	11.203	75.065	88.901	12.815	21.399	78.293	91.804
	2	25.499	77.465	89.414	14.769	22.509	86.921	102.555
	3	31.422	85.618	94.223	39.514	41.030	90.751	102.682
	4	46.947	86.141	99.258	41.021	49.378	95.231	114.025
$\pi/2$	1	5.6344	18.757	28.875	13.570	21.848	28.754	39.274
	2	11.724	25.491	32.998	16.635	23.679	45.505	54.963
	3	15.825	41.360	48.575	22.320	27.634	53.806	63.642
	4	27.761	42.306	49.078	37.420	40.797	62.373	72.898
$3\pi/4$	1	3.5787	7.1550	12.138	14.598	22.466	26.289	31.315
	2	4.9181	18.335	21.376	15.008	22.664	26.570	34.544
	3	8.3994	19.207	23.673	22.782	28.235	40.285	49.556
	4	13.994	20.658	23.820	26.321	30.282	50.600	59.527
π	1	2.4902	3.1939	6.0606	14.731	22.517	23.135	29.297
	2	2.6064	9.6122	12.938	14.960	22.668	25.209	30.625
	3	5.4335	10.087	13.014	21.990	27.370	31.561	36.692
	4	7.4475	17.262	17.967	23.496	28.851	35.752	43.198
$5\pi4$	1	1.5952	1.4971	3.3390	14.805	22.570	22.463	28.530
	2	1.8119	5.5228	7.4940	14.959	22.659	23.343	28.671
	3	3.8722	5.6869	7.8281	21.241	26.976	27.562	34.163
	4	4.4292	11.566	13.214	23.912	28.954	31.578	35.014

Table 5.15 gives the first four frequency parameters Ω of a three-layered, cross-ply [0°/90°/0°] open cylindrical shell with various boundary conditions and circumferential included angles (θ_0). The material properties and geometric dimensions of the shell are: $E_2 = 15E_2$, $R = 1$ m, $L/R = 2$, $h/R = 0.1$. Five different circumferential included angles, i.e., $\theta_0 = \pi/4$, $\pi/2$, $3\pi/4$, π and $5\pi/4$, are considered in the calculation. These results may serve as benchmark values for future researches. It is obvious from the tables that the increase of the circumferential included angle will results in decreases in the frequency parameters. Meanwhile, we can see that an open shell with higher constraining rigidity will have higher vibration frequency parameters. For the sake of enhancing our understanding of vibration behaviors of the open cylindrical and conical shells, the first three mode shapes for the shell with CCCC boundary conditions are present in Fig. 5.15, which is constructed in three-dimension views.

Table 5.16 lists the lowest four frequency parameters Ω for a two-layered, cross-ply [0°/90°] open cylindrical shell with various boundary conditions and thickness-to-radius ratios. The open cylindrical shell under consideration having radius

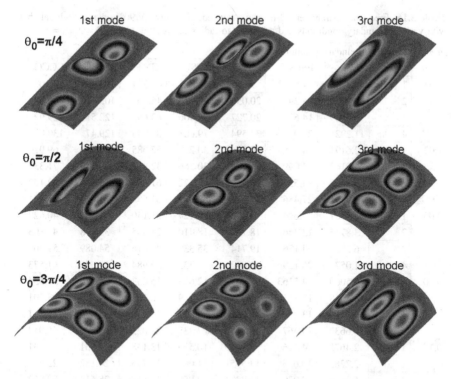

Fig. 5.15 Mode shapes for a CCCC restrained [0°/90°/0°] laminated open cylindrical shell

$R = 1$ m, length-to-radius ratio of 2 and circumferential included angel of π. Five thickness-to-radius ratios, i.e., $h/R = 0.01, 0.02, 0.05, 0.1$ and 0.15 are used in the analysis. It is interesting to see that the frequency parameters of the shell decrease with thickness-to-radius ratio increases. Furthermore, by comparing results of the shell with FSFS (or FCFC) boundary conditions with those of SFSF (or CFCF) case, it is obvious that fixation on a curved boundary will results in higher frequency parameters than on a straight boundary.

As the final numerical example, Table 5.17 shows the lowest five frequency parameters Ω of a three-layered [0°/ϑ/0°] open cylindrical shell with different boundary conditions and fiber orientations. Four different lamination schemes, i.e., $\vartheta = 0°, 30°, 60°$ and $90°$ are performed in the calculation. The shell parameters with following material and geometry properties are used in the investigation: $R = 1$ m, $L/R = 2$, $h/R = 0.05$, $E_1/E_2 = 15$. It is interesting to see that the maximum fundamental frequency parameters for different boundary conditions are obtained at different fiber orientations. For FFFF, FSFS and FCFC boundary conditions, the corresponding maximum fundamental frequency parameters occur at $\vartheta = 90°$. These

Table 5.16 Frequency parameters Ω of a two-layered, cross-ply $[0°/90°]$ open cylindrical shell with various boundary conditions and thickness-to-radius ratios ($R = 1$ m, $L/R = 2$, $\theta_0 = \pi$)

h/R	Mode	Boundary conditions						
		FFFF	FSFS	FCFC	SFSF	CFCF	SSSS	CCCC
0.01	1	2.7758	4.8157	9.4618	45.087	46.724	104.38	106.04
	2	3.9626	14.292	20.056	45.222	46.904	105.72	106.88
	3	6.5409	14.680	20.727	79.406	83.653	122.51	129.92
	4	11.292	28.036	34.594	79.608	83.945	129.47	130.07
0.02	1	2.6191	4.7869	9.3262	30.237	32.685	68.450	69.976
	2	3.9162	14.099	19.714	30.366	32.818	69.339	71.337
	3	6.3503	14.556	20.430	55.422	61.528	85.308	92.157
	4	11.239	27.486	33.621	55.517	61.688	91.626	93.121
0.05	1	2.5501	4.6945	9.1067	18.124	21.140	39.402	40.629
	2	3.8715	13.446	18.767	18.410	21.505	39.478	42.805
	3	6.2003	14.126	19.714	35.339	36.116	54.089	59.261
	4	11.057	25.129	29.092	36.435	39.984	58.676	61.573
0.10	1	2.5060	4.5163	8.6298	12.651	15.653	24.724	27.534
	2	3.8048	12.113	16.377	13.299	16.324	27.269	28.391
	3	6.0116	13.250	18.066	21.833	22.930	35.093	39.281
	4	10.663	19.467	19.878	27.237	28.023	42.300	46.914
0.15	1	2.4675	4.3166	8.0369	10.539	13.198	19.651	21.131
	2	3.7283	10.655	13.535	11.002	13.716	20.182	22.567
	3	5.8170	12.279	14.937	17.058	18.066	28.612	32.653
	4	10.188	14.864	16.173	19.937	20.858	32.926	37.089

Table 5.17 Frequency parameters Ω of a three-layered, $[0°/\vartheta/0°]$ open cylindrical shell with various boundary conditions and fiber orientations ($R = 1$ m, $L/R = 2$, $h/R = 0.05$)

ϑ	Mode	Boundary conditions						
		FFFF	FSFS	FCFC	SFSF	CFCF	SSSS	CCCC
0°	1	2.1221	2.6128	5.0303	19.047	28.564	32.137	39.663
	2	2.5086	7.9443	10.978	19.186	28.650	33.589	40.141
	3	5.2843	9.1689	11.820	30.775	37.686	39.470	47.388
	4	6.1129	15.962	18.954	32.796	39.484	44.132	50.542
30°	1	2.1451	2.6571	5.1170	21.693	30.077	41.030	47.136
	2	2.7402	8.0809	11.174	21.710	30.089	41.159	47.779
	3	5.6245	9.6128	12.314	38.942	44.967	49.744	55.636
	4	6.2681	16.249	19.961	40.521	46.410	50.257	57.472
60°	1	2.2825	2.9658	5.7048	20.985	29.273	39.105	45.667
	2	2.8505	9.0178	12.470	21.060	29.317	40.405	46.919
	3	5.6029	10.197	13.296	37.684	43.651	48.884	55.513
	4	7.1056	18.165	21.597	38.109	43.870	51.548	59.705
90°	1	2.5305	3.2222	6.2084	19.347	28.270	34.420	41.995
	2	2.6181	9.8038	13.573	19.611	28.430	36.452	42.414
	3	5.5442	10.496	13.889	32.218	38.778	43.661	52.435
	4	7.5371	19.500	22.541	36.199	42.098	51.308	57.262

are obtained at $\vartheta = 30°$ for SFSF, CFCF, SSSS and CCCC boundary conditions. In addition, CFCF yields the minimum fundamental frequency parameter at $\vartheta = 90°$ while it is obtained at $\vartheta = 0°$ for other boundary conditions.

Chapter 6
Conical Shells

Conical shells are another special type of shells of revolution. The middle surface of a conical shell is generated by revolving a straight line (generator line) around an axis that is not paralleled to the line itself. Conical shells can have different geometrical shapes. This chapter is organizationally limited to conical shells (both the closed shells and the open ones) having circular cross-sections. In this type of conical shells, the generator line rotates about a fixed axis and results in a constant vertex half-angle angle (φ) with respect to the axis. Specially, the vertex half-angle angle may be equal to zero or 90° ($\pi/2$). In the first case, the cylindrical shells discussed in Chap. 5 will be obtained. Thus, cylindrical shells can be viewed as a special type of the conical shells, and conical shells have all the classifying parameters of the cylindrical shells (Leissa 1973). For the second case, the generator line is vertical to the axis, and the special case of circular plates is obtained.

Laminated conical shells are also one of the important structural components which are widely used in naval vessels, missiles, spacecrafts and other cutting-edge engineering fields. The vibration analysis of them is often required and has always been one important research subject in these fields (Civalek 2006, 2007, 2013; Lam et al. 2002; Ng et al. 2003; Shu 1996; Tong 1993, 1994a, b; Tornabene 2011; Viswanathan et al. 2012; Wu and Lee 2011; Wu and Wu 2000). However, comparing with the cylindrical shells and the circular plates, relatively little literature is available regarding the conical shells due to the fact that the conical coordinate system is function of the meridional direction and the equations of motion for conical shells consist of a set of partial differential equations with variable coefficients.

This chapter is focused on vibration analysis of laminated conical shells with general boundary conditions. Equations of conical shells on the basis of the classical shell theory (CST) and the shear deformation shell theory (SDST) are presented in the first and second sections, respectively, by substituting the proper Lamé parameters of conical shells in the general shell equations (see Chap. 1). Then, in the framework of SDST, numerous vibration results of laminated closed and open conical shells with different boundary conditions, lamination schemes and geometry parameters are given in the third and fourth sections by using the modified Fourier series and the weak form solution procedure.

© Science Press, Beijing and Springer-Verlag Berlin Heidelberg 2015
G. Jin et al., *Structural Vibration*, DOI 10.1007/978-3-662-46364-2_6

Fig. 6.1 Geometry notations
and coordinate system of
conical shells

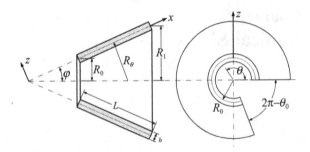

As shown in Fig. 6.1, a general laminated open conical shell with length L, vertex half-angle angle φ, total thickness h, circumferential included angle θ_0, small edge radius R_0 and large edge radius R_1 is selected as the analysis model. The middle surface of the conical shell where an orthogonal coordinate system (x, θ and z) is fixed is taken as the reference surface, in which the x co-ordinate is measured along the generator of the cone starting at the vertex and the θ and z co-ordinates are taken in the circumferential and radial directions, respectively. The middle surface displacements of the conical shell in the x, θ and z directions are denoted by u, v and w, respectively. The conical shell is assumed to be composed of arbitrary number of liner orthotropic laminas which are bonded together rigidly. The mean radius of the conical shell at any point x along its length can be written as:

$$R = x \sin \varphi \tag{6.1}$$

Considering the conical shell in Fig. 6.1 and its conical coordinate system, the coordinates, characteristics of the Lamé parameters and radii of curvatures are:

$$\alpha = x, \quad \beta = \theta, \quad A = 1, \quad B = x \sin \varphi, \quad R_\alpha = \infty, \quad R_\beta = x \tan \varphi \tag{6.2}$$

6.1 Fundamental Equations of Thin Laminated Conical Shells

Closed conical shells can be defined as a special case of open conical shells having circumferential included angle of 2π (360°). We will first derive the fundamental equations for thin open conical shells. The equations are formulated for the general dynamic analysis by substituting Eq. (6.2) into the general classical shell equations developed in Sect. 1.2. It can be readily specialized to the static and free vibration analysis.

6.1.1 *Kinematic Relations*

Substituting Eq. (6.2) into Eq. (1.7), the middle surface strains and curvature changes of conical shells can be specialized from those of general thin shells. They are formed in terms of the middle surface displacements as:

$$
\begin{aligned}
\varepsilon_x^0 &= \frac{\partial u}{\partial x} & \chi_x &= -\frac{\partial^2 w}{\partial x^2} \\
\varepsilon_\theta^0 &= \frac{\partial v}{xs\partial\theta} + \frac{u}{x} + \frac{w}{xt} & \chi_\theta &= \frac{c\partial v}{x^2 s^2 \partial\theta} - \frac{\partial^2 w}{x^2 s^2 \partial\theta^2} - \frac{\partial w}{x\partial x} \\
\gamma_{x\theta}^0 &= \frac{\partial v}{\partial x} + \frac{\partial u}{xs\partial\theta} - \frac{v}{x} & \chi_{x\theta} &= \frac{\partial v}{xt\partial x} - \frac{2v}{x^2 t} - \frac{2\partial^2 w}{xs\partial x\partial\theta} + \frac{2\partial w}{x^2 s\partial\theta}
\end{aligned}
\tag{6.3}
$$

$$
\text{and}\quad s = \sin\varphi \qquad c = \cos\varphi \qquad t = \tan\varphi
$$

where ε_x^0, ε_θ^0 and $\gamma_{x\theta}^0$ denote the normal and shear middle surface strains. χ_x, χ_θ and $\chi_{x\theta}$ are the corresponding curvature and twist changes. Then, the state of strain at an arbitrary point in the kth layer of a laminated conical shell can be written as:

$$
\begin{aligned}
\varepsilon_x &= \varepsilon_x^0 + z\chi_x \\
\varepsilon_\theta &= \varepsilon_\theta^0 + z\chi_\theta \\
\gamma_{x\theta} &= \gamma_{x\theta}^0 + z\chi_{x\theta}
\end{aligned}
\tag{6.4}
$$

where $Z_{k+1} < z < Z_k$. And Z_{k+1} and Z_k denote the distances from the top surface and bottom surface of the layer to the referenced middle surface, respectively.

6.1.2 *Stress-Strain Relations and Stress Resultants*

For laminated conical shells made of composite layers, the well-known stress-strain relations are given as in Eq. (5.4). It should be noted that the materials considered in this chapter are restricted to conical orthotropy. Substituting Eq. (6.2) into Eq. (1.14), the force and moment resultants of the conical shell can be obtained in terms of the middle surface strains and curvature changes as

$$
\begin{bmatrix} N_x \\ N_\theta \\ N_{x\theta} \\ M_x \\ M_\theta \\ M_{x\theta} \end{bmatrix} =
\begin{bmatrix}
A_{11} & A_{12} & A_{16} & B_{11} & B_{12} & B_{16} \\
A_{12} & A_{22} & A_{26} & B_{12} & B_{22} & B_{26} \\
A_{16} & A_{26} & A_{66} & B_{16} & B_{26} & B_{66} \\
B_{11} & B_{12} & B_{16} & D_{11} & D_{12} & D_{16} \\
B_{12} & B_{22} & B_{26} & D_{12} & D_{22} & D_{26} \\
B_{16} & B_{26} & B_{66} & D_{16} & D_{26} & D_{66}
\end{bmatrix}
\begin{bmatrix} \varepsilon_x^0 \\ \varepsilon_\theta^0 \\ \gamma_{x\theta}^0 \\ \chi_x \\ \chi_\theta \\ \chi_{x\theta} \end{bmatrix}
\tag{6.5}
$$

where N_x, N_θ and $N_{x\theta}$ are the normal and shear force resultants and M_x, M_θ and $M_{x\theta}$ denote the bending and twisting moment resultants. The stiffness coefficients A_{ij}, B_{ij}, and D_{ij} are written as in Eq. (1.15).

6.1.3 Energy Functions

The strain energy function of thin laminated conical shells during vibration can be written in terms of the middle surface strains, curvature changes and stress resultants as:

$$U_s = \frac{1}{2} \int_x \int_\theta \left\{ \begin{array}{c} N_x \varepsilon_x^0 + N_\theta \varepsilon_\theta^0 + N_{x\theta} \gamma_{x\theta}^0 \\ + M_x \chi_x + M_\theta \chi_\theta + M_{x\theta} \chi_{x\theta} \end{array} \right\} xs d\theta dx \tag{6.6}$$

Substituting Eqs. (6.3) and (6.5) into Eq. (6.6), the strain energy functions of the shells can be written in terms of middle surface displacements. The corresponding kinetic energy (T) of the conical shells during vibration can be written as:

$$T = \frac{1}{2} \int_x \int_\theta I_0 \left\{ \left(\frac{\partial u}{\partial t}\right)^2 + \left(\frac{\partial v}{\partial t}\right)^2 + \left(\frac{\partial w}{\partial t}\right)^2 \right\} xs d\theta dx \tag{6.7}$$

where the inertia term I_0 is the same as in Eq. (1.19). Suppose q_x, q_θ and q_z are the external loads in the x, θ and z directions, respectively. Thus, the external work can be expressed as:

$$W_e = \int_x \int_\theta \{q_x u + q_\theta v + q_z w\} R dx d\theta \tag{6.8}$$

The same as usual, the general boundary conditions of a conical shell are implemented by using the artificial spring boundary technique. Letting symbols k_ψ^u, k_ψ^v, k_ψ^w and K_ψ^w ($\psi = x_0$, θ_0, x_1 and θ_1) to indicate the stiffness of the boundary springs at the boundaries $R = R_0$, $\theta = 0$, $R = R_1$ and $\theta = \theta_0$, respectively, thus, the deformation strain energy stored in the boundary springs (U_{sp}) during vibration can be defined as:

$$U_{sp} = \frac{1}{2} \int_\theta \left\{ \begin{array}{c} x[k_{x0}^u u^2 + k_{x0}^v v^2 + k_{x0}^w w^2 + K_{x0}^x (\partial w/\partial x)]|_{R=R_0} \\ + x[k_{x1}^u u^2 + k_{x1}^v v^2 + k_{x1}^w w^2 + K_{x1}^x (\partial w/\partial x)]|_{R=R_1} \end{array} \right\} s d\theta$$

$$+ \frac{1}{2} \int_x \left\{ \begin{array}{c} [k_{\theta0}^u u^2 + k_{\theta0}^v v^2 + k_{\theta0}^w w^2 + K_{\theta0}^x (\partial w/xs\partial\theta)]|_{\theta=0} \\ + [k_{\theta1}^u u^2 + k_{\theta1}^v v^2 + k_{\theta1}^w w^2 + K_{\theta1}^x (\partial w/xs\partial\theta)]|_{\theta=\theta_0} \end{array} \right\} dx \tag{6.9}$$

6.1.4 Governing Equations and Boundary Conditions

Substituting Eq. (6.2) into Eq. (1.28) and then simplifying the expressions, the resulting governing equations for conical shells are:

$$\frac{\partial N_x}{\partial x} + \frac{N_x - N_\theta}{x} + \frac{\partial N_{x\theta}}{xs\partial\theta} + q_x = I_0 \frac{\partial^2 u}{\partial t^2}$$

$$\frac{\partial N_{x\theta}}{\partial x} + \frac{\partial N_\theta}{xs\partial\theta} + \frac{2N_{x\theta}}{x} + \frac{Q_\theta}{xt} + q_\theta = I_0 \frac{\partial^2 v}{\partial t^2} \qquad (6.10)$$

$$-\frac{N_\theta}{xt} + \frac{\partial Q_x}{\partial x} + \frac{Q_x}{x} + \frac{\partial Q_\theta}{xs\partial\theta} + q_z = I_0 \frac{\partial^2 w}{\partial t^2}$$

where Q_x and Q_θ are defined as

$$Q_x = \frac{\partial M_x}{\partial x} + \frac{\partial M_{x\theta}}{xs\partial\theta} + \frac{M_x - M_\theta}{x}$$

$$Q_\theta = \frac{\partial M_\theta}{xs\partial\theta} + \frac{\partial M_{x\theta}}{\partial x} + \frac{2M_{x\theta}}{x} \qquad (6.11)$$

Substituting Eqs. (6.3), (6.5) and (6.11) into Eq. (6.10) yields the governing equations in terms of displacements as:

$$\left(\begin{bmatrix} L_{11} & L_{12} & L_{13} \\ L_{21} & L_{22} & L_{23} \\ L_{31} & L_{32} & L_{33} \end{bmatrix} - \omega^2 \begin{bmatrix} -I_0 & 0 & 0 \\ 0 & -I_0 & 0 \\ 0 & 0 & -I_0 \end{bmatrix} \right) \begin{bmatrix} u \\ v \\ w \end{bmatrix} = \begin{bmatrix} -p_x \\ -p_\theta \\ -p_z \end{bmatrix} \qquad (6.12)$$

The coefficients of the linear operator L_{ij} are written as

$$L_{11} = \frac{\partial}{\partial x}\left(A_{11}\frac{\partial}{\partial x} + \frac{A_{12}}{x} + \frac{A_{16}}{xs}\frac{\partial}{\partial\theta} \right) + \frac{1}{xs}\frac{\partial}{\partial\theta}\left(A_{16}\frac{\partial}{\partial x} + \frac{A_{26}}{x} + \frac{A_{66}}{xs}\frac{\partial}{\partial\theta} \right)$$

$$+ \frac{1}{x}\left((A_{11} - A_{12})\frac{\partial}{\partial x} + \frac{A_{12} - A_{22}}{x} + \frac{A_{16} - A_{26}}{xs}\frac{\partial}{\partial\theta} \right)$$

$$L_{12} = \frac{\partial}{\partial x}\left(\frac{A_{12}}{xs}\frac{\partial}{\partial\theta} + A_{16}\left(\frac{\partial}{\partial x} - \frac{1}{x}\right) + \frac{B_{12}c}{x^2 s^2}\frac{\partial}{\partial\theta} + \frac{B_{16}}{xt}\left(\frac{\partial}{\partial x} - \frac{2}{x}\right) \right)$$

$$+ \frac{1}{x}\left(\frac{A_{12}}{xs}\frac{\partial}{\partial\theta} + A_{16}\left(\frac{\partial}{\partial x} - \frac{1}{x}\right) + \frac{B_{12}c}{x^2 s^2}\left(\frac{c\partial v}{x^2 s^2 \partial\theta}\right) + \frac{B_{16}}{xt}\left(\frac{\partial}{\partial x} - \frac{2}{x}\right) \right)$$

$$- \frac{1}{x}\left(\frac{A_{22}}{xs}\frac{\partial}{\partial\theta} + A_{26}\left(\frac{\partial}{\partial x} - \frac{1}{x}\right) + \frac{B_{22}c}{x^2 s^2}\left(\frac{c\partial v}{x^2 s^2 \partial\theta}\right) + \frac{B_{26}}{xt}\left(\frac{\partial}{\partial x} - \frac{2}{x}\right) \right)$$

$$+ \frac{1}{xs}\frac{\partial}{\partial\theta}\left(\frac{A_{26}}{xs}\frac{\partial}{\partial\theta} + A_{66}\left(\frac{\partial}{\partial x} - \frac{1}{x}\right) + \frac{B_{26}c}{x^2 s^2}\frac{\partial}{\partial\theta} + \frac{B_{66}}{xt}\left(\frac{\partial}{\partial x} - \frac{2}{x}\right) \right)$$

$$L_{13} = \frac{\partial}{\partial x}\left(\frac{A_{12}}{xt} - B_{11}\frac{\partial^2}{\partial x^2} - \frac{B_{12}}{x}\left(\frac{\partial^2}{xs^2\partial\theta^2} + \frac{\partial}{\partial x}\right) - \frac{2B_{16}}{xs}\left(\frac{\partial^2}{\partial x\partial\theta} - \frac{\partial}{x\partial\theta}\right) \right)$$

$$+ \frac{1}{x}\left(\frac{A_{12}}{xt} - B_{11}\frac{\partial^2}{\partial x^2} - \frac{B_{12}}{x}\left(\frac{\partial^2}{xs^2\partial\theta^2} + \frac{\partial}{\partial x}\right) - \frac{2B_{16}}{xs}\left(\frac{\partial^2}{\partial x\partial\theta} - \frac{\partial}{x\partial\theta}\right) \right)$$

$$- \frac{1}{x}\left(\frac{A_{22}}{xt} - B_{12}\frac{\partial^2}{\partial x^2} - \frac{B_{22}}{x}\left(\frac{\partial^2}{xs^2\partial\theta^2} + \frac{\partial}{\partial x}\right) - \frac{2B_{26}}{xs}\left(\frac{\partial^2}{\partial x\partial\theta} - \frac{\partial}{x\partial\theta}\right) \right)$$

$$+ \frac{1}{xs}\frac{\partial}{\partial\theta}\left(\frac{A_{16}}{xt} - B_{16}\frac{\partial^2}{\partial x^2} - \frac{B_{26}}{x}\left(\frac{\partial^2}{xs^2\partial\theta^2} + \frac{\partial}{\partial x}\right) - \frac{2B_{66}}{xs}\left(\frac{\partial^2}{\partial x\partial\theta} - \frac{\partial}{x\partial\theta}\right) \right)$$

$$L_{21} = \frac{\partial}{\partial x}\left(A_{16}\frac{\partial}{\partial x} + \frac{A_{26}}{x} + \frac{A_{66}}{xs}\frac{\partial}{\partial\theta}\right) + \frac{2}{x}\left(A_{16}\frac{\partial}{\partial x} + \frac{A_{26}}{x} + \frac{A_{66}}{xs}\frac{\partial}{\partial\theta}\right)$$

$$+ \frac{1}{xs}\frac{\partial}{\partial\theta}\left(A_{12}\frac{\partial}{\partial x} + \frac{A_{22}}{x} + \frac{A_{26}}{xs}\frac{\partial}{\partial\theta}\right) + \frac{c}{x^2s^2}\frac{\partial}{\partial\theta}\left(B_{12}\frac{\partial}{\partial x} + \frac{B_{22}}{x} + \frac{B_{26}}{xs}\frac{\partial}{\partial\theta}\right)$$

$$+ \frac{1}{xt}\frac{\partial}{\partial x}\left(B_{16}\frac{\partial}{\partial x} + \frac{B_{26}}{x} + \frac{B_{66}}{xs}\frac{\partial}{\partial\theta}\right) + \frac{2}{x^2t}\left(B_{16}\frac{\partial}{\partial x} + \frac{B_{26}}{x} + \frac{B_{66}}{xs}\frac{\partial}{\partial\theta}\right)$$

$$L_{22} = \frac{\partial}{\partial x}\left(\frac{A_{26}}{xs}\frac{\partial}{\partial\theta} + A_{66}\left(\frac{\partial}{\partial x} - \frac{1}{x}\right) + \frac{B_{26}c}{x^2s^2}\frac{\partial}{\partial\theta} + \frac{B_{66}}{xt}\left(\frac{\partial}{\partial x} - \frac{2}{x}\right)\right)$$

$$+ \frac{2}{x}\left(\frac{A_{26}}{xs}\frac{\partial}{\partial\theta} + A_{66}\left(\frac{\partial}{\partial x} - \frac{1}{x}\right) + \frac{B_{26}c}{x^2s^2}\frac{\partial}{\partial\theta} + \frac{B_{66}}{xt}\left(\frac{\partial}{\partial x} - \frac{2}{x}\right)\right)$$

$$+ \frac{1}{xs}\frac{\partial}{\partial\theta}\left(\frac{A_{22}}{xs}\frac{\partial}{\partial\theta} + A_{26}\left(\frac{\partial}{\partial x} - \frac{1}{x}\right) + \frac{B_{22}c}{x^2s^2}\frac{\partial}{\partial\theta} + \frac{B_{26}}{xt}\left(\frac{\partial}{\partial x} - \frac{2}{x}\right)\right)$$

$$+ \frac{c}{x^2s^2}\frac{\partial}{\partial\theta}\left(\frac{B_{22}}{xs}\frac{\partial}{\partial\theta} + B_{26}\left(\frac{\partial}{\partial x} - \frac{1}{x}\right) + \frac{D_{22}c}{x^2s^2}\frac{\partial}{\partial\theta} + \frac{D_{26}}{xt}\left(\frac{\partial}{\partial x} - \frac{2}{x}\right)\right)$$

$$+ \frac{1}{xt}\frac{\partial}{\partial x}\left(\frac{B_{26}}{xs}\frac{\partial}{\partial\theta} + B_{66}\left(\frac{\partial}{\partial x} - \frac{1}{x}\right) + \frac{D_{26}c}{x^2s^2}\frac{\partial}{\partial\theta} + \frac{D_{66}}{xt}\left(\frac{\partial}{\partial x} - \frac{2}{x}\right)\right)$$

$$+ \frac{2}{x^2t}\left(\frac{B_{26}}{xs}\frac{\partial}{\partial\theta} + B_{66}\left(\frac{\partial}{\partial x} - \frac{1}{x}\right) + \frac{D_{26}c}{x^2s^2}\frac{\partial}{\partial\theta} + \frac{D_{66}}{xt}\left(\frac{\partial}{\partial x} - \frac{2}{x}\right)\right)$$

$$L_{23} = \frac{\partial}{\partial x}\left(\frac{A_{26}}{xt} - B_{16}\frac{\partial^2}{\partial x^2} - \frac{B_{26}}{x}\left(\frac{\partial^2}{xs^2\partial\theta^2} + \frac{\partial}{\partial x}\right) - \frac{2B_{66}}{xs}\left(\frac{\partial^2}{\partial x\partial\theta} - \frac{\partial w}{x\partial\theta}\right)\right)$$

$$+ \frac{2}{x}\left(\frac{A_{26}}{xt} - B_{16}\frac{\partial^2}{\partial x^2} - \frac{B_{26}}{x}\left(\frac{\partial^2}{xs^2\partial\theta^2} + \frac{\partial}{\partial x}\right) - \frac{2B_{66}}{xs}\left(\frac{\partial^2}{\partial x\partial\theta} - \frac{\partial w}{x\partial\theta}\right)\right)$$

$$+ \frac{1}{xs}\frac{\partial}{\partial\theta}\left(\frac{A_{22}}{xt} - B_{12}\frac{\partial^2}{\partial x^2} - \frac{B_{22}}{x}\left(\frac{\partial^2}{xs^2\partial\theta^2} + \frac{\partial}{\partial x}\right) - \frac{2B_{26}}{xs}\left(\frac{\partial^2}{\partial x\partial\theta} - \frac{\partial w}{x\partial\theta}\right)\right)$$

$$+ \frac{c}{x^2s^2}\frac{\partial}{\partial\theta}\left(\frac{B_{22}}{xt} - D_{12}\frac{\partial^2}{\partial x^2} - \frac{D_{22}}{x}\left(\frac{\partial^2}{xs^2\partial\theta^2} + \frac{\partial}{\partial x}\right) - \frac{2D_{26}}{xs}\left(\frac{\partial^2}{\partial x\partial\theta} - \frac{\partial w}{x\partial\theta}\right)\right)$$

$$+ \frac{1}{xt}\frac{\partial}{\partial x}\left(\frac{B_{26}}{xt} - D_{16}\frac{\partial^2}{\partial x^2} - \frac{D_{26}}{x}\left(\frac{\partial^2}{xs^2\partial\theta^2} + \frac{\partial}{\partial x}\right) - \frac{2D_{66}}{xs}\left(\frac{\partial^2}{\partial x\partial\theta} - \frac{\partial w}{x\partial\theta}\right)\right)$$

$$+ \frac{2}{x^2t}\left(\frac{B_{26}}{xt} - D_{16}\frac{\partial^2}{\partial x^2} - \frac{D_{26}}{x}\left(\frac{\partial^2}{xs^2\partial\theta^2} + \frac{\partial}{\partial x}\right) - \frac{2D_{66}}{xs}\left(\frac{\partial^2}{\partial x\partial\theta} - \frac{\partial w}{x\partial\theta}\right)\right)$$

$$L_{31} = \frac{-1}{xt}\left(A_{12}\frac{\partial}{\partial x} + \frac{A_{22}}{x} + \frac{A_{26}}{xs}\frac{\partial}{\partial\theta}\right)$$

$$+\frac{\partial}{\partial x}\left(\begin{array}{l}\frac{\partial}{\partial x}\left(B_{11}\frac{\partial}{\partial x} + \frac{B_{12}}{x} + \frac{B_{16}}{xs}\frac{\partial}{\partial\theta}\right) + \frac{1}{xs}\frac{\partial}{\partial\theta}\left(B_{16}\frac{\partial}{\partial x} + \frac{B_{26}}{x} + \frac{B_{66}}{xs}\frac{\partial}{\partial\theta}\right)\\[4pt] +\frac{1}{x}\left(B_{11}\frac{\partial}{\partial x} + \frac{B_{12}}{x} + \frac{B_{16}}{xs}\frac{\partial}{\partial\theta}\right) - \frac{1}{x}\left(B_{12}\frac{\partial}{\partial x} + \frac{B_{22}}{x} + \frac{B_{26}}{xs}\frac{\partial}{\partial\theta}\right)\end{array}\right)$$

$$+\frac{1}{x}\left(\begin{array}{l}\frac{\partial}{\partial x}\left(B_{11}\frac{\partial}{\partial x} + \frac{B_{12}}{x} + \frac{B_{16}}{xs}\frac{\partial}{\partial\theta}\right) + \frac{1}{xs}\frac{\partial}{\partial\theta}\left(B_{16}\frac{\partial}{\partial x} + \frac{B_{26}}{x} + \frac{B_{66}}{xs}\frac{\partial}{\partial\theta}\right)\\[4pt] +\frac{1}{x}\left(B_{11}\frac{\partial}{\partial x} + \frac{B_{12}}{x} + \frac{B_{16}}{xs}\frac{\partial}{\partial\theta}\right) - \frac{1}{x}\left(B_{12}\frac{\partial}{\partial x} + \frac{B_{22}}{x} + \frac{B_{26}}{xs}\frac{\partial}{\partial\theta}\right)\end{array}\right)$$

$$+\frac{1}{xs}\frac{\partial}{\partial\theta}\left(\begin{array}{l}\frac{1}{xs}\frac{\partial}{\partial\theta}\left(B_{12}\frac{\partial}{\partial x} + \frac{B_{22}}{x} + \frac{B_{26}}{xs}\right) + \frac{\partial}{\partial x}\left(B_{16}\frac{\partial}{\partial x} + \frac{B_{26}}{x} + \frac{B_{66}}{xs}\frac{\partial}{\partial\theta}\right)\\[4pt] +\frac{2}{x}\left(B_{16}\frac{\partial}{\partial x} + \frac{B_{26}}{x} + \frac{B_{66}}{xs}\frac{\partial}{\partial\theta}\right)\end{array}\right)$$

$$L_{32} = \frac{-1}{xt}\left(\frac{A_{22}}{xs}\frac{\partial}{\partial\theta} + A_{26}\left(\frac{\partial}{\partial x} - \frac{1}{x}\right) + \frac{B_{22}c}{x^2s^2}\frac{\partial}{\partial\theta} + \frac{B_{26}}{xt}\left(\frac{\partial}{\partial x} - \frac{2}{x}\right)\right)$$

$$+\frac{\partial}{\partial x}\left(\begin{array}{l}\frac{\partial}{\partial x}\left(\frac{B_{12}}{xs}\frac{\partial}{\partial\theta} + B_{16}\left(\frac{\partial}{\partial x} - \frac{1}{x}\right) + \frac{D_{12}c}{x^2s^2}\frac{\partial}{\partial\theta} + \frac{D_{16}}{xt}\left(\frac{\partial}{\partial x} - \frac{2}{x}\right)\right)\\[4pt] +\frac{1}{xs}\frac{\partial}{\partial\theta}\left(\frac{B_{26}}{xs}\frac{\partial}{\partial\theta} + B_{66}\left(\frac{\partial}{\partial x} - \frac{1}{x}\right) + \frac{D_{26}c}{x^2s^2}\frac{\partial}{\partial\theta} + \frac{D_{66}}{xt}\left(\frac{\partial}{\partial x} - \frac{2}{x}\right)\right)\\[4pt] +\frac{1}{x}\left(\frac{B_{12}}{xs}\frac{\partial}{\partial\theta} + B_{16}\left(\frac{\partial}{\partial x} - \frac{1}{x}\right) + \frac{D_{12}c}{x^2s^2}\frac{\partial}{\partial\theta} + \frac{D_{16}}{xt}\left(\frac{\partial}{\partial x} - \frac{2}{x}\right)\right)\\[4pt] -\frac{1}{x}\left(\frac{B_{22}}{xs}\frac{\partial}{\partial\theta} + B_{26}\left(\frac{\partial}{\partial x} - \frac{1}{x}\right) + \frac{D_{22}c}{x^2s^2}\frac{\partial}{\partial\theta} + \frac{D_{26}}{xt}\left(\frac{\partial}{\partial x} - \frac{2}{x}\right)\right)\end{array}\right)$$

$$+\frac{1}{x}\left(\begin{array}{l}\frac{\partial}{\partial x}\left(\frac{B_{12}}{xs}\frac{\partial}{\partial\theta} + B_{16}\left(\frac{\partial}{\partial x} - \frac{1}{x}\right) + \frac{D_{12}c}{x^2s^2}\frac{\partial}{\partial\theta} + \frac{D_{16}}{xt}\left(\frac{\partial}{\partial x} - \frac{2}{x}\right)\right)\\[4pt] +\frac{1}{xs}\frac{\partial}{\partial\theta}\left(\frac{B_{26}}{xs}\frac{\partial}{\partial\theta} + B_{66}\left(\frac{\partial}{\partial x} - \frac{1}{x}\right) + \frac{D_{26}c}{x^2s^2}\frac{\partial}{\partial\theta} + \frac{D_{66}}{xt}\left(\frac{\partial}{\partial x} - \frac{2}{x}\right)\right)\\[4pt] +\frac{1}{x}\left(\frac{B_{12}}{xs}\frac{\partial}{\partial\theta} + B_{16}\left(\frac{\partial}{\partial x} - \frac{1}{x}\right) + \frac{D_{12}c}{x^2s^2}\frac{\partial}{\partial\theta} + \frac{D_{16}}{xt}\left(\frac{\partial}{\partial x} - \frac{2}{x}\right)\right)\\[4pt] -\frac{1}{x}\left(\frac{B_{22}}{xs}\frac{\partial}{\partial\theta} + B_{26}\left(\frac{\partial}{\partial x} - \frac{1}{x}\right) + \frac{D_{22}c}{x^2s^2}\frac{\partial}{\partial\theta} + \frac{D_{26}}{xt}\left(\frac{\partial}{\partial x} - \frac{2}{x}\right)\right)\end{array}\right)$$

$$+\frac{1}{xs}\frac{\partial}{\partial\theta}\left(\begin{array}{l}\frac{1}{xs}\frac{\partial}{\partial\theta}\left(\frac{B_{22}}{xs}\frac{\partial}{\partial\theta} + B_{26}\left(\frac{\partial}{\partial x} - \frac{1}{x}\right) + \frac{D_{22}c}{x^2s^2}\frac{\partial}{\partial\theta} + \frac{D_{26}}{xt}\left(\frac{\partial}{\partial x} - \frac{2}{x}\right)\right)\\[4pt] +\frac{\partial}{\partial x}\left(\frac{B_{26}}{xs}\frac{\partial}{\partial\theta} + B_{66}\left(\frac{\partial}{\partial x} - \frac{1}{x}\right) + \frac{D_{26}c}{x^2s^2}\frac{\partial}{\partial\theta} + \frac{D_{66}}{xt}\left(\frac{\partial}{\partial x} - \frac{2}{x}\right)\right)\\[4pt] +\frac{2}{x}\left(\frac{B_{26}}{xs}\frac{\partial}{\partial\theta} + B_{66}\left(\frac{\partial}{\partial x} - \frac{1}{x}\right) + \frac{D_{26}c}{x^2s^2}\frac{\partial}{\partial\theta} + \frac{D_{66}}{xt}\left(\frac{\partial}{\partial x} - \frac{2}{x}\right)\right)\end{array}\right)$$

$$L_{33} = \frac{-1}{xt}\left(\frac{A_{22}}{xt} - B_{12}\frac{\partial^2}{\partial x^2} + \frac{B_{22}}{x}\left(\frac{\partial^2}{xs^2\partial\theta^2} + \frac{\partial}{\partial x}\right) - \frac{B_{26}}{xs}\left(\frac{2\partial^2}{\partial x\partial\theta} - \frac{2\partial}{x\partial\theta}\right)\right)$$

$$+\frac{\partial}{\partial x}\left(\begin{array}{l}\frac{\partial}{\partial x}\left(\frac{B_{12}}{xt} - D_{11}\frac{\partial^2}{\partial x^2} + \frac{D_{12}}{x}\left(\frac{\partial^2}{xs^2\partial\theta^2} + \frac{\partial}{\partial x}\right) - \frac{D_{16}}{xs}\left(\frac{2\partial^2}{\partial x\partial\theta} - \frac{2\partial}{x\partial\theta}\right)\right)\\[4pt] +\frac{1}{xs}\frac{\partial}{\partial\theta}\left(\frac{B_{26}}{xt} - D_{16}\frac{\partial^2}{\partial x^2} + \frac{D_{26}}{x}\left(\frac{\partial^2}{xs^2\partial\theta^2} + \frac{\partial}{\partial x}\right) - \frac{D_{66}}{xs}\left(\frac{2\partial^2}{\partial x\partial\theta} - \frac{2\partial}{x\partial\theta}\right)\right)\\[4pt] +\frac{1}{x}\left(\frac{B_{12}}{xt} - D_{11}\frac{\partial^2}{\partial x^2} + \frac{D_{12}}{x}\left(\frac{\partial^2}{xs^2\partial\theta^2} + \frac{\partial}{\partial x}\right) - \frac{D_{16}}{xs}\left(\frac{2\partial^2}{\partial x\partial\theta} - \frac{2\partial}{x\partial\theta}\right)\right)\\[4pt] -\frac{1}{x}\left(\frac{B_{22}}{xt} - D_{12}\frac{\partial^2}{\partial x^2} + \frac{D_{22}}{x}\left(\frac{\partial^2}{xs^2\partial\theta^2} + \frac{\partial}{\partial x}\right) - \frac{D_{26}}{xs}\left(\frac{2\partial^2}{\partial x\partial\theta} - \frac{2\partial}{x\partial\theta}\right)\right)\end{array}\right)$$

$$L_{33} = \frac{-1}{xt}\left(\frac{A_{22}}{xt} - B_{12}\frac{\partial^2}{\partial x^2} + \frac{B_{22}}{x}\left(\frac{\partial^2}{xs^2\partial\theta^2} + \frac{\partial}{\partial x}\right) - \frac{B_{26}}{xs}\left(\frac{2\partial^2}{\partial x\partial\theta} - \frac{2\partial}{x\partial\theta}\right)\right)$$

$$+ \frac{\partial}{\partial x}\left(\begin{array}{l} \frac{\partial}{\partial x}\left(\frac{B_{12}}{xt} - D_{11}\frac{\partial^2}{\partial x^2} + \frac{D_{12}}{x}\left(\frac{\partial^2}{xs^2\partial\theta^2} + \frac{\partial}{\partial x}\right) - \frac{D_{16}}{xs}\left(\frac{2\partial^2}{\partial x\partial\theta} - \frac{2\partial}{x\partial\theta}\right)\right) \\[4pt] + \frac{1}{xs}\frac{\partial}{\partial\theta}\left(\frac{B_{26}}{xt} - D_{16}\frac{\partial^2}{\partial x^2} + \frac{D_{26}}{x}\left(\frac{\partial^2}{xs^2\partial\theta^2} + \frac{\partial}{\partial x}\right) - \frac{D_{66}}{xs}\left(\frac{2\partial^2}{\partial x\partial\theta} - \frac{2\partial}{x\partial\theta}\right)\right) \\[4pt] + \frac{1}{x}\left(\frac{B_{12}}{xt} - D_{11}\frac{\partial^2}{\partial x^2} + \frac{D_{12}}{x}\left(\frac{\partial^2}{xs^2\partial\theta^2} + \frac{\partial}{\partial x}\right) - \frac{D_{16}}{xs}\left(\frac{2\partial^2}{\partial x\partial\theta} - \frac{2\partial}{x\partial\theta}\right)\right) \\[4pt] - \frac{1}{x}\left(\frac{B_{22}}{xt} - D_{12}\frac{\partial^2}{\partial x^2} + \frac{D_{22}}{x}\left(\frac{\partial^2}{xs^2\partial\theta^2} + \frac{\partial}{\partial x}\right) - \frac{D_{26}}{xs}\left(\frac{2\partial^2}{\partial x\partial\theta} - \frac{2\partial}{x\partial\theta}\right)\right) \end{array}\right)$$

$$+ \frac{1}{x}\left(\begin{array}{l} \frac{\partial}{\partial x}\left(\frac{B_{12}}{xt} - D_{11}\frac{\partial^2}{\partial x^2} + \frac{D_{12}}{x}\left(\frac{\partial^2}{xs^2\partial\theta^2} + \frac{\partial}{\partial x}\right) - \frac{D_{16}}{xs}\left(\frac{2\partial^2}{\partial x\partial\theta} - \frac{2\partial}{x\partial\theta}\right)\right) \\[4pt] + \frac{1}{xs}\frac{\partial}{\partial\theta}\left(\frac{B_{26}}{xt} - D_{16}\frac{\partial^2}{\partial x^2} + \frac{D_{26}}{x}\left(\frac{\partial^2}{xs^2\partial\theta^2} + \frac{\partial}{\partial x}\right) - \frac{D_{66}}{xs}\left(\frac{2\partial^2}{\partial x\partial\theta} - \frac{2\partial}{x\partial\theta}\right)\right) \\[4pt] + \frac{1}{x}\left(\frac{B_{12}}{xt} - D_{11}\frac{\partial^2}{\partial x^2} + \frac{D_{12}}{x}\left(\frac{\partial^2}{xs^2\partial\theta^2} + \frac{\partial}{\partial x}\right) - \frac{D_{16}}{xs}\left(\frac{2\partial^2}{\partial x\partial\theta} - \frac{2\partial}{x\partial\theta}\right)\right) \\[4pt] - \frac{1}{x}\left(\frac{B_{22}}{xt} - D_{12}\frac{\partial^2}{\partial x^2} + \frac{D_{22}}{x}\left(\frac{\partial^2}{xs^2\partial\theta^2} + \frac{\partial}{\partial x}\right) - \frac{D_{26}}{xs}\left(\frac{2\partial^2}{\partial x\partial\theta} - \frac{2\partial}{x\partial\theta}\right)\right) \end{array}\right)$$

$$+ \frac{1}{xs}\frac{\partial}{\partial\theta}\left(\begin{array}{l} \frac{1}{xs}\frac{\partial}{\partial\theta}\left(\frac{B_{22}}{xt} - D_{12}\frac{\partial^2}{\partial x^2} + \frac{D_{22}}{x}\left(\frac{\partial^2}{xs^2\partial\theta^2} + \frac{\partial}{\partial x}\right) - \frac{D_{26}}{xs}\left(\frac{2\partial^2}{\partial x\partial\theta} - \frac{2\partial}{x\partial\theta}\right)\right) \\[4pt] + \frac{\partial}{\partial x}\left(\frac{B_{26}}{xt} - D_{16}\frac{\partial^2}{\partial x^2} + \frac{D_{26}}{x}\left(\frac{\partial^2}{xs^2\partial\theta^2} + \frac{\partial}{\partial x}\right) - \frac{D_{66}}{xs}\left(\frac{2\partial^2}{\partial x\partial\theta} - \frac{2\partial}{x\partial\theta}\right)\right) \\[4pt] + \frac{2}{x}\left(\frac{B_{26}}{xt} - D_{16}\frac{\partial^2}{\partial x^2} + \frac{D_{26}}{x}\left(\frac{\partial^2}{xs^2\partial\theta^2} + \frac{\partial}{\partial x}\right) - \frac{D_{66}}{xs}\left(\frac{2\partial^2}{\partial x\partial\theta} - \frac{2\partial}{x\partial\theta}\right)\right) \end{array}\right) \tag{6.13}$$

The above equations show the level of complexity in solving the governing equations of a general laminated conical shell. When a conical shell is laminated symmetrically with respect to its middle surface, the constants B_{ij} equal to zero, and, hence the equations are much simplified. Substituting Eq. (6.2) into Eqs. (1.29) and (1.30), the general boundary conditions of thin conical shells are:

$$R = R_0: \begin{cases} N_x - k_{x0}^u u = 0 \\ N_{x\theta} + \frac{cM_{x\theta}}{R_0} - k_{x0}^v v = 0 \\ Q_x + \frac{\partial M_{x\theta}}{R_0 \partial\theta} - k_{x0}^w w = 0 \\ -M_x - K_{x0}^w \frac{\partial w}{\partial x} = 0 \end{cases} \qquad R = R_1: \begin{cases} N_x + k_{x1}^u u = 0 \\ N_{x\theta} + \frac{cM_{x\theta}}{R_1} + k_{x1}^v v = 0 \\ Q_x + \frac{\partial M_{x\theta}}{R_1 \partial\theta} + k_{x1}^w w = 0 \\ -M_x + K_{x1}^w \frac{\partial w}{\partial x} = 0 \end{cases}$$

$$\theta = 0: \begin{cases} N_{x\theta} - k_{\theta 0}^u u = 0 \\ N_\theta + \frac{cM_\theta}{R_0} - k_{\theta 0}^v v = 0 \\ Q_\theta + \frac{\partial M_{x\theta}}{\partial x} - k_{\theta 0}^w w = 0 \\ -M_\theta - K_{\theta 0}^w \frac{\partial w}{R_0 \partial\theta} = 0 \end{cases} \qquad \theta = \theta_0: \begin{cases} N_{x\theta} + k_{\theta 1}^u u = 0 \\ N_\theta + \frac{cM_\theta}{R_1} + k_{\theta 1}^v v = 0 \\ Q_\theta + \frac{\partial M_{x\theta}}{\partial x} + k_{\theta 1}^w w = 0 \\ -M_\theta + K_{\theta 1}^w \frac{\partial w}{R_1 \partial\theta} = 0 \end{cases} \tag{6.14}$$

Alternately, the governing equations and boundary conditions of thin laminated conical shells can be obtained by the Hamilton's principle in the same manner as described in Sect. 1.2.4.

Table 6.1 Possible classical boundary conditions for thin laminated conical shells at boundary $R = R_0$

Boundary type	Conditions
Free boundary conditions	
F	$N_x = N_{x\theta} + \frac{cM_{x\theta}}{R_0} = Q_x + \frac{\partial M_{x\theta}}{R_0\partial\theta} = M_x = 0$
F2	$u = N_{x\theta} + \frac{cM_{x\theta}}{R_0} = Q_x + \frac{\partial M_{x\theta}}{R_0\partial\theta} = M_x = 0$
F3	$N_x = v = Q_x + \frac{\partial M_{x\theta}}{R_0\partial\theta} = M_x = 0$
F4	$u = v = Q_x + \frac{\partial M_{x\theta}}{R_0\partial\theta} = M_x = 0$
Simply supported boundary conditions	
S	$u = v = w = M_x = 0$
SD	$N_x = v = w = M_x = 0$
S3	$u = N_{x\theta} + \frac{cM_{x\theta}}{R_0} = w = M_x = 0$
S4	$N_x = N_{x\theta} + \frac{cM_{x\theta}}{R_0} = w = M_x = 0$
Clamped boundary conditions	
C	$u = v = w = \frac{\partial w}{\partial x} = 0$
C2	$N_x = v = w = \frac{\partial w}{\partial x} = 0$
C3	$u = N_{x\theta} + \frac{cM_{x\theta}}{R_0} = w = \frac{\partial w}{\partial x} = 0$
C4	$N_x = N_{x\theta} + \frac{cM_{x\theta}}{R_0} = w = \frac{\partial w}{\partial x} = 0$

Thin conical shells can have up to 12 possible classical boundary conditions at each edge. This yields a numerous combinations of boundary conditions, particular for open conical shells. The possible combinations for each classical boundary conditions at boundary $R = R_0$ are given in Table 6.1. Similar boundary conditions can be obtained for the other three boundaries (i.e., $R = R_1$, $\theta = 0$ and $\theta = \theta_0$).

6.2 Fundamental Equations of Thick Laminated Conical Shells

The fundamental equations of thin laminated conical shells presented in the previous section are based on the CST and are applicable only when the total thickness of a shell is smaller than 1/20 of the smallest of the wave lengths and/or radii of curvature (Qatu 2004) due to the fact that both shear deformation and rotary inertia are neglected in the formulation. Fundamental equations of thick laminated conical shells will be derived in this section. As usual, the equations that follow are a specialization of the general first-order shear deformation shell theory (see Sect. 1.3) to those of thick laminated conical shells.

6.2.1 *Kinematic Relations*

On the basis of the assumptions of the SDST, the displacement field of a conical shell is expressed in terms of the middle surface displacements and rotation components as

$$
\begin{aligned}
U(x, \theta, z) &= u(x, \theta) + z\phi_x \\
V(x, \theta, z) &= v(x, \theta) + z\phi_\theta \\
W(x, \theta, z) &= w(x, \theta)
\end{aligned}
\tag{6.15}
$$

where u, v and w are the middle surface displacements of the shell in the x, θ and z directions, respectively, and ϕ_x and ϕ_θ represent the rotations of the transverse normal respect to θ- and x-axes.

Specializing Eqs. (1.33) and (1.34) to those of conical shells, the normal and shear strains at any point of the shell space can be defined as:

$$
\begin{aligned}
\varepsilon_x &= \varepsilon_x^0 + z\chi_x \\
\varepsilon_\theta &= \frac{1}{(1 + z/R_\theta)} \left(\varepsilon_\theta^0 + z\chi_\theta \right) \\
\gamma_{x\theta} &= \left(\gamma_{x\theta}^0 + z\chi_{x\theta} \right) + \frac{1}{(1 + z/R_\theta)} \left(\gamma_{\theta x}^0 + z\chi_{\theta x} \right) \\
\gamma_{xz} &= \gamma_{xz}^0 \\
\gamma_{\theta z} &= \frac{\gamma_{\theta z}^0}{(1 + z/R_\theta)}
\end{aligned}
\tag{6.16}
$$

where ε_x^0, ε_θ^0, $\gamma_{x\theta}^0$ and $\gamma_{\theta x}^0$ are the normal and shear strains. χ_x, χ_θ, $\chi_{x\theta}$ and $\chi_{\theta x}$ denote the curvature and twist changes; γ_{xz}^0 and $\gamma_{\theta z}^0$ represent the transverse shear strains. They are defined in terms of the middle surface displacements and rotation components as:

$$
\begin{aligned}
\varepsilon_x^0 &= \frac{\partial u}{\partial x} & \chi_x &= \frac{\partial \phi_x}{\partial x} \\
\varepsilon_\theta^0 &= \frac{\partial v}{xs\partial\theta} + \frac{u}{x} + \frac{w}{xt} & \chi_\theta &= \frac{\partial \phi_\theta}{xs\partial\theta} + \frac{\phi_x}{x} \\
\gamma_{x\theta}^0 &= \frac{\partial v}{\partial x} & \chi_{x\theta} &= \frac{\partial \phi_\theta}{\partial x} \\
\gamma_{\theta x}^0 &= \frac{\partial u}{xs\partial\theta} - \frac{v}{x} & \chi_{\theta x} &= \frac{\partial \phi_x}{xs\partial\theta} - \frac{\phi_\theta}{x} \\
\gamma_{xz}^0 &= \frac{\partial w}{\partial x} + \phi_x & \gamma_{\theta z}^0 &= \frac{\partial w}{xs\partial\theta} - \frac{v}{xt} + \phi_\theta
\end{aligned}
\tag{6.17}
$$

6.2.2 Stress-Strain Relations and Stress Resultants

The stress-strain relations derived earlier for the thick cylindrical shells [i.e., Eq. (5.20)] are applicable for the conical shells. Substituting Eq. (6.2) into (1.40), the force and moment resultants of a thick conical shell can be obtained by following integration operation:

$$
\begin{bmatrix} N_x \\ N_{x\theta} \\ Q_x \end{bmatrix} = \int_{-h/2}^{h/2} \begin{bmatrix} \sigma_x \\ \tau_{x\theta} \\ \tau_{xz} \end{bmatrix} \left(1 + \tfrac{z}{R_\theta}\right) dz \quad
\begin{bmatrix} N_\theta \\ N_{\theta x} \\ Q_\theta \end{bmatrix} = \int_{-h/2}^{h/2} \begin{bmatrix} \sigma_\theta \\ \tau_{\theta x} \\ \tau_{\theta z} \end{bmatrix} dz
$$

$$
\begin{bmatrix} M_x \\ M_{x\theta} \end{bmatrix} = \int_{-h/2}^{h/2} \begin{bmatrix} \sigma_x \\ \tau_{x\theta} \end{bmatrix} \left(1 + \tfrac{z}{R_\theta}\right) z \, dz \quad
\begin{bmatrix} M_\theta \\ M_{\theta x} \end{bmatrix} = \int_{-h/2}^{h/2} \begin{bmatrix} \sigma_\theta \\ \tau_{\theta x} \end{bmatrix} z \, dz
\tag{6.18}
$$

Since the radius of curvature R_θ is a function of the x coordinate, thus, the resulting stiffness parameters will be functions of the x coordinate. This will results in much complexity in equations of thick conical shells. Qatu (2004) suggested taking the average curvature of the conical shells when using these equations. Figures 3.7 and 3.8 showed that the effects of the deepness term z/R on the frequency parameters of an extremely deep, unsymmetrically laminated curved beam ($\theta_0 = 286.48°$) with thickness-to-radius ratio $h/R = 0.1$ is very small and the maximum effect is less than 0.41 % for the worse case. It is proposed here to neglect the effects of the deepness term z/R_β. In many prior researches, the effects of the deepness term z/R_β are often neglected (for example, Jin et al. 2013b, 2014a; Qu et al. 2013a, b; Ye et al. 2014b). Neglecting the effects of the deepness term, the force and moment resultants of a thick conical shell can be rewritten as:

$$
\begin{bmatrix} N_x \\ N_\theta \\ N_{x\theta} \\ N_{\theta x} \\ M_x \\ M_\theta \\ M_{x\theta} \\ M_{\theta x} \end{bmatrix} =
\begin{bmatrix}
A_{11} & A_{12} & A_{16} & A_{16} & B_{11} & B_{12} & B_{16} & B_{16} \\
A_{12} & A_{22} & A_{26} & A_{26} & B_{12} & B_{22} & B_{26} & B_{26} \\
A_{16} & A_{26} & A_{66} & A_{66} & B_{16} & B_{26} & B_{66} & B_{66} \\
A_{16} & A_{26} & A_{66} & A_{66} & B_{16} & B_{26} & B_{66} & B_{66} \\
B_{11} & B_{12} & B_{16} & B_{16} & D_{11} & D_{12} & D_{16} & D_{16} \\
B_{12} & B_{22} & B_{26} & B_{26} & D_{12} & D_{22} & D_{26} & D_{26} \\
B_{16} & B_{26} & B_{66} & B_{66} & D_{16} & D_{26} & D_{66} & D_{66} \\
B_{16} & B_{26} & B_{66} & B_{66} & D_{16} & D_{26} & D_{66} & D_{66}
\end{bmatrix}
\begin{bmatrix} \varepsilon_x^0 \\ \varepsilon_\theta^0 \\ \gamma_{x\theta}^0 \\ \gamma_{\theta x}^0 \\ \chi_x \\ \chi_\theta \\ \chi_{x\theta} \\ \chi_{\theta x} \end{bmatrix}
\tag{6.19a}
$$

$$
\begin{bmatrix} Q_\theta \\ Q_x \end{bmatrix} =
\begin{bmatrix} A_{44} & A_{45} \\ A_{45} & A_{55} \end{bmatrix}
\begin{bmatrix} \gamma_{\theta z}^0 \\ \gamma_{xz}^0 \end{bmatrix}
\tag{6.19b}
$$

The stiffness coefficients A_{ij}, B_{ij} and D_{ij} are given as in Eq. (1.43).

6.2.3 Energy Functions

The strain energy (U_s) of thick conical shells during vibration can be defined in terms of the middle surface strains and curvature changes and stress resultants as

$$U_s = \frac{1}{2}\int_x \int_\theta \left\{ \begin{array}{l} N_x \varepsilon_x^0 + N_\theta \varepsilon_\theta^0 + N_{x\theta}\gamma_{x\theta}^0 + N_{\theta x}\gamma_{\theta x}^0 + M_x \chi_x \\ +M_\theta \chi_\theta + M_{x\theta}\chi_{x\theta} + M_{\theta x}\chi_{\theta x} + Q_\theta \gamma_{\theta z} + Q_x \gamma_{xz} \end{array} \right\} xsd\theta dx \quad (6.20)$$

Substituting Eqs. (6.17) and (6.19a, 6.19b) into Eq. (6.20), the strain energy of the shell can be expressed in terms of the middle surface displacements (u, v, w) and rotation components (ϕ_x, ϕ_θ) as:

$$U_s = \frac{1}{2}\int_x \int_\theta \left\{ \begin{array}{l} A_{11}\left(\frac{\partial u}{\partial x}\right)^2 + 2A_{12}\left(\frac{\partial v}{xs\partial\theta}+\frac{u}{x}+\frac{w}{xt}\right)\left(\frac{\partial u}{\partial x}\right) \\ +2A_{16}\left(\frac{\partial v}{\partial x}+\frac{\partial u}{xs\partial\theta}-\frac{v}{x}\right)\left(\frac{\partial u}{\partial x}\right)+A_{22}\left(\frac{\partial v}{xs\partial\theta}+\frac{u}{x}+\frac{w}{xt}\right)^2 \\ +2A_{26}\left(\frac{\partial v}{\partial x}+\frac{\partial u}{xs\partial\theta}-\frac{v}{x}\right)\left(\frac{\partial v}{xs\partial\theta}+\frac{u}{x}+\frac{w}{xt}\right) \\ +A_{66}\left(\frac{\partial v}{\partial x}+\frac{\partial u}{xs\partial\theta}-\frac{v}{x}\right)^2+A_{44}\left(\frac{\partial w}{xs\partial\theta}-\frac{v}{xt}+\phi_\theta\right)^2 \\ +2A_{45}\left(\frac{\partial w}{xs\partial\theta}-\frac{v}{xt}+\phi_\theta\right)\left(\frac{\partial w}{\partial x}+\phi_x\right)+A_{55}\left(\frac{\partial w}{\partial x}+\phi_x\right)^2 \end{array} \right\} xsd\theta dx$$

$$+\int_x \int_\theta \left\{ \begin{array}{l} B_{11}\left(\frac{\partial\phi_x}{\partial x}\right)\left(\frac{\partial u}{\partial x}\right)+B_{12}\left(\frac{\partial\phi_\theta}{xs\partial\theta}+\frac{\phi_x}{x}\right)\left(\frac{\partial u}{\partial x}\right) \\ +B_{16}\left(\frac{\partial\phi_\theta}{\partial x}+\frac{\partial\phi_x}{xs\partial\theta}-\frac{\phi_\theta}{x}\right)\left(\frac{\partial u}{\partial x}\right)+B_{12}\left(\frac{\partial\phi_x}{\partial x}\right) \\ \times\left(\frac{\partial v}{xs\partial\theta}+\frac{u}{x}+\frac{w}{xt}\right)+B_{22}\left(\frac{\partial\phi_\theta}{xs\partial\theta}+\frac{\phi_x}{x}\right) \\ \times\left(\frac{\partial v}{xs\partial\theta}+\frac{u}{x}+\frac{w}{xt}\right)+B_{26}\left(\frac{\partial\phi_\theta}{\partial x}+\frac{\partial\phi_x}{xs\partial\theta}-\frac{\phi_\theta}{x}\right) \\ \times\left(\frac{\partial v}{xs\partial\theta}+\frac{u}{x}+\frac{w}{xt}\right)+B_{16}\left(\frac{\partial\phi_x}{\partial x}\right)\left(\frac{\partial v}{\partial x}+\frac{\partial u}{xs\partial\theta}-\frac{v}{x}\right) \\ +B_{26}\left(\frac{\partial\phi_\theta}{xs\partial\theta}+\frac{\phi_x}{x}\right)\left(\frac{\partial v}{\partial x}+\frac{\partial u}{xs\partial\theta}-\frac{v}{x}\right) \\ +B_{66}\left(\frac{\partial\phi_\theta}{\partial x}+\frac{\partial\phi_x}{xs\partial\theta}-\frac{\phi_\theta}{x}\right)\left(\frac{\partial v}{\partial x}+\frac{\partial u}{xs\partial\theta}-\frac{v}{x}\right) \end{array} \right\} xsd\theta dx$$

$$+\frac{1}{2}\int_x \int_\theta \left\{ \begin{array}{l} D_{11}\left(\frac{\partial\phi_x}{\partial x}\right)^2 + 2D_{12}\left(\frac{\partial\phi_\theta}{xs\partial\theta}+\frac{\phi_x}{x}\right)\left(\frac{\partial\phi_x}{\partial x}\right) \\ +2D_{16}\left(\frac{\partial\phi_\theta}{\partial x}+\frac{\partial\phi_x}{xs\partial\theta}-\frac{\phi_\theta}{x}\right)\left(\frac{\partial\phi_x}{\partial x}\right) \\ +D_{22}\left(\frac{\partial\phi_\theta}{xs\partial\theta}+\frac{\phi_x}{x}\right)^2+2D_{26}\left(\frac{\partial\phi_\theta}{\partial x}+\frac{\partial\phi_x}{xs\partial\theta}-\frac{\phi_\theta}{x}\right) \\ \times\left(\frac{\partial\phi_\theta}{xs\partial\theta}+\frac{\phi_x}{x}\right)+D_{66}\left(\frac{\partial\phi_\theta}{\partial x}+\frac{\partial\phi_x}{xs\partial\theta}-\frac{\phi_\theta}{x}\right)^2 \end{array} \right\} xsd\theta dx$$

$$(6.21)$$

and the kinetic energy (T) function can be written as:

$$T = \frac{1}{2}\int_x \int_\theta \left\{ \begin{array}{l} I_0\left(\frac{\partial u}{\partial t}\right)^2 + 2I_1\frac{\partial u}{\partial t}\frac{\partial \phi_x}{\partial t} + I_2\left(\frac{\partial \phi_x}{\partial t}\right)^2 + I_0\left(\frac{\partial v}{\partial t}\right)^2 \\ + 2I_1\frac{\partial v}{\partial t}\frac{\partial \phi_\theta}{\partial t} + I_2\left(\frac{\partial \phi_\theta}{\partial t}\right)^2 + I_0\left(\frac{\partial w}{\partial t}\right)^2 \end{array} \right\} xsd\theta dx \qquad (6.22)$$

where the inertia terms are given as in Eq. (1.52).

Suppose the shell is subjected to external forces q_x, q_θ and q_z (in the x, θ and z directions, respectively) and external couples m_x and m_θ (in the middle surface), thus, the work done by the external forces and moments is written as

$$W_e = \int_x \int_\theta \{q_x u + q_\theta v + q_z w + m_x \phi_x + m_\theta \phi_\theta\} xsd\theta dx \qquad (6.23)$$

Using the artificial spring boundary technique similar to that described earlier, let k_ψ^u, k_ψ^v, k_ψ^w, K_ψ^x and $K_\psi^\theta (\psi = x_0, \theta_0, x_1$ and $\theta_1)$ to represent the rigidities (per unit length) of the boundary springs at the boundaries $R = R_0$, $\theta = 0$, $R = R_1$ and $\theta = \theta_0$, respectively. Therefore, the deformation strain energy (U_{sp}) of the boundary springs during vibration is:

$$U_{sp} = \frac{1}{2}\int_\theta \left\{ \begin{array}{l} x\left[k_{x0}^u u^2 + k_{x0}^v v^2 + k_{x0}^w w^2 + K_{x0}^x \phi_x^2 + K_{x0}^\theta \phi_\theta^2\right]|_{R=R_0} \\ +x\left[k_{x1}^u u^2 + k_{x1}^v v^2 + k_{x1}^w w^2 + K_{x1}^x \phi_x^2 + K_{x1}^\theta \phi_\theta^2\right]|_{R=R_1} \end{array} \right\} xsd\theta$$
$$+ \frac{1}{2}\int_x \left\{ \begin{array}{l} \left[k_{\theta0}^u u^2 + k_{\theta0}^v v^2 + k_{\theta0}^w w^2 + K_{\theta0}^x \phi_x^2 + K_{\theta0}^\theta \phi_\theta^2\right]|_{\theta=0} \\ + \left[k_{\theta1}^u u^2 + k_{\theta1}^v v^2 + k_{\theta1}^w w^2 + K_{\theta1}^x \phi_x^2 + K_{\theta1}^\theta \phi_\theta^2\right]|_{\theta=\theta_0} \end{array} \right\} dx \qquad (6.24)$$

6.2.4 Governing Equations and Boundary Conditions

The governing equations and boundary conditions of thick laminated conical shells can be obtained by the Hamilton's principle in the same manner as described in Sect. 1.2.4. Alternately, it can be specialized from those of general thick shells by substituting Eq. (6.2) into Eq. (1.59). According to Eq. (1.59), we have (after being divided by $B = xs$)

$$\frac{\partial N_x}{\partial x} + \frac{N_x - N_\theta}{x} + \frac{\partial N_{\theta x}}{xs\partial\theta} + q_x = I_0\frac{\partial^2 u}{\partial t^2} + I_1\frac{\partial^2\phi_x}{\partial t^2}$$

$$\frac{\partial N_{x\theta}}{\partial x} + \frac{\partial N_\theta}{xs\partial\theta} + \frac{N_{x\theta} + N_{\theta x}}{x} + \frac{Q_\theta}{xt} + q_\theta = I_0\frac{\partial^2 v}{\partial t^2} + I_1\frac{\partial^2\phi_\theta}{\partial t^2}$$

$$-\frac{N_\theta}{xt} + \frac{\partial Q_x}{\partial x} + \frac{Q_x}{x} + \frac{\partial Q_\theta}{xs\partial\theta} + q_z = I_0\frac{\partial^2 w}{\partial t^2} \qquad (6.25)$$

$$\frac{\partial M_x}{\partial x} + \frac{M_x - M_\theta}{x} + \frac{\partial M_{\theta x}}{xs\partial\theta} - Q_x + m_x = I_1\frac{\partial^2 u}{\partial t^2} + I_2\frac{\partial^2\phi_x}{\partial t^2}$$

$$\frac{\partial M_{x\theta}}{\partial x} + \frac{\partial M_\theta}{xs\partial\theta} + \frac{M_{x\theta} + M_{\theta x}}{x} - Q_\theta + m_\theta = I_1\frac{\partial^2 v}{\partial t^2} + I_2\frac{\partial^2\phi_\theta}{\partial t^2}$$

Substituting Eqs. (6.17) and (6.19a, 6.19b) into above equation, the governing equations can be written in terms of displacements as

$$\left(\begin{bmatrix} L_{11} & L_{12} & L_{13} & L_{14} & L_{15} \\ L_{21} & L_{22} & L_{23} & L_{24} & L_{25} \\ L_{31} & L_{32} & L_{33} & L_{34} & L_{35} \\ L_{41} & L_{42} & L_{43} & L_{44} & L_{45} \\ L_{51} & L_{52} & L_{53} & L_{54} & L_{55} \end{bmatrix} + \omega^2\begin{bmatrix} I_0 & 0 & 0 & I_1 & 0 \\ 0 & I_0 & 0 & 0 & I_1 \\ 0 & 0 & I_0 & 0 & 0 \\ I_1 & 0 & 0 & I_2 & 0 \\ 0 & I_1 & 0 & 0 & I_2 \end{bmatrix}\right)\begin{bmatrix} u \\ v \\ w \\ \phi_x \\ \phi_\theta \end{bmatrix} = \begin{bmatrix} -p_x \\ -p_y \\ -p_z \\ -m_x \\ -m_\theta \end{bmatrix} \quad (6.26)$$

The coefficients of the linear operator L_{ij} are given as

$$L_{11} = \frac{\partial}{\partial x}\left(A_{11}\frac{\partial}{\partial x} + \frac{A_{12}}{x} + \frac{A_{16}}{xs}\frac{\partial}{\partial\theta}\right) + \frac{1}{x}\left(A_{11}\frac{\partial}{\partial x} + \frac{A_{12}}{x} + \frac{A_{16}}{xs}\frac{\partial}{\partial\theta}\right)$$
$$-\frac{1}{x}\left(A_{12}\frac{\partial}{\partial x} + \frac{A_{22}}{x} + \frac{A_{26}}{xs}\frac{\partial}{\partial\theta}\right) + \frac{1}{xs}\frac{\partial}{\partial\theta}\left(A_{16}\frac{\partial}{\partial x} + \frac{A_{26}}{x} + \frac{A_{66}}{xs}\frac{\partial}{\partial\theta}\right)$$

$$L_{12} = \frac{\partial}{\partial x}\left(\frac{A_{12}}{xs}\frac{\partial}{\partial\theta} + A_{16}\left(\frac{\partial}{\partial x} - \frac{1}{x}\right)\right) + \frac{1}{x}\left(\frac{A_{12}}{xs}\frac{\partial}{\partial\theta} + A_{16}\left(\frac{\partial}{\partial x} - \frac{1}{x}\right)\right)$$
$$-\frac{1}{x}\left(\frac{A_{22}}{xs}\frac{\partial}{\partial\theta} + A_{26}\left(\frac{\partial}{\partial x} - \frac{1}{x}\right)\right) + \frac{1}{xs}\frac{\partial}{\partial\theta}\left(\frac{A_{26}}{xs}\frac{\partial}{\partial\theta} + A_{66}\left(\frac{\partial}{\partial x} - \frac{1}{x}\right)\right)$$

$$L_{13} = \frac{\partial}{\partial x}\left(\frac{A_{12}}{xt}\right) + \frac{1}{x}\left(\frac{A_{12}}{xt}\right) - \frac{1}{x}\left(\frac{A_{22}}{xt}\right) + \frac{1}{xs}\frac{\partial}{\partial\theta}\left(\frac{A_{26}}{xt}\right)$$

$$L_{14} = \frac{\partial}{\partial x}\left(B_{11}\frac{\partial}{\partial x} + \frac{B_{12}}{x} + \frac{B_{16}}{xs}\frac{\partial}{\partial\theta}\right) + \frac{1}{x}\left(B_{11}\frac{\partial}{\partial x} + \frac{B_{12}}{x} + \frac{B_{16}}{xs}\frac{\partial}{\partial\theta}\right)$$
$$-\frac{1}{x}\left(B_{12}\frac{\partial}{\partial x} + \frac{B_{22}}{x} + \frac{B_{26}}{xs}\frac{\partial}{\partial\theta}\right) + \frac{1}{xs}\frac{\partial}{\partial\theta}\left(B_{16}\frac{\partial}{\partial x} + \frac{B_{26}}{x} + \frac{B_{66}}{xs}\frac{\partial}{\partial\theta}\right)$$

$$L_{15} = \frac{\partial}{\partial x}\left(\frac{B_{12}}{xs}\frac{\partial}{\partial\theta} + B_{16}\left(\frac{\partial}{\partial x} - \frac{1}{x}\right)\right) + \frac{1}{x}\left(\frac{B_{12}}{xs}\frac{\partial}{\partial\theta} + B_{16}\left(\frac{\partial}{\partial x} - \frac{1}{x}\right)\right)$$
$$-\frac{1}{x}\left(\frac{B_{22}}{xs}\frac{\partial}{\partial\theta} + B_{26}\left(\frac{\partial}{\partial x} - \frac{1}{x}\right)\right) + \frac{1}{xs}\frac{\partial}{\partial\theta}\left(\frac{B_{22}}{xs}\frac{\partial}{\partial\theta} + B_{26}\left(\frac{\partial}{\partial x} - \frac{1}{x}\right)\right)$$

$$L_{21} = \frac{1}{xs}\frac{\partial}{\partial\theta}\left(A_{12}\frac{\partial}{\partial x} + \frac{A_{22}}{x} + \frac{A_{26}}{xs}\frac{\partial}{\partial\theta}\right) + \frac{\partial}{\partial x}\left(A_{16}\frac{\partial}{\partial x} + \frac{A_{26}}{x} + \frac{A_{66}}{xs}\frac{\partial}{\partial\theta}\right)$$
$$+\frac{2}{x}\left(A_{16}\frac{\partial}{\partial x} + \frac{A_{26}}{x} + \frac{A_{66}}{xs}\frac{\partial}{\partial\theta}\right)$$

$$L_{22} = \frac{1}{xs}\frac{\partial}{\partial\theta}\left(\frac{A_{22}}{xs}\frac{\partial}{\partial\theta} + A_{26}\left(\frac{\partial}{\partial x} - \frac{1}{x}\right)\right) + \frac{\partial}{\partial x}\left(\frac{A_{26}}{xs}\frac{\partial}{\partial\theta} + A_{66}\left(\frac{\partial}{\partial x} - \frac{1}{x}\right)\right)$$

$$+ \frac{2}{x}\left(\frac{A_{26}}{xs}\frac{\partial}{\partial\theta} + A_{66}\left(\frac{\partial}{\partial x} - \frac{1}{x}\right)\right) - \frac{A_{44}}{x^2 t^2}$$

$$L_{23} = \frac{1}{xs}\frac{\partial}{\partial\theta}\left(\frac{A_{22}}{xt}\right) + \frac{\partial}{\partial x}\left(\frac{A_{26}}{xt}\right) + \frac{2A_{26}}{x^2 t} + \frac{A_{44}}{x^2 st}\frac{\partial}{\partial\theta} + \frac{A_{45}}{xt}\frac{\partial}{\partial x}$$

$$L_{24} = \frac{1}{xs}\frac{\partial}{\partial\theta}\left(B_{12}\frac{\partial}{\partial x} + \frac{B_{22}}{x} + \frac{B_{26}}{xs}\frac{\partial}{\partial\theta}\right) + \frac{\partial}{\partial x}\left(B_{16}\frac{\partial}{\partial x} + \frac{B_{26}}{x} + \frac{B_{66}}{xs}\frac{\partial}{\partial\theta}\right)$$

$$+ \frac{2}{x}\left(B_{16}\frac{\partial}{\partial x} + \frac{B_{26}}{x} + \frac{B_{66}}{xs}\frac{\partial}{\partial\theta}\right) + \frac{A_{45}}{xt}$$

$$L_{25} = \frac{1}{xs}\frac{\partial}{\partial\theta}\left(\frac{B_{22}}{xs}\frac{\partial}{\partial\theta} + B_{26}\left(\frac{\partial}{\partial x} - \frac{1}{x}\right)\right) + \frac{\partial}{\partial x}\left(\frac{B_{26}}{xs}\frac{\partial}{\partial\theta} + B_{66}\left(\frac{\partial}{\partial x} - \frac{1}{x}\right)\right)$$

$$+ \frac{2}{x}\left(\frac{B_{26}}{xs}\frac{\partial}{\partial\theta} + B_{66}\left(\frac{\partial}{\partial x} - \frac{1}{x}\right)\right) + \frac{A_{44}}{xt}$$

$$L_{31} = -\frac{A_{12}}{xt}\frac{\partial}{\partial x} - \frac{A_{22}}{x^2 t} - \frac{cA_{26}}{x^2 s^2}\frac{\partial}{\partial\theta}$$

$$L_{32} = \frac{-1}{xt}\left(\frac{A_{22}}{xs}\frac{\partial}{\partial\theta} + A_{26}\left(\frac{\partial}{\partial x} - \frac{1}{x}\right)\right) - \frac{\partial}{\partial x}\left(\frac{A_{45}}{xt}\right) - \frac{A_{45}}{x^2 t} - \frac{cA_{44}}{x^2 s^2}\frac{\partial}{\partial\theta}$$

$$L_{33} = \frac{-A_{22}}{x^2 t^2} + \frac{\partial}{\partial x}\left(\frac{A_{45}}{xs}\frac{\partial}{\partial\theta} + A_{55}\frac{\partial}{\partial x}\right) + \frac{1}{x}\left(\frac{A_{45}}{xs}\frac{\partial}{\partial\theta} + A_{55}\frac{\partial}{\partial x}\right)$$

$$+ \frac{1}{xs}\frac{\partial}{\partial\theta}\left(\frac{A_{44}}{xs}\frac{\partial}{\partial\theta} + A_{45}\frac{\partial}{\partial x}\right)$$

$$L_{34} = \frac{-1}{xt}\left(B_{12}\frac{\partial}{\partial x} + \frac{B_{22}}{x} + \frac{B_{26}}{xs}\frac{\partial}{\partial\theta}\right) + A_{55}\frac{\partial}{\partial x} + \frac{A_{55}}{x} + \frac{A_{45}}{xs}\frac{\partial}{\partial\theta}$$

$$L_{35} = \frac{-1}{xt}\left(\frac{B_{22}}{xs}\frac{\partial}{\partial\theta} + B_{26}\left(\frac{\partial}{\partial x} - \frac{1}{x}\right)\right) + A_{45}\frac{\partial}{\partial x} + \frac{A_{45}}{x} + \frac{A_{44}}{xs}\frac{\partial}{\partial\theta}$$

$$L_{41} = \frac{\partial}{\partial x}\left(B_{11}\frac{\partial}{\partial x} + \frac{B_{12}}{x} + \frac{B_{16}}{xs}\frac{\partial}{\partial\theta}\right) + \frac{1}{x}\left(B_{11}\frac{\partial}{\partial x} + \frac{B_{12}}{x} + \frac{B_{16}}{xs}\frac{\partial}{\partial\theta}\right)$$

$$- \frac{1}{x}\left(B_{12}\frac{\partial}{\partial x} + \frac{B_{22}}{x} + \frac{B_{26}}{xs}\frac{\partial}{\partial\theta}\right) + \frac{1}{xs}\frac{\partial}{\partial\theta}\left(B_{16}\frac{\partial}{\partial x} + \frac{B_{26}}{x} + \frac{B_{66}}{xs}\frac{\partial}{\partial\theta}\right)$$

$$L_{42} = \frac{\partial}{\partial x}\left(\frac{B_{12}}{xs}\frac{\partial}{\partial\theta} + B_{16}\left(\frac{\partial}{\partial x} - \frac{1}{x}\right)\right) + \frac{1}{x}\left(\frac{B_{12}}{xs}\frac{\partial}{\partial\theta} + B_{16}\left(\frac{\partial}{\partial x} - \frac{1}{x}\right)\right)$$

$$- \frac{1}{x}\left(\frac{B_{22}}{xs}\frac{\partial}{\partial\theta} + B_{26}\left(\frac{\partial}{\partial x} - \frac{1}{x}\right)\right) + \frac{1}{xs}\frac{\partial}{\partial\theta}\left(\frac{B_{26}}{xs}\frac{\partial}{\partial\theta} + B_{66}\left(\frac{\partial}{\partial x} - \frac{1}{x}\right)\right) + \frac{A_{45}}{xt}$$

$$L_{43} = \frac{\partial}{\partial x}\left(\frac{B_{12}}{xt}\right) + \frac{1}{x}\left(\frac{B_{12}}{xt}\right) - \frac{1}{x}\left(\frac{B_{22}}{xt}\right) + \frac{1}{xs}\frac{\partial}{\partial \theta}\left(\frac{B_{26}}{xt}\right)$$

$$- \left(\frac{A_{45}}{xs}\frac{\partial}{\partial \theta} + A_{55}\frac{\partial}{\partial x}\right)$$

$$L_{44} = \frac{\partial}{\partial x}\left(D_{11}\frac{\partial}{\partial x} + \frac{D_{12}}{x} + \frac{D_{16}}{xs}\frac{\partial}{\partial \theta}\right) + \frac{1}{x}\left(D_{11}\frac{\partial}{\partial x} + \frac{D_{12}}{x} + \frac{D_{16}}{xs}\frac{\partial}{\partial \theta}\right)$$

$$- \frac{1}{x}\left(D_{12}\frac{\partial}{\partial x} + \frac{D_{22}}{x} + \frac{D_{26}}{xs}\frac{\partial}{\partial \theta}\right)$$

$$+ \frac{1}{xs}\frac{\partial}{\partial \theta}\left(D_{16}\frac{\partial}{\partial x} + \frac{D_{26}}{x} + \frac{D_{66}}{xs}\frac{\partial}{\partial \theta}\right) - A_{55}$$

$$L_{45} = \frac{\partial}{\partial x}\left(\frac{D_{12}}{xs}\frac{\partial}{\partial \theta} + D_{16}\left(\frac{\partial}{\partial x} - \frac{1}{x}\right)\right) + \frac{1}{x}\left(\frac{D_{12}}{xs}\frac{\partial}{\partial \theta} + D_{16}\left(\frac{\partial}{\partial x} - \frac{1}{x}\right)\right)$$

$$- \frac{1}{x}\left(\frac{D_{22}}{xs}\frac{\partial}{\partial \theta} + D_{26}\left(\frac{\partial}{\partial x} - \frac{1}{x}\right)\right)$$

$$+ \frac{1}{xs}\frac{\partial}{\partial \theta}\left(\frac{D_{22}}{xs}\frac{\partial}{\partial \theta} + D_{26}\left(\frac{\partial}{\partial x} - \frac{1}{x}\right)\right) - A_{45}$$

$$L_{51} = \frac{1}{xs}\frac{\partial}{\partial \theta}\left(B_{12}\frac{\partial}{\partial x} + \frac{B_{22}}{x} + \frac{B_{26}}{xs}\frac{\partial}{\partial \theta}\right) + \frac{\partial}{\partial x}\left(B_{16}\frac{\partial}{\partial x} + \frac{B_{26}}{x} + \frac{B_{66}}{xs}\frac{\partial}{\partial \theta}\right) \quad (6.27)$$

$$+ \frac{2}{x}\left(B_{16}\frac{\partial}{\partial x} + \frac{B_{26}}{x} + \frac{B_{66}}{xs}\frac{\partial}{\partial \theta}\right)$$

$$L_{52} = \frac{1}{xs}\frac{\partial}{\partial \theta}\left(\frac{B_{22}}{xs}\frac{\partial}{\partial \theta} + B_{26}\left(\frac{\partial}{\partial x} - \frac{1}{x}\right)\right) + \frac{\partial}{\partial x}\left(\frac{B_{26}}{xs}\frac{\partial}{\partial \theta} + B_{66}\left(\frac{\partial}{\partial x} - \frac{1}{x}\right)\right)$$

$$+ \frac{2}{x}\left(\frac{B_{26}}{xs}\frac{\partial}{\partial \theta} + B_{66}\left(\frac{\partial}{\partial x} - \frac{1}{x}\right)\right) + \frac{A_{44}}{xt}$$

$$L_{53} = \frac{1}{xs}\frac{\partial}{\partial \theta}\left(\frac{B_{22}}{xt}\right) + \frac{\partial}{\partial x}\left(\frac{B_{26}}{xt}\right) + \frac{2B_{26}}{x^2 t} - \left(\frac{A_{44}}{xs}\frac{\partial}{\partial \theta} + A_{45}\frac{\partial}{\partial x}\right)$$

$$L_{54} = \frac{1}{xs}\frac{\partial}{\partial \theta}\left(D_{12}\frac{\partial}{\partial x} + \frac{D_{22}}{x} + \frac{D_{26}}{xs}\frac{\partial}{\partial \theta}\right) + \frac{\partial}{\partial x}\left(D_{16}\frac{\partial}{\partial x} + \frac{D_{26}}{x} + \frac{D_{66}}{xs}\frac{\partial}{\partial \theta}\right)$$

$$+ \frac{2}{x}\left(D_{16}\frac{\partial}{\partial x} + \frac{D_{26}}{x} + \frac{D_{66}}{xs}\frac{\partial}{\partial \theta}\right) - A_{45}$$

$$L_{55} = \frac{1}{xs}\frac{\partial}{\partial \theta}\left(\frac{D_{22}}{xs}\frac{\partial}{\partial \theta} + D_{26}\left(\frac{\partial}{\partial x} - \frac{1}{x}\right)\right) + \frac{\partial}{\partial x}\left(\frac{D_{26}}{xs}\frac{\partial}{\partial \theta} + D_{66}\left(\frac{\partial}{\partial x} - \frac{1}{x}\right)\right)$$

$$+ \frac{2}{x}\left(\frac{D_{26}}{xs}\frac{\partial}{\partial \theta} + D_{66}\left(\frac{\partial}{\partial x} - \frac{1}{x}\right)\right) - A_{44}$$

These equations are proven useful when exact solutions are desired.

According to Eqs. (1.60) and (1.61), the general boundary conditions of thick conical shells are:

$$R = R_0: \begin{cases} N_x - k_{x0}^u u = 0 \\ N_{x\theta} - k_{x0}^v v = 0 \\ Q_x - k_{x0}^w w = 0 \\ M_x - K_{x0}^x \phi_x = 0 \\ M_{x\theta} - K_{x0}^\theta \phi_\theta = 0 \end{cases} \quad R = R_1: \begin{cases} N_x + k_{x1}^u u = 0 \\ N_{x\theta} + k_{x1}^v v = 0 \\ Q_x + k_{x1}^w w = 0 \\ M_x + K_{x1}^x \phi_x = 0 \\ M_{x\theta} + K_{x1}^\theta \phi_\theta = 0 \end{cases}$$

$$\theta = 0: \begin{cases} N_{\theta x} - k_{\theta 0}^u u = 0 \\ N_\theta - k_{\theta 0}^v v = 0 \\ Q_\theta - k_{\theta 0}^w w = 0 \\ M_{\theta x} - K_{\theta 0}^x \phi_x = 0 \\ M_\theta - K_{\theta 0}^\theta \phi_\theta = 0 \end{cases} \quad \theta = \theta_0: \begin{cases} N_{\theta x} + k_{\theta 1}^u u = 0 \\ N_\theta + k_{\theta 1}^v v = 0 \\ Q_\theta + k_{\theta 1}^w w = 0 \\ M_{\theta x} + K_{\theta 1}^x \phi_x = 0 \\ M_\theta + K_{\theta 1}^\theta \phi_\theta = 0 \end{cases} \quad (6.28)$$

Thick conical shells can have up to 24 possible classical boundary conditions at each boundary (see Table 6.2) which leads a high number of combinations of boundary conditions. The classification of the classical boundary conditions shown in Table 1.3 for general thick shells is applicable for thick conical shells.

Table 6.2 shows the importance of developing an accurate, robust and efficient method which is capable of simplifying solution algorithms, reducing model input data and universally dealing with various boundary conditions. This difficulty can be overcome by using the artificial spring boundary technique. In this chapter, we mainly consider four typical boundary conditions which are frequently encountered in practices, i.e., F, SD, S and C boundary conditions. Taking edge $R = R_0$ for example, the corresponding spring rigidities for the four classical boundary conditions are given as in Eq. (5.37).

6.3 Vibration of Laminated Closed Conical Shells

In this section, we consider vibrations of laminated closed conical shells. The open ones will then be treated in the later section of this chapter. It is commonly believed an exact solution is available only for laminated closed conical shells having cross-ply lamination schemes and shear diaphragm boundary conditions at both ends. Tong (1993, 1994a) obtained such solutions for thin and thick conical shells. In this chapter, in the framework of SDST, accurate vibration solutions for laminated conical shells with general boundary conditions, lamination schemes and different geometry parameters will be presented by using the modified Fourier series and weak form solution procedure. For a laminated closed conical shell, there exists two boundaries, i.e., $R = R_0$ and $R = R_1$. Thus, a two-letter string is employed to denote the boundary conditions of the shell, such as F-C identifies the shell with

Table 6.2 Possible classical boundary conditions for thick conical shells at each boundary of $R = $ constant

Boundary type	Conditions
Free boundary conditions	
F	$N_x = N_{x\theta} = Q_x = M_x = M_{x\theta} = 0$
F2	$u = N_{x\theta} = Q_x = M_x = M_{x\theta} = 0$
F3	$N_x = v = Q_x = M_x = M_{x\theta} = 0$
F4	$u = v = Q_x = M_x = M_{x\theta} = 0$
F5	$N_x = N_{x\theta} = Q_x = M_x = \phi_\theta = 0$
F6	$u = N_{x\theta} = Q_x = M_x = \phi_\theta = 0$
F7	$N_x = v = Q_x = M_x = \phi_\theta = 0$
F8	$u = v = Q_x = M_x = \phi_\theta = 0$
Simply supported boundary conditions	
S	$u = v = w = M_x = \phi_\theta = 0$
SD	$N_x = v = w = M_x = \phi_\theta = 0$
S3	$u = N_{x\theta} = w = M_x = \phi_\theta = 0$
S4	$N_x = N_{x\theta} = w = M_x = \phi_\theta = 0$
S5	$u = v = w = M_x = M_{x\theta} = 0$
S6	$N_x = v = w = M_x = M_{x\theta} = 0$
S7	$u = N_{x\theta} = w = M_x = M_{x\theta} = 0$
S8	$N_x = N_{x\theta} = w = M_x = M_{x\theta} = 0$
Clamped boundary conditions	
C	$u = v = w = \phi_x = \phi_\theta = 0$
C2	$N_x = v = w = \phi_x = \phi_\theta = 0$
C3	$u = N_{x\theta} = w = \phi_x = \phi_\theta = 0$
C4	$N_x = N_{x\theta} = w = \phi_x = \phi_\theta = 0$
C5	$u = v = w = \phi_x = M_{x\theta} = 0$
C6	$N_x = v = w = \phi_x = M_{x\theta} = 0$
C7	$u = N_{x\theta} = w = \phi_x = M_{x\theta} = 0$
C8	$N_x = N_{x\theta} = w = \phi_x = M_{x\theta} = 0$

completely free and clamped boundary conditions at the edges $R = R_0$ and $R = R_1$, respectively. Unless otherwise stated, the non-dimensional frequency parameter $\Omega = \omega R_1 \sqrt{\rho h / A_{11}}$ is used in the subsequent analysis and laminated conical shells under consideration are assumed to be composed of composite layers having following material properties: $E_2 = 10$ GPa, $E_1/E_2 = $ open, $\mu_{12} = 0.25$, $G_{12} = G_{13} = 0.6E_2$, $G_{23} = 0.5E_2$, $\rho = 1{,}500$ kg/m^3.

Considering the circumferential symmetry of closed conical shells, each displacement/rotation component of a closed conical shell is expanded as a 1-D modified Fourier series of the following form through Fourier decomposition of the circumferential wave motion:

$$u(x, \theta) = \sum_{m=0}^{M} \sum_{n=0}^{N} A_{mn} \cos \lambda_m x \cos n\theta + \sum_{l=1}^{2} \sum_{n=0}^{N} a_{ln} P_l(x) \cos n\theta$$

$$v(x, \theta) = \sum_{m=0}^{M} \sum_{n=0}^{N} B_{mn} \cos \lambda_m x \sin n\theta + \sum_{l=1}^{2} \sum_{n=0}^{N} b_{ln} P_l(x) \sin n\theta$$

$$w(x, \theta) = \sum_{m=0}^{M} \sum_{n=0}^{N} C_{mn} \cos \lambda_m x \cos n\theta + \sum_{l=1}^{2} \sum_{n=0}^{N} c_{ln} P_l(x) \cos n\theta \qquad (6.29)$$

$$\phi_x(x, \theta) = \sum_{m=0}^{M} \sum_{n=0}^{N} D_{mn} \cos \lambda_m x \cos n\theta + \sum_{l=1}^{2} \sum_{n=0}^{N} d_{ln} P_l(x) \cos n\theta$$

$$\phi_\theta(x, \theta) = \sum_{m=0}^{M} \sum_{n=0}^{N} E_{mn} \cos \lambda_m x \sin n\theta + \sum_{l=1}^{2} \sum_{n=0}^{N} e_{ln} P_l(x) \sin n\theta$$

where $\lambda_m = m\pi/L$. Similarly, n represents the circumferential wave number of the corresponding mode. It should be note that n is a non-negative integer. Interchanging of sin $n\theta$ and cos $n\theta$ in Eq. (6.29), another set of free vibration modes (anti-symmetric modes) can be obtained. It is obvious that each displacement and rotation component in the FSDT displacement field is required to have up to the second derivatives [see Eq. (6.27)]. Thus, two auxiliary polynomial functions P_l (x) are introduced in each displacement expression to remove all the discontinuities potentially associated with the first-order derivatives at the boundaries. These auxiliary functions are defined as in Eq. (5.39).

6.3.1 Convergence Studies and Result Verification

Table 6.3 shows the convergence study of the natural frequencies (Hz) for a single-layered conical shell with F–F and C–C boundary conditions. The geometric and material constants of the shell are: $R_0 = 1$ m, $L = 2$ m, $h = 0.1$ m, $\varphi = 45°$, $E_1/E_2 = 15$. Five truncation schemes (i.e. $M = 11 - 15$ and $N = 10$) are performed in the study. The table shows the present solutions converge fast. The maximum differences between the '11 × 10' and '15 × 10' form results for the F–F and C–C boundary conditions are less than 0.025 and 0.004 %, respectively. In addition, comparing with Table 5.3, we can find that the convergence of the solutions for the cylindrical shells is better than the conical ones. Furthermore, the C–C solutions converge faster than those of F–F boundary conditions. Unless otherwise stated, the truncated number of the displacement expressions will be uniformly selected as $M = 15$ in the following discussions.

To further validate the accuracy and reliability of current method, the current solutions are compared with those reported by other researchers by the subsequent numerical examples. Table 6.4 lists the comparison of the fundamental dimensionless frequencies Ω for cross-ply conical shells with different boundary

Table 6.3 Convergence of the natural frequencies (Hz) for a single-layered conical shell with F–F and C–C boundary conditions ($R_0 = 1$ m, $L = 2$ m, $h = 0.1$ m, $\varphi = 45°$, $E_1/E_2 = 15$)

B.C.	M	Mode number					
		1	2	3	4	5	6
F–F	11	1.3975	10.045	25.970	40.249	46.792	71.916
	12	1.3972	10.045	25.969	40.243	46.791	71.913
	13	1.3972	10.045	25.969	40.242	46.790	71.913
	14	1.3971	10.045	25.969	40.239	46.790	71.912
	15	1.3971	10.045	25.969	40.239	46.790	71.912
C–C	11	247.80	252.15	253.50	266.46	269.13	281.76
	12	247.80	252.15	253.50	266.47	269.13	281.76
	13	247.80	252.14	253.50	266.46	269.13	281.76
	14	247.80	252.14	253.50	266.46	269.13	281.76
	15	247.79	252.14	253.50	266.45	269.13	281.76

Table 6.4 Comparison of the fundamental frequency parameters Ω for two cross-ply conical shells with various thickness-radius ratios ($R_0 = 0.75$ m, $L = 0.5$ m, $\varphi = 30°$)

B.C.	Layout	Method	h/R_1				
			0.01	0.03	0.05	0.07	0.09
SD–SD	$[0°/90°]$	CST (Shu 1996)	0.1799	0.2397	0.2841	0.3277	0.3680
		FSDT (Wu and Lee 2011)	0.1759	0.2320	0.2710	0.3061	0.3358
		Present	0.1759	0.2320	0.2710	0.3061	0.3358
	$[0°/90°]_{10}$	CST (Shu 1996)	0.1976	0.2669	0.3304	0.3873	0.4321
		FSDT (Wu and Lee 2011)	0.1958	0.2607	0.3134	0.3544	0.3832
		Present	0.1958	0.2607	0.3134	0.3544	0.3832
C–C	$[0°/90°]$	CST (Shu 1996)	0.2986	0.6210	0.9331	1.2344	1.5206
		FSDT (Wu and Lee 2011)	0.3045	0.5834	0.7967	0.9476	1.0457
		Present	0.2966	0.5835	0.7971	0.9480	1.0466
	$[0°/90°]_{10}$	CST (Shu 1996)	0.3771	0.8578	1.3361	1.5730	1.5735
		FSDT (Wu and Lee 2011)	0.3720	0.7509	0.9797	1.1025	1.1759
		Present	0.3720	0.7509	0.9799	1.1031	1.1770

conditions and thickness-to-large edge radius ratios (h/R_1). Two types of lamination schemes, i.e., $[0°/90°]$ and $[0°/90°]_{10}$, are examined. The SD–SD and C–C boundary conditions are considered in the comparisons. The thickness-to-large edge radius ratio h/R_1 is varied from 0.01 to 0.09 by a step of 0.02, corresponding to thin to moderately thick conical shells. The material constants and geometry parameters of the shell are: $E_2 = 10$ GPa, $E_1/E_2 = 15$, $\mu_{12} = 0.25$, $G_{12} = 0.5E_2$, $G_{13} = 0.3846E_2$,

Table 6.5 Comparison of the frequency parameters Ω for a cross-ply $[0°/90°]_{10}$ conical shell with various boundary conditions ($R_0 = 0.75$ m, $L = 0.5$ m, $\varphi = 30°$, $m = 1$)

n	FSDT (Qu et al. 2013a)			Present		
	S–F	S–S	S–C	S–F	S–S	S–C
0	0.6527	1.2919	1.4065	0.6528	1.2919	1.4065
1	0.4016	1.0903	1.1977	0.4016	1.0903	1.1976
2	0.2711	0.9789	1.1020	0.2712	0.9789	1.1020
3	0.2941	0.9692	1.0945	0.2941	0.9692	1.0944
4	0.4315	1.0312	1.1488	0.4315	1.0312	1.1487
5	0.6203	1.1490	1.2545	0.6203	1.1490	1.2545
6	0.8315	1.3076	1.4007	0.8315	1.3076	1.4006
7	1.0532	1.4936	1.5757	1.0532	1.4936	1.5756
8	1.2796	1.6968	1.7700	1.2796	1.6968	1.7700
9	1.5078	1.9105	1.9768	1.5078	1.9105	1.9767

$G_{23} = 0.3846E_2$, $R_0 = 0.75$ m, $L = 0.5$ m and $\varphi = 30°$. The comparisons are performed between the present results and the FSDT solutions reported by Wu and Lee (2011) and CST solutions published by Shu (1996). The comparisons in the table show a excellent agreement between the present results and those reported by Wu and Lee (2011). It is obvious that the discrepancies are negligible and not exceed 0.012 % for the worst case. The comparisons validate the high accuracy of the modified Fourier series method in predicting vibrations of composite conical shells.

In Table 6.5, the first longitudinal mode ($m = 1$) frequency parameters Ω for the $[0°/90°]_{10}$ conical shell given in Table 6.4 with three set of classical boundary conditions, i.e., S–F, S–S and S–C are presented. The lowest ten circumference wave numbers (i.e., $n = 1$–10) are considered in the calculation. The results reported by Qu et al. (2013a) based on the first-order shear deformation theory are also included in the table. A consistent agreement of the present results and the referential data can be seen from the table. The discrepancies are very small and less than 0.015 % for the worst case. The small discrepancies in the results may be attributed to the different solution approaches were used in the literature.

6.3.2 Laminated Closed Conical Shells with General Boundary Conditions

Table 6.6 shows the lowest four frequency parameters Ω for a two-layered, $[0°/90°]$ laminated conical shell with various thickness-to-small edge radius ratios (h/R_0). The material properties and geometric constants of the layers of the shell are: $E_1/E_2 = 15$, $R_0 = 1$ m, $L/R_0 = 2$, $\varphi = 45°$. Six sets of classical boundary combinations, i.e., F–S, S–F, F–C, C–F, S–S and C–C and five kinds of thickness-to-small edge radius ratios ($h/R_0 = 0.01, 0.02, 0.05, 0.1$ and 0.15) are included in the table. It can be seen from the

Table 6.6 Frequency parameters Ω for a [0°/90°] laminated conical shell with various boundary conditions and thickness-to-small edge radius ratios ($E_1/E_2 = 15$, $R_0 = 1$ m, $L/R_0 = 2$, $\varphi = 45°$)

h/R_0	Mode	Boundary conditions					
		F–S	S–F	F–C	C–F	S–S	C–C
0.01	1	0.1188	0.0415	0.1189	0.0416	0.1364	0.1373
	2	0.1198	0.0454	0.1198	0.0454	0.1399	0.1411
	3	0.1291	0.0455	0.1293	0.0456	0.1417	0.1426
	4	0.1365	0.0546	0.1366	0.0548	0.1527	0.1540
0.02	1	0.1559	0.0543	0.1559	0.0543	0.1786	0.1825
	2	0.1581	0.0613	0.1584	0.0614	0.1802	0.1836
	3	0.1741	0.0620	0.1759	0.0621	0.1961	0.2001
	4	0.1867	0.0775	0.1867	0.0777	0.1977	0.2022
0.05	1	0.2192	0.0754	0.2208	0.0761	0.2538	0.2699
	2	0.2302	0.0914	0.2347	0.0914	0.2728	0.2900
	3	0.2697	0.0922	0.2779	0.0939	0.2766	0.2921
	4	0.2758	0.1268	0.2811	0.1270	0.3256	0.3435
0.10	1	0.2896	0.0988	0.3000	0.1034	0.3408	0.3835
	2	0.2988	0.1061	0.3064	0.1072	0.3746	0.4140
	3	0.3560	0.1584	0.3788	0.1585	0.3862	0.4277
	4	0.4472	0.1585	0.4604	0.1626	0.4593	0.4981
0.15	1	0.3269	0.1070	0.3414	0.1142	0.4197	0.4852
	2	0.3640	0.1396	0.3863	0.1412	0.4300	0.4966
	3	0.4527	0.1595	0.4735	0.1672	0.4992	0.5575
	4	0.4783	0.2249	0.5173	0.2249	0.5760	0.6327

table that the frequency parameters of the shell increase as the thickness-to-small edge radius ratio increases. In addition, it is obviously that the boundary conditions have a conspicuous effect on the vibration frequencies. Increasing the restraint stiffness always results in increments of the frequency parameters.

Table 6.7 lists the lowest four frequency parameters Ω for conical shells with different angle-ply lamination schemes and boundary conditions. Single-layered lamination [0°], antisymmetric laminations [0°/45°] and [0°/45°/0°/45°] and symmetric lamination [0°/45°/0°] are used in the calculation. The geometric and material constants of these shells are: $E_1/E_2 = 15$, $R_0 = 1$ m, $L/R_0 = 2$, $h/R_0 = 0.1$, $\varphi = 45°$. As can be observed from Table 6.7, frequency parameters Ω of the [0°/45°/0°/45°] shell are larger than those of the other three lamination schemes and the minimum frequency parameter for each mode in all the boundary condition occurs at those of single-layered lamination [0°]. In order to enhance our understanding of the vibration behaviors of laminated conical shells, some selected mode shapes and their corresponding frequency parameters $\Omega_{n,m}$ for the [0°/45°/0°/45°] laminated conical shell with F–C boundary condition are plotted in Fig. 6.2, where n and m denote the circumferential wave number and longitudinal mode number. These

Table 6.7 Frequency parameters Ω for laminated conical shells with various lamination schemes and boundary conditions ($E_1/E_2 = 15$, $R_0 = 1$ m, $L/R_0 = 2$, $h/R_0 = 0.1$, $\varphi = 45°$)

Structure element	Mode	Boundary conditions					
		F–S	S–F	F–C	C–F	S–S	C–C
[0°]	1	0.1652	0.0611	0.2024	0.0793	0.2399	0.3751
	2	0.1818	0.0674	0.2143	0.0855	0.2496	0.3817
	3	0.1946	0.0793	0.2366	0.0914	0.2534	0.3837
	4	0.2449	0.1063	0.2766	0.1165	0.2809	0.4033
[0°/45°]	1	0.2890	0.0903	0.3020	0.0972	0.3813	0.4632
	2	0.3219	0.1177	0.3453	0.1219	0.3919	0.4780
	3	0.3612	0.1185	0.3750	0.1254	0.4268	0.5002
	4	0.3995	0.1745	0.4371	0.1762	0.4588	0.5427
[0°/45°/0°]	1	0.2480	0.0898	0.2853	0.1088	0.3389	0.4710
	2	0.2592	0.1016	0.3010	0.1147	0.3476	0.4775
	3	0.3073	0.1150	0.3443	0.1328	0.3623	0.4882
	4	0.3171	0.1380	0.3681	0.1447	0.3873	0.5077
[0°/45°/0°/45°]	1	0.3180	0.1104	0.3427	0.1197	0.4224	0.5190
	2	0.3676	0.1306	0.3977	0.1420	0.4355	0.5270
	3	0.3728	0.1451	0.4013	0.1494	0.4645	0.5582
	4	0.4646	0.2090	0.5102	0.2107	0.4992	0.5814

mode shapes are constructed by means of considering the displacement field in Eq. (6.29) after solving the eigenvalue problem.

The vertex half-angle angle (φ) is a key parameter of a conical shell. The cylindrical shells and annular plates considered in previous chapters can be seen as the special cases of conical shells with zero and 90° ($\pi/2$) vertex half-angle angles, respectively. In the following example, influence of vertex half-angle angle (φ) on the vibration characteristics of laminated conical shells is investigated. Table 6.8 shows the lowest four frequency parameters Ω for a three-layered, cross-ply [0°/90°/0°] conical shell with different boundary conditions. The material constants and geometric parameters are the same as the previous example (see Table 6.7) except that the vertex half-angle angle (φ) of the shell is changed from $\varphi = 15°$ to $\varphi = 75°$ by a step of 15°. The table shows that the shell with C–F and S–F boundary conditions yields lower frequency parameters than those of F–C and F–S boundary conditions. The similar observation can be seen in Tables 6.6 and 6.7. The second observation made here is that the shell with C–F and S–F boundary conditions gave closer results. In addition, we can see that the maximum frequency parameters of the shell with F–S, S–F, F–C, C–F, S–S and C–C boundary conditions occur at $\varphi = 45°$, 15°, 60°, 15°, 45° and 75°, respectively.

Effects of the lamination layer number on the frequency parameters of conical shells are investigated as well. In Fig. 6.3, variation of the lowest three dimensionless frequencies Ω of a $[0°/\vartheta]_n$ layered conical shell with F–C boundary condition against the number of layers n are depicted, respectively (where $n = 1$ means

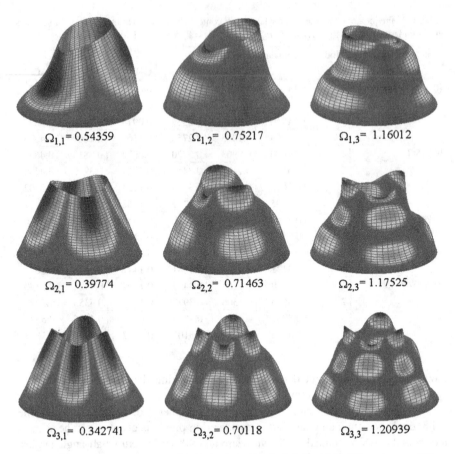

$\Omega_{1,1} = 0.54359$ $\Omega_{1,2} = 0.75217$ $\Omega_{1,3} = 1.16012$

$\Omega_{2,1} = 0.39774$ $\Omega_{2,2} = 0.71463$ $\Omega_{2,3} = 1.17525$

$\Omega_{3,1} = 0.342741$ $\Omega_{3,2} = 0.70118$ $\Omega_{3,3} = 1.20939$

Fig. 6.2 Mode shapes for a [0°/45°/0°/45°] laminated conical shell with F–C boundary condition ($\Omega_{n,m}$, $n = 1$–3, $m = 1$–3)

a single-layered shell; $n = 2$ means a two-layered shell, i.e. [0°/ϑ], and so forth). Three lamination schemes, i.e. $\vartheta = 30°$, 45° and 60° are considered in the investigation. The layers of the shells are of equal thickness and made from the same material with follow proprieties: $E_1/E_2 = 15$, $R_0 = 1$ m, $L/R_0 = 2$, $h/R_0 = 0.1$, $\varphi = 45°$. As clearly observed from Fig. 6.3, the frequency parameters of the shell increases rapidly and may reaches their crest around $n = 18$, and beyond this range, the frequency parameters remain unchanged (ignore the fluctuation). The fluctuation on the curves curves curved may be due to the fact that the shell are symmetrically laminated when n is an odd number, and n equal to an even number means the shells are unsymmetrically laminated.

The following numerical analysis is conducted to laminated conical shells with elastic boundary conditions. As what treated in previous chapter, the following two typical uniform elastic boundary conditions are considered in the subsequent analysis (taking edge $R = R_0$ for example):

Table 6.8 Frequency parameters Ω for a $[0°/90°/0°]$ laminated conical shells with different boundary conditions and vertex half-angle angles ($E_1/E_2 = 15$, $R_0 = 1$ m, $L/R_0 = 2$, $h/R_0 = 0.1$)

φ	Mode	Boundary conditions					
		F–S	S–F	F–C	C–F	S–S	C–C
15°	1	0.1504	0.0942	0.1671	0.1051	0.2360	0.3182
	2	0.1582	0.0946	0.1742	0.1087	0.2571	0.3350
	3	0.2094	0.1384	0.2270	0.1434	0.2719	0.3420
	4	0.2466	0.1483	0.2561	0.1568	0.3202	0.3861
30°	1	0.1936	0.0878	0.2226	0.1043	0.2841	0.3996
	2	0.2200	0.0942	0.2455	0.1139	0.2939	0.4078
	3	0.2421	0.1214	0.2761	0.1297	0.3318	0.4310
	4	0.3244	0.1513	0.3630	0.1635	0.3451	0.4474
45°	1	0.2165	0.0791	0.2588	0.1006	0.3066	0.4575
	2	0.2457	0.0843	0.2830	0.1099	0.3159	0.4648
	3	0.2625	0.1092	0.3136	0.1208	0.3506	0.4852
	4	0.3402	0.1394	0.4048	0.1563	0.3628	0.4991
60°	1	0.2131	0.0658	0.2699	0.0931	0.3031	0.4898
	2	0.2226	0.0678	0.2724	0.0970	0.3216	0.5019
	3	0.2683	0.0992	0.3355	0.1132	0.3269	0.5034
	4	0.3414	0.1105	0.3858	0.1332	0.3708	0.5359
75°	1	0.1582	0.0414	0.2200	0.0788	0.2806	0.4994
	2	0.1928	0.0566	0.2612	0.0841	0.2860	0.5026
	3	0.2107	0.0633	0.2626	0.0943	0.3129	0.5179
	4	0.2648	0.0920	0.3442	0.1072	0.3176	0.5213

E^1: the transverse direction is elastically restrained ($w \neq 0$, $u = v = \phi_x = \phi_\theta = 0$), i.e., $k_w = \Gamma$;

E^2: the rotation is elastically restrained ($\phi_x \neq 0$, $u = v = w = \phi_\theta = 0$), i.e., $K_x = \Gamma$.

Table 6.9 shows the lowest three frequency parameters Ω of a two-layered $[0°/90°]$ conical shell with different restrain parameters Γ and vertex half-angle angles. The vertex half-angle angle φ varies from 0° to 90° by a step of 30°. The shell parameters used are $E_1/E_2 = 15$, $R_0 = 1$ m, $L/R_0 = 2$, $h/R_0 = 0.05$. The shell is clamped at the edge of $R = R_1$ and with elastic boundary conditions at the other edge. The table shows that when the vertex half-angle angle $\varphi = 0°$, 30° and 60°, increasing restraint rigidities in the transverse and rotation directions have very limited effects on the frequency parameters of the shell. When the restrained rigidity parameter Γ is varied from $10^{-2} \times D$ to $10^4 \times D$, the corresponding maximum differences of the lowest frequency parameters for the shell with E^1–C and E^2–C boundary conditions are less than 1.36, 2.42, 0.85 % and 3.32, 4.45, 1.35 %, respectively. However, the similar frequency parameter differences reach 177, 60, 16 % and 21.4, 22. 3, 18.3 % for the shell with vertex half-angle angle of $\varphi = 90°$.

Fig. 6.3 Variation of the frequency parameters Ω versus the number of layers n for a $[0°/\vartheta]_n$ layered conical shell: **a** $\vartheta = 30°$; **b** $\vartheta = 45°$; **c** $\vartheta = 60°$

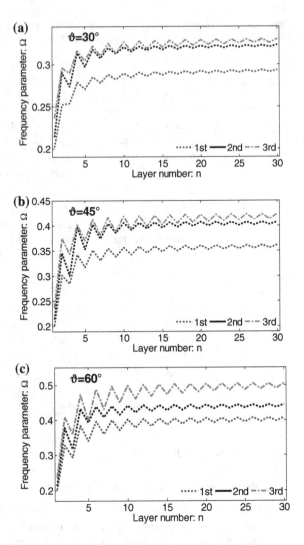

Table 6.10 shows the similar studies for the shell with F boundary conditions at the edge of $R = R_1$. The table also reveals that when the vertex half-angle angle $\varphi = 0°$, 30° and 60°, increasing the restraint rigidities in the transverse and rotation directions have very limited effects on the frequency parameters of the shell. In addition, it can be seen that the change of the restraint rigidity parameter Γ has large effects on the frequency parameters of all conical shells with vertex half-angle angle $\varphi = 90°$. Increasing the rigidity parameter Γ from $10^{-2} \times D$ to $10^4 \times D$ increases the frequency parameters by almost 651.4, 22.5, 36.5 % and 101.2, 54.6, 94.6 % for the E^1–F and E^2–F boundary conditions, respectively.

Table 6.9 Frequency parameters Ω for a two-layered [0°/90°] conical shell with different restrain parameters Γ and vertex half-angle angles ($R_0 = 1$ m, $L/R_0 = 2$, $h/R_0 = 0.05$, $E_1/E_2 = 15$)

φ	Γ	E^1–C			E^2–C		
		1	2	3	1	2	3
0°	$10^{-2} \times D$	0.1614	0.1800	0.1960	0.1610	0.1797	0.1955
	$10^{-1} \times D$	0.1614	0.1800	0.1960	0.1610	0.1798	0.1956
	$10^{0} \times D$	0.1614	0.1800	0.1960	0.1616	0.1802	0.1961
	$10^{1} \times D$	0.1615	0.1801	0.1961	0.1627	0.1809	0.1973
	$10^{2} \times D$	0.1619	0.1804	0.1965	0.1630	0.1812	0.1976
	$10^{3} \times D$	0.1627	0.1810	0.1973	0.1631	0.1812	0.1977
	$10^{4} \times D$	0.1630	0.1812	0.1977	0.1631	0.1812	0.1977
30°	$10^{-2} \times D$	0.2555	0.2732	0.2831	0.2533	0.2706	0.2822
	$10^{-1} \times D$	0.2555	0.2732	0.2832	0.2535	0.2708	0.2822
	$10^{0} \times D$	0.2556	0.2732	0.2831	0.2548	0.2726	0.2828
	$10^{1} \times D$	0.2557	0.2734	0.2832	0.2572	0.2760	0.2838
	$10^{2} \times D$	0.2564	0.2745	0.2834	0.2580	0.2771	0.2841
	$10^{3} \times D$	0.2576	0.2765	0.2840	0.2581	0.2772	0.2842
	$10^{4} \times D$	0.2581	0.2771	0.2841	0.2581	0.2772	0.2842
60°	$10^{-2} \times D$	0.2532	0.2577	0.2837	0.2485	0.2527	0.2813
	$10^{-1} \times D$	0.2533	0.2577	0.2838	0.2488	0.2532	0.2815
	$10^{0} \times D$	0.2532	0.2576	0.2839	0.2512	0.2565	0.2825
	$10^{1} \times D$	0.2534	0.2582	0.2834	0.2554	0.2622	0.2845
	$10^{2} \times D$	0.2548	0.2608	0.2843	0.2566	0.2638	0.2851
	$10^{3} \times D$	0.2563	0.2634	0.2850	0.2567	0.2640	0.2851
	$10^{4} \times D$	0.2567	0.2640	0.2851	0.2567	0.2640	0.2851
90°	$10^{-2} \times D$	0.0592	0.1036	0.1434	0.1351	0.1360	0.1406
	$10^{-1} \times D$	0.0614	0.1010	0.1436	0.1370	0.1379	0.1422
	$10^{0} \times D$	0.0793	0.1067	0.1439	0.1476	0.1489	0.1515
	$10^{1} \times D$	0.1364	0.1402	0.1521	0.1607	0.1628	0.1632
	$10^{2} \times D$	0.1608	0.1629	0.1633	0.1636	0.1659	0.1660
	$10^{3} \times D$	0.1637	0.1659	0.1660	0.1640	0.1663	0.1663
	$10^{4} \times D$	0.1640	0.1663	0.1663	0.1640	0.1663	0.1664

6.4   Vibration of Laminated Open Conical Shells

Laminated open conical shells can be obtained by cutting a segment of the laminated closed conical shells. For an open conical shell, the assumption of whole periodic wave numbers in the circumferential direction is inappropriate, and thus, a set of complete two-dimensional analysis is required and resort must be made to a full two-dimensional solution scheme. This forms a major deterrent so that the analyses of deep open conical shells have not been widely available. Among those available, Bardell et al. (1999) studied the vibration of a general three-layer conical

Table 6.10 Frequency parameters Ω for a two-layered $[0°/90°]$ conical shell with different restrain parameters Γ and vertex half-angle angles ($R_0 = 1$ m, $L/R_0 = 2$, $h/R_0 = 0.05$, $E_1/E_2 = 15$)

φ	Γ	E^1–F			E^2–F		
		1	2	3	1	2	3
0°	$10^{-2} \times D$	0.0904	0.0909	0.1403	0.0909	0.0912	0.1404
	$10^{-1} \times D$	0.0904	0.0909	0.1403	0.0909	0.0912	0.1404
	$10^{0} \times D$	0.0904	0.0909	0.1403	0.0909	0.0912	0.1404
	$10^{1} \times D$	0.0904	0.0909	0.1403	0.0909	0.0912	0.1404
	$10^{2} \times D$	0.0905	0.0910	0.1403	0.0909	0.0912	0.1404
	$10^{3} \times D$	0.0908	0.0912	0.1404	0.0909	0.0912	0.1404
	$10^{4} \times D$	0.0909	0.0912	0.1404	0.0909	0.0912	0.1404
30°	$10^{-2} \times D$	0.0831	0.1018	0.1025	0.0841	0.1027	0.1031
	$10^{-1} \times D$	0.0831	0.1018	0.1025	0.0841	0.1027	0.1031
	$10^{0} \times D$	0.0831	0.1019	0.1025	0.0841	0.1027	0.1033
	$10^{1} \times D$	0.0832	0.1019	0.1025	0.0842	0.1027	0.1037
	$10^{2} \times D$	0.0835	0.1024	0.1026	0.0843	0.1027	0.1038
	$10^{3} \times D$	0.0841	0.1027	0.1034	0.0843	0.1027	0.1038
	$10^{4} \times D$	0.0842	0.1027	0.1037	0.0843	0.1027	0.1038
60°	$10^{-2} \times D$	0.0617	0.0705	0.0801	0.0620	0.0710	0.0803
	$10^{-1} \times D$	0.0617	0.0705	0.0801	0.0620	0.0712	0.0803
	$10^{0} \times D$	0.0617	0.0706	0.0801	0.0625	0.0722	0.0803
	$10^{1} \times D$	0.0618	0.0709	0.0801	0.0631	0.0737	0.0804
	$10^{2} \times D$	0.0625	0.0724	0.0802	0.0632	0.0742	0.0804
	$10^{3} \times D$	0.0631	0.0739	0.0803	0.0633	0.0742	0.0804
	$10^{4} \times D$	0.0632	0.0742	0.0804	0.0633	0.0742	0.0804
90°	$10^{-2} \times D$	0.0030	0.0196	0.0232	0.0110	0.0156	0.0163
	$10^{-1} \times D$	0.0091	0.0197	0.0233	0.0127	0.0172	0.0175
	$10^{0} \times D$	0.0203	0.0224	0.0233	0.0182	0.0208	0.0250
	$10^{1} \times D$	0.0216	0.0236	0.0304	0.0216	0.0235	0.0306
	$10^{2} \times D$	0.0221	0.0240	0.0316	0.0221	0.0240	0.0316
	$10^{3} \times D$	0.0222	0.0240	0.0317	0.0222	0.0240	0.0317
	$10^{4} \times D$	0.0222	0.0240	0.0317	0.0222	0.0240	0.0317

sandwich panel based on the h-p version finite element method. Chern and Chao (2000) made a general survey and comparison for variety of simply supported shallow spherical, cylindrical, plate and saddle panels in rectangular planform. Lee et al. (2002) and Hu et al. (2002) reported the vibration characteristics of twisted cantilevered conical composite shells. Also, vibration of cantilevered laminated composite shallow conical shells was presented by Lim et al. (1998), etc.

In this section, we consider free vibration of laminated deep open conical shells. As was done previously for laminated open cylindrical shells, regardless of boundary conditions, each displacement and rotation component of the open conical shells under consideration is expanded as a two-dimensional modified Fourier series as Eq. (5.43). The similar non-dimensional parameter $\Omega = \omega R_1 \sqrt{\rho h / A_{11}}$ is used in the calculations. And unless otherwise stated, the layers of open conical shells under consideration are made of composite material with following properties: $E_2 = 10$ GPa, $E_1/E_2 =$ open, $\mu_{12} = 0.25$, $G_{12} = G_{13} = G_{23} = 0.5E_2$, $\rho = 1,500$ kg/m^3. For an open conical shell, there exits four boundaries, i.e., $R = R_0$, $R = R_1$, $\theta = 0$ and $\theta = \theta_0$. Similarly, the boundary condition of an open conical shell is represented by a four-letter character, such as FCSC identify the shell with F, C, S and C boundary conditions at boundaries $R = R_0$, $\theta = 0$, $R = R_1$ and $\theta = \theta_0$, respectively.

6.4.1 Convergence Studies and Result Verification

Tables 6.11 shows the convergence studies of the lowest six frequency parameters Ω for a three-layered, [0°/90°/0°] deep open conical shell with FFFF and CCCC boundary conditions, respectively. The material and geometry constants of the shell are: $E_1/E_2 = 15$, $R_0 = 1$ m, $L/R_0 = 2$, $h/R_0 = 0.1$, $\varphi = \pi/4$, $\theta_0 = \pi$. The zero frequency parameters corresponding to the rigid body modes of the shell with FFFF boundary conditions are omitted from the results. Excellent convergence of frequencies can be observed in the table. Furthermore, the convergence of the FFFF solutions is faster than those of CCCC boundary conditions.

Table 6.12 shows the comparison of the non-dimensional frequency parameters $\Omega = \omega L^2 \sqrt{\rho h / D}$ of SSSS and CCCC supported, [0°/90°/0°] laminated shallow and deep open conical shells with results provided by Ye et al. (2014b) based on the conjunction of Ritz method and the first-order shear deformable shell theory. The

Table 6.11 Convergence of the frequency parameters Ω of a [0°/90°/0°] laminated open conical shell with FFFF and CCCC boundary conditions ($R_0 = 1$ m, $L/R_0 = 2$, $h/R_0 = 0.1$, $\theta_0 = \pi$, $E_1/E_2 = 15$)

B.C.	$M \times N$	Mode number					
		1	2	3	4	5	6
FFFF	14 × 14	0.0181	0.0325	0.0485	0.0711	0.0886	0.1268
	15 × 15	0.0181	0.0325	0.0485	0.0711	0.0885	0.1268
	16 × 16	0.0181	0.0325	0.0485	0.0711	0.0885	0.1268
	17 × 17	0.0181	0.0325	0.0485	0.0711	0.0885	0.1268
	18 × 18	0.0181	0.0325	0.0484	0.0711	0.0885	0.1267
CCCC	14 × 14	0.4588	0.4616	0.5151	0.5259	0.5991	0.6611
	15 × 15	0.4588	0.4615	0.5150	0.5258	0.5987	0.6610
	16 × 16	0.4588	0.4615	0.5149	0.5257	0.5987	0.6602
	17 × 17	0.4588	0.4615	0.5149	0.5257	0.5986	0.6601
	18 × 18	0.4588	0.4615	0.5148	0.5257	0.5986	0.6597

Table 6.12 Comparison of frequency parameters $\Omega = \omega L^2 \sqrt{\rho h / D}$ for [0°/90°/0°] laminated shallow and deep open conical shells with SSSS and CCCC boundary conditions ($R_0 = 1$ m, $L/R_0 = 2$, $h/R_0 = 0.1$, $\varphi = \pi/4$, $E_1/E_2 = 15$)

θ_0	Mode	Ye et al. (2014b)		Present		Difference (%)	
		SSSS	CCCC	SSSS	CCCC	SSSS	CCCC
45°	1	29.953	37.831	29.956	37.831	0.010	0.002
	2	32.542	44.176	32.543	44.177	0.003	0.002
	3	50.005	58.363	50.006	58.364	0.003	0.001
	4	52.410	67.406	52.415	67.408	0.010	0.002
90°	1	16.702	24.103	16.704	24.104	0.012	0.002
	2	19.810	27.671	19.814	27.673	0.019	0.008
	3	29.910	37.898	29.918	37.900	0.027	0.003
	4	33.347	39.299	33.349	39.304	0.005	0.014
135°	1	15.281	22.565	15.284	22.566	0.019	0.004
	2	16.536	22.888	16.537	22.889	0.010	0.004
	3	19.943	26.779	19.949	26.782	0.027	0.010
	4	22.977	30.186	22.989	30.197	0.049	0.035
180°	1	14.820	21.855	14.823	21.856	0.023	0.002
	2	15.129	21.983	15.129	21.985	0.005	0.010
	3	17.444	24.519	17.454	24.525	0.058	0.022
	4	19.367	25.038	19.375	25.042	0.038	0.018
225°	1	14.571	21.526	14.572	21.527	0.006	0.002
	2	14.588	21.667	14.592	21.670	0.028	0.012
	3	16.470	23.078	16.480	23.078	0.065	0.002
	4	16.915	23.431	16.916	23.438	0.003	0.030

shell parameters used in the comparison are the same as those for Table 6.11. Five different circumferential included angles, i.e., $\theta_0 = 45°$, 90°, 135°, 180° and 225°, corresponding to shallow to deep open conical shells are performed in the comparison. It is clearly evident that the present solutions are in a good agreement with the referential data, although different admissible displacement functions were employed by Ye et al. (2014b). The differences between the two results are very small, and do not exceed 0.065 % for the worst case.

6.4.2   Laminated Open Conical Shells with General Boundary Conditions

Some further numerical results for laminated open conical shells with different boundary conditions and shell parameters, such as geometric properties, lamination schemes are given in the subsequent discussions.

Isotropic open conical shells are a special case of the laminated ones. Vibration results of these shells with general boundary conditions are rare in the literature as well. Therefore, Table 6.13 shows the first four frequency parameters Ω of an isotropic ($E = 210$ GPa, $\mu_{12} = 0.3$, $\rho = 7{,}800$ kg/m^3) open conical shell with different types of boundary conditions (FFFC, FFCC, FCCC, CCCF, CCFF and CFFF) and thickness-to-small edge radius ratios (h/R_0). The geometry parameters used in the analysis are: $R_0 = 1$ m, $L = 2$ m, $\theta_0 = 90°$, $\varphi = 30°$. Four different thickness-to-small edge radius ratios, i.e., $h/R_0 = 0.01, 0.02, 0.05$ and 0.1 are used in the calculation. The table shows that the frequency parameters of the shell tend to increase with thickness-to-small edge radius ratio increases. This is true due to the stiffness of an isotropic open conical shell increases with thickness increases. Furthermore, it is interesting to find that the frequency parameters of the shell with cantilever boundary conditions in the curved edge are higher than those of straight edges.

The first six mode shapes for the shell with thickness-to-small edge radius ratio $h/R_0 = 0.01$ and FCCC boundary condition are given in Fig. 6.4. These 3-D view mode shapes serve to enhance our understanding of the vibratory characteristics of the open conical shell. Table 6.14 shows the similar studies for a two-layered [0°/ 90°] open conical shell. The layers of the shell are made of composite material with following properties: $E_1/E_2 = 15$. Table 6.14 also shows that the frequency parameters of laminated open conical shells increase with thickness-to-small edge radius ratio increases.

Table 6.13 Frequency parameters Ω of an isotropic open conical shell with various boundary conditions and thickness-to-radius ratios ($R_0 = 1$ m, $L = 2$ m, $\theta_0 = 90°$, $\varphi = 30°$)

h/R_0	Mode	Boundary conditions					
		FFFC	FFCC	FCCC	CCCF	CCFF	CFFF
0.01	1	0.0037	0.0590	0.2277	0.0996	0.0263	0.0240
	2	0.0100	0.1212	0.2537	0.1907	0.0738	0.0281
	3	0.0172	0.1788	0.3233	0.2479	0.0995	0.0622
	4	0.0345	0.2172	0.3411	0.2769	0.1060	0.0908
0.02	1	0.0074	0.0872	0.3525	0.1501	0.0393	0.0360
	2	0.0198	0.1837	0.3722	0.3004	0.1052	0.0393
	3	0.0342	0.2540	0.4800	0.3473	0.1571	0.1018
	4	0.0681	0.3293	0.5113	0.4079	0.1707	0.1143
0.05	1	0.0184	0.1470	0.5520	0.2695	0.0682	0.0510
	2	0.0489	0.3335	0.6682	0.5239	0.1861	0.0820
	3	0.0842	0.4356	0.8394	0.5964	0.2820	0.1307
	4	0.1629	0.5654	0.9446	0.6877	0.2874	0.2357
0.10	1	0.0366	0.2299	0.8543	0.4357	0.1096	0.0817
	2	0.0953	0.5164	0.9780	0.7747	0.2542	0.1080
	3	0.1652	0.6100	1.2614	0.9499	0.4405	0.2062
	4	0.2996	0.8463	1.2658	1.0000	0.4729	0.4047

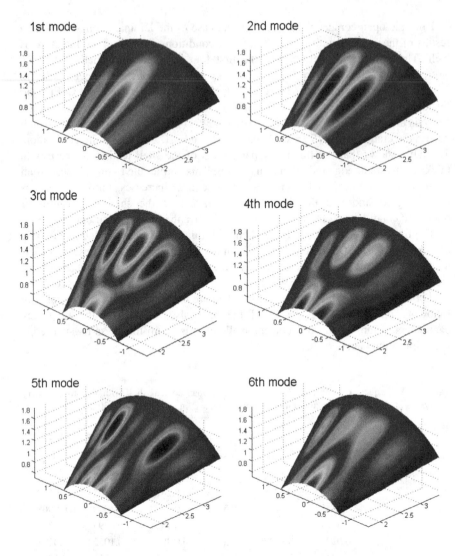

Fig. 6.4 Mode shapes for an isotropic open conical shell with FCCC boundary conditions

Table 6.14 Frequency parameters Ω of a two-layered [0°/90°] open conical shell with various boundary conditions and thickness-to-radius ratios ($R_0 = 1$ m, $L = 2$ m, $\theta_0 = 90°$, $\varphi = 30°$, $E_1/E_2 = 15$)

h/R_0	Mode	Boundary conditions					
		FFFC	FFCC	FCCC	CCCF	CCFF	CFFF
0.01	1	0.0024	0.0412	0.1344	0.0628	0.0187	0.0173
	2	0.0052	0.0765	0.1466	0.1147	0.0498	0.0200
	3	0.0102	0.1158	0.1899	0.1394	0.0616	0.0438
	4	0.0202	0.1307	0.1906	0.1556	0.0698	0.0577
0.02	1	0.0047	0.0578	0.1979	0.0906	0.0267	0.0256
	2	0.0102	0.1123	0.2039	0.1760	0.0684	0.0262
	3	0.0201	0.1552	0.2693	0.1926	0.0942	0.0654
	4	0.0399	0.1934	0.2941	0.2173	0.1112	0.0758
0.05	1	0.0118	0.0914	0.3067	0.1583	0.0441	0.0336
	2	0.0254	0.1990	0.3915	0.2830	0.1135	0.0524
	3	0.0491	0.2533	0.4713	0.3519	0.1591	0.0824
	4	0.0953	0.3186	0.5573	0.3917	0.1742	0.1454
0.10	1	0.0234	0.1377	0.4786	0.2556	0.0690	0.0468
	2	0.0498	0.3080	0.6516	0.4231	0.1442	0.0742
	3	0.0946	0.3353	0.6981	0.5271	0.2666	0.1182
	4	0.1735	0.4728	0.8905	0.5919	0.2746	0.2194

At the end of this section, the influence of circumferential included angle θ_0 on the frequency parameters of laminated open conical shells is investigated. A two-layered [0°/90°] open conical shell with small edge radius $R_0 = 1$ m, length $L = 2$ m, thickness $h = 0.05$ m is investigated. The layers of the shell are made of composite material having orthotropy ratio $E_1/E_2 = 15$. Three different vertex half-angle angles, i.e., $\varphi = 0°$, 45° and 90°, corresponding to open cylindrical shell, general open conical shell and sectorial plate are performed in the investigation. Figure 6.5 shows the variation of the lowest three frequency parameters Ω of the shell with SSSS boundary condition against the circumferential included angle θ_0. As observed from the figure, the frequency parameter traces of the shell decline when the circumferential included angle θ_0 is varied from 30° to 330° by a step of 15°. It is attributed to the stiffness of the shell decreases with circumferential included angle θ_0 increases. Furthermore, it is obvious that the effect of the

Fig. 6.5 Variation of frequency parameters Ω versus θ_0 for [0°/90°] laminated open conical shells with SSSS boundary condition: **a** $\varphi = 0°$; **b** $\varphi = 45°$; **c** $\varphi = 90°$

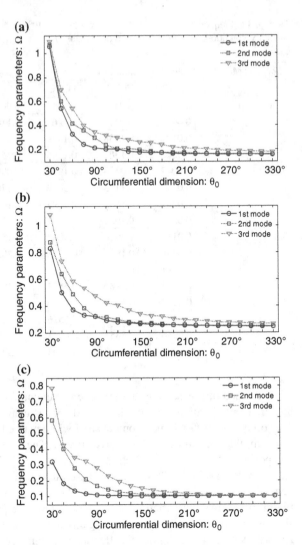

circumferential included angle θ_0 is much higher for open cylindrical, conical shells and sectorial plates with lower circumferential included angle. Changing θ_0 from 30° to 90° results in frequency parameters that are more than five times lower. Conversely, increasing θ_0 from 90° to 330° yields less than 30 % decrement of the frequency parameters.

Fig. 6.6 Variation of
frequency parameters Ω
versus θ_0 for [0°/90°]
laminated open conical shells
with CCCC boundary
condition: **a** $\varphi = 0$°; **b** $\varphi = 45$°;
c $\varphi = 90$°

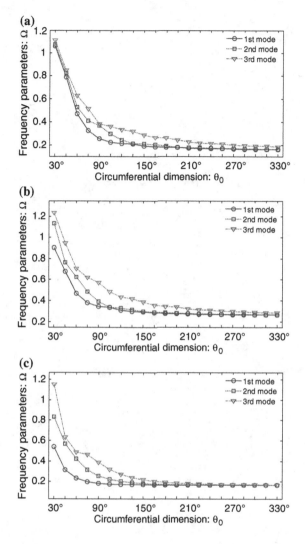

Figure 6.6 shows the similar studies for the conical shell with CCCC boundary
condition. The similar observations can be seen in this figure.

Chapter 7
Spherical Shells

The cylindrical and conical shells considered in Chaps. 5 and 6 are special cases of shells of revolution. Spherical shells are another special case of shells of revolution. A spherical shell is a doubly-curved shell characterized by a middle surface generated by the rotation of a circular cure line segment (generator) about a fixed axis. If the axis of rotation along the diameter of the circle of the line segment, a spherical shell with constant curvature in the meridional and circumferential directions will be resulted and the two radii of curvature are equal. It is noticeable that the spherical shells are very stiff for both in-plane and bending loads due to the curvature of the middle surface, which is also a reason for the analysis difficulties of the shells, especially the exact three-dimensional elasticity (3-D) analysis. The spherical shells can be closed and open. If the generator rotates less than one full revolution about the axis, the spherical shell is open and has four boundaries. If further, the generator rotates one full revolution about the axis and the proper continuity conditions are satisfied along the junction line, a closed spherical shell results, which has only two edges.

Laminated spherical shells composed of advance composite material layers are another important structural components widely used in an increasing number of engineering structures to satisfy special functional requirements. Understanding the vibration characteristics of these structural elements is particularly important for engineers to design suitable structures with low vibration and noise radiation characteristics. However, the literature of vibration analysis of laminated spherical shells is limited. Among those available, Qatu (2004) presented the fundamental equations of thin and thick spherical shells. Gautham and Ganesan (1997) dealt with the free vibration characteristics of isotropic and laminated orthotropic spherical caps using semi-analytical shell finite element. Sai Ram and Sreedhar Babu (2002) applied an eight-node degenerated isoparametric shell element method to investigate the free vibration of composite spherical shell cap with and without a cutout. Qu et al. (2013a) developed a unified formulation for vibration analysis of composite laminated cylindrical, conical and spherical shells including shear deformation and rotary inertia, in which a modified variational principle in conjunction with a multi-segment partitioning technique is employed to derive the formulation based on the first-order shear deformation theory. Jin et al. (2014a) investigated vibrations of laminated spherical shells with general boundary conditions. Free vibration of

© Science Press, Beijing and Springer-Verlag Berlin Heidelberg 2015
G. Jin et al., *Structural Vibration*, DOI 10.1007/978-3-662-46364-2_7

simply supported laminated spherical panels with random material properties is reported by Singh et al. (2001) based on the high-order shear deformation shallow theory. Ye et al. (2014b, c) developed a unified Chebyshev–Ritz formulation for the vibration analysis of thin and thick laminated open cylindrical, conical and spherical shells with arbitrary boundary conditions, etc.

The emphasis in this chapter is the closed and deep open spherical shells composed of layers having spherical orthotropy. Fundamental equations of thin (CST) and thick (SDST) spherical shells are presented in the first and second sections, respectively, including the strain-displacement relations, force and moment resultants, energy functions, governing equations and boundary condition equations. These equations are obtained by substituting the proper Lamé parameters of spherical shells in the general shell equations. On the basis of shear deformation shell theory (SDST) and modified Fourier series, numerous vibration results of thin and thick laminated closed spherical shells with different boundary conditions, lamination schemes and geometry parameters are given in the third section by applying the weak form solution procedure. Vibration of deep open laminated spherical shells will then be treated in the latest section of this chapter.

Closed spherical shells can be viewed as a special case of open spherical shells having circumferential included angles of 2π (or 360°) and the proper continuity conditions are satisfied along the junction line. For the sake of brevity, a general laminated open spherical shell with mean radius R and total thickness h is selected as the analysis model. As shown in Fig. 7.1, the geometry and dimensions of the shell are defined with respect to the coordinates φ, θ and z along the meridional, circumferential and radial directions which is located in the middle surface of the shell. The shell domain is bounded by $\varphi_0 < \varphi < \varphi_1$, $0 < \theta < \theta_0$ and $h/2 < z < h/2$. The middle surface displacements of the spherical shell in the φ, θ and z directions are denoted by u, v and w, respectively.

Consider the spherical shell in Fig. 7.1 and its spherical coordinate system, the coordinates, characteristics of the Lamé parameters and radii of curvatures are:

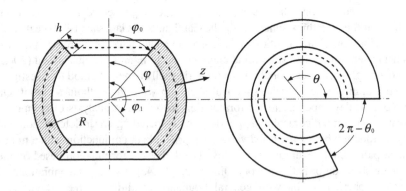

Fig. 7.1 Geometry notations and coordinate system of spherical shells

$$\alpha = \varphi, \quad \beta = \theta, \quad A = R, \quad B = R\sin\varphi, \quad R_\alpha = R_\beta = R \qquad (7.1)$$

The equations of spherical shells are a specialization of the general shell equations given in Chap. 1 by substituting Eq. (7.1) into such shell equations.

7.1 Fundamental Equations of Thin Laminated Spherical Shells

We will first derive the fundamental equations for thin laminated spherical shells by substituting Eq. (7.1) into the general thin shell equations developed in Sect. 1.2. The equations are formulated for the general dynamic analysis. It can be readily specialized to static (letting frequency ω equal to zero) and free vibration (neglecting the external load) analysis.

7.1.1 Kinematic Relations

Substituting Eq. (7.1) into Eq. (1.7) yields the middle surface strains and curvature changes for a thin spherical shell in terms of middle surface displacements as:

$$\varepsilon_\varphi^0 = \frac{1}{R}\left(\frac{\partial u}{\partial\varphi} + w\right) \quad \varepsilon_\theta^0 = \frac{1}{R}\left(\frac{u}{t} + \frac{\partial v}{s\partial\theta} + w\right)$$

$$\gamma_{\varphi\theta}^0 = \frac{1}{R}\left(\frac{\partial u}{s\partial\theta} + \frac{\partial v}{\partial\varphi} - \frac{v}{t}\right) \quad \chi_\varphi = \frac{1}{R^2}\left(\frac{\partial u}{\partial\varphi} - \frac{\partial^2 w}{\partial\varphi^2}\right)$$

$$\chi_\theta = \frac{1}{R^2}\left(\frac{u}{t} + \frac{\partial v}{s\partial\theta} - \frac{\partial^2 w}{s^2\partial\theta^2} - \frac{\partial w}{t\partial\varphi}\right) \qquad (7.2)$$

$$\chi_{\varphi\theta} = \frac{1}{R^2}\left(\frac{\partial u}{s\partial\theta} + \frac{\partial v}{\partial\varphi} - \frac{v}{t} - \frac{2\partial^2 w}{s\partial\varphi\partial\theta} + \frac{\partial w}{st\partial\theta}\right)$$

and $s = \sin\varphi \quad c = \cos\varphi \quad t = \tan\varphi$

where ε_φ^0, ε_θ^0 and $\gamma_{\varphi\theta}^0$ are the middle surface normal and shear strains. χ_φ, χ_θ and $\chi_{\varphi\theta}$ denote the middle surface curvature and twist changes. Then, the state of strain at an arbitrary point in the kth layer of the thin spherical shell can be defined as:

$$\varepsilon_\varphi = \varepsilon_\varphi^0 + z\chi_\varphi$$

$$\varepsilon_\theta = \varepsilon_\theta^0 + z\chi_\theta \qquad (7.3)$$

$$\gamma_{\varphi\theta} = \gamma_{\varphi\theta}^0 + z\chi_{\varphi\theta}$$

where $Z_{k+1} < z < Z_k$ (Z_{k+1} and Z_k denote the distances from the top surface and bottom surface of the layer to the referenced middle surface, respectively).

7.1.2 Stress-Strain Relations and Stress Resultants

According to the generalized Hooke's law, the corresponding stress-strain relations in the kth layer of the thin spherical shells are:

$$\left\{ \begin{array}{c} \sigma_\varphi \\ \sigma_\theta \\ \tau_{\varphi\theta} \end{array} \right\}_k = \left[\begin{array}{ccc} \overline{Q}_{11}^k & \overline{Q}_{12}^k & \overline{Q}_{16}^k \\ \overline{Q}_{12}^k & \overline{Q}_{22}^k & \overline{Q}_{26}^k \\ \overline{Q}_{16}^k & \overline{Q}_{26}^k & \overline{Q}_{66}^k \end{array} \right] \left\{ \begin{array}{c} \varepsilon_\varphi \\ \varepsilon_\theta \\ \gamma_{\varphi\theta} \end{array} \right\}_k \tag{7.4}$$

where σ_φ and σ_θ are the normal stresses in the φ and θ directions, $\tau_{\varphi\theta}$ is the shear stress. The constants $\overline{Q}_{ij}^k (i, j = 1, 2, 6)$ represent the elastic properties of the material of the layer. They are written as in Eq. (1.12). It should be noted that the orthotropy treated here is spherical orthotropy.

Integrating the stresses over the shell thickness or substituting Eq. (7.1) into Eq. (1.14), the force and moment resultants of thin spherical shells can be obtained in terms of the middle surface strains and curvature changes as

$$\left[\begin{array}{c} N_\varphi \\ N_\theta \\ N_{\varphi\theta} \\ M_\varphi \\ M_\theta \\ M_{\varphi\theta} \end{array} \right] = \left[\begin{array}{cccccc} A_{11} & A_{12} & A_{16} & B_{11} & B_{12} & B_{16} \\ A_{12} & A_{22} & A_{26} & B_{12} & B_{22} & B_{26} \\ A_{16} & A_{26} & A_{66} & B_{16} & B_{26} & B_{66} \\ B_{11} & B_{12} & B_{16} & D_{11} & D_{12} & D_{16} \\ B_{12} & B_{22} & B_{26} & D_{12} & D_{22} & D_{26} \\ B_{16} & B_{26} & B_{66} & D_{16} & D_{26} & D_{66} \end{array} \right] \left[\begin{array}{c} \varepsilon_\varphi^0 \\ \varepsilon_\theta^0 \\ \gamma_{\varphi\theta}^0 \\ \chi_\varphi \\ \chi_\theta \\ \chi_{\varphi\theta} \end{array} \right] \tag{7.5}$$

where N_φ and N_θ are the normal force resultants in the φ and θ directions, $N_{\varphi\theta}$ is the corresponding shear force resultants. M_φ, M_θ and $M_{\varphi\theta}$ denote the bending and twisting moment resultants. The stiffness coefficients A_{ij}, B_{ij}, and D_{ij} are written as in (1.15).

7.1.3 Energy Functions

The strain energy (U_s) of thin laminated spherical shells during vibration can be defined as:

$$U_s = \frac{1}{2} \int_\varphi \int_\theta \left\{ \begin{array}{c} N_\varphi \varepsilon_\varphi^0 + N_\theta \varepsilon_\theta^0 + N_{\varphi\theta} \gamma_{\varphi\theta}^0 \\ + M_\varphi \chi_\varphi + M_\theta \chi_\theta + M_{\varphi\theta} \chi_{\varphi\theta} \end{array} \right\} R^2 s d\theta d\varphi \tag{7.6}$$

The corresponding kinetic energy function is:

$$T = \frac{1}{2} \int_\varphi \int_\theta I_0 \left\{ \left(\frac{\partial u}{\partial t}\right)^2 + \left(\frac{\partial v}{\partial t}\right)^2 + \left(\frac{\partial w}{\partial t}\right)^2 \right\} R^2 s d\theta d\varphi \qquad (7.7)$$

where the inertia term I_0 is defined as in Eq. (1.19). Suppose q_φ, q_θ and q_z are the external loads in the φ, θ and z directions, respectively. Thus, the external work can be expressed as:

$$W_e = \int_\varphi \int_\theta \{ q_\varphi u + q_\theta v + q_z w \} R^2 s d\theta d\varphi \qquad (7.8)$$

The same as usual, the general boundary conditions of a thin spherical shell are implemented by using the artificial spring boundary technique (see Sect. 1.2.3). Letting k_ψ^u, k_ψ^v, k_ψ^w and K_ψ^w ($\psi = \varphi 0$, $\theta 0$, $\varphi 1$ and $\theta 1$) to indicate the stiffness of the boundary springs at the boundaries $\varphi = \varphi_0$, $\theta = 0$, $\varphi = \varphi_1$ and $\theta = \theta_0$, respectively, therefore, the deformation strain energy stored in the boundary springs (U_{sp}) during vibration can be defined as:

$$U_{sp} = \frac{1}{2} \int_\theta \left\{ \begin{array}{l} \left[k_{\varphi 0}^u u^2 + k_{\varphi 0}^v v^2 + k_{\varphi 0}^w w^2 + K_{\varphi 0}^w (\partial w / R \partial \varphi) \right] |_{\varphi = \varphi_0} \\ + \left[k_{\varphi 1}^u u^2 + k_{\varphi 1}^v v^2 + k_{\varphi 1}^w w^2 + K_{\varphi 1}^w (\partial w / R \partial \varphi) \right] |_{\varphi = \varphi_1} \end{array} \right\} R s d\theta$$
$$+ \frac{1}{2} \int_\varphi \left\{ \begin{array}{l} \left[k_{\theta 0}^u u^2 + k_{\theta 0}^v v^2 + k_{\theta 0}^w w^2 + K_{\theta 0}^w (\partial w / R \partial \varphi) \right] |_{\theta = 0} \\ + \left[k_{\theta 1}^u u^2 + k_{\theta 1}^v v^2 + k_{\theta 1}^w w^2 + K_{\theta 1}^w (\partial w / R \partial \varphi) \right] |_{\theta = \theta_0} \end{array} \right\} R d\varphi \qquad (7.9)$$

The energy functions presented here can be applied to derive the governing equations and boundary conditions of thin laminated spherical shells or search the approximate solutions of practical problems.

7.1.4 Governing Equations and Boundary Conditions

Substituting Eq. (7.1) into Eq. (1.28) and then simplifying the expressions, the resulting governing equations of thin laminated spherical shells are:

$$\frac{\partial N_\varphi}{\partial \varphi} + \frac{N_\varphi}{t} + \frac{\partial N_{\varphi\theta}}{s \partial \theta} - \frac{N_\theta}{t} + Q_\varphi + R q_\varphi = R I_0 \frac{\partial^2 u}{\partial t^2}$$

$$\frac{\partial N_{\varphi\theta}}{\partial \varphi} + \frac{2 N_{\varphi\theta}}{t} + \frac{\partial N_\theta}{s \partial \theta} + Q_\theta + R q_\theta = R I_0 \frac{\partial^2 v}{\partial t^2} \qquad (7.10)$$

$$- (N_\varphi + N_\theta) + \frac{\partial Q_\varphi}{\partial \varphi} + \frac{Q_\varphi}{t} + \frac{\partial Q_\theta}{s \partial \theta} + R q_z = R I_0 \frac{\partial^2 w}{\partial t^2}$$

where

$$Q_\varphi = \frac{\partial M_\varphi}{R\partial\varphi} + \frac{M_\varphi}{Rt} - \frac{M_\theta}{Rt} + \frac{\partial M_{\varphi\theta}}{Rs\partial\theta}$$

$$Q_\theta = \frac{\partial M_{\varphi\theta}}{R\partial\varphi} + \frac{2M_{\varphi\theta}}{Rt} + \frac{\partial M_\theta}{Rs\partial\theta} \tag{7.11}$$

According to Eqs. (1.29) and (1.30), the boundary conditions of thin laminated spherical shells are:

$$\varphi = \varphi_0 : \begin{cases} N_\varphi + \frac{M_\varphi}{R} - k^u_{\varphi 0} u = 0 \\ N_{\varphi\theta} + \frac{M_{\varphi\theta}}{R} - k^v_{\varphi 0} v = 0 \\ Q_\varphi + \frac{\partial M_{\varphi\theta}}{RS\partial\theta} - k^w_{\varphi 0} w = 0 \\ -M_\varphi - K^w_{\varphi 0}\frac{\partial w}{R\partial\varphi} = 0 \end{cases} \qquad \varphi = \varphi_1 : \begin{cases} N_\varphi + \frac{M_\varphi}{R} + k^u_{\varphi 1} u = 0 \\ N_{\varphi\theta} + \frac{M_{\varphi\theta}}{R} + k^v_{\varphi 1} v = 0 \\ Q_\varphi + \frac{\partial M_{\varphi\theta}}{Rs\partial\theta} + k^w_{\varphi 1} w = 0 \\ -M_\varphi + K^w_{\varphi 1}\frac{\partial w}{R\partial\varphi} = 0 \end{cases}$$

$$\theta = 0 : \begin{cases} N_{\varphi\theta} + \frac{M_{\varphi\theta}}{R} - k^u_{\theta 0} u = 0 \\ N_\theta + \frac{M_\theta}{R} - k^v_{\theta 0} v = 0 \\ Q_\theta + \frac{\partial M_{\varphi\theta}}{R\partial\varphi} - k^w_{\theta 0} w = 0 \\ -M_\theta - K^w_{\theta 0}\frac{\partial w}{Rs\partial\theta} = 0 \end{cases} \qquad \theta = \theta_0 : \begin{cases} N_{\varphi\theta} + \frac{M_{\varphi\theta}}{R} + k^u_{\theta 1} u = 0 \\ N_\theta + \frac{M_\theta}{R} + k^v_{\theta 1} v = 0 \\ Q_\theta + \frac{\partial M_{\varphi\theta}}{R\partial\varphi} + k^w_{\theta 1} w = 0 \\ -M_\theta + K^w_{\theta 1}\frac{\partial w}{Rs\partial\theta} = 0 \end{cases} \tag{7.12}$$

For thin spherical shells, each boundary can have up to 12 possible classical boundary conditions. The possible classical boundary conditions for boundaries $\varphi = $ constant are given in Table 7.1, similar boundary conditions can be obtained for boundaries $\theta = $ constant.

Table 7.1 Possible classical boundary conditions for thin spherical shells at each boundary of $\varphi = $ constant

Boundary type	Conditions
Free boundary conditions	
F	$N_\varphi + \frac{M_\varphi}{R} = N_{\varphi\theta} + \frac{M_{\varphi\theta}}{R} = Q_\varphi + \frac{\partial M_{\varphi\theta}}{Rs\partial\theta} = M_\varphi = 0$
F2	$u = N_{\varphi\theta} + \frac{M_{\varphi\theta}}{R} = Q_\varphi + \frac{\partial M_{\varphi\theta}}{Rs\partial\theta} = M_\varphi = 0$
F3	$N_\varphi + \frac{M_\varphi}{R} = v = Q_\varphi + \frac{\partial M_{\varphi\theta}}{Rs\partial\theta} = M_\varphi = 0$
F4	$u = v = Q_\varphi + \frac{\partial M_{\varphi\theta}}{Rs\partial\theta} = M_\varphi = 0$
Simply supported boundary conditions	
S	$u = v = w = M_\varphi = 0$
SD	$N_\varphi + \frac{M_\varphi}{R} = v = w = M_\varphi = 0$
S3	$u = N_{\varphi\theta} + \frac{M_{\varphi\theta}}{R} = w = M_\varphi = 0$
S4	$N_\varphi + \frac{M_\varphi}{R} = N_{\varphi\theta} + \frac{M_{\varphi\theta}}{R} = w = M_\varphi = 0$
Clamped boundary conditions	
C	$u = v = w = \frac{\partial w}{\partial\varphi} = 0$
C2	$N_\varphi + \frac{M_\varphi}{R} = v = w = \frac{\partial w}{\partial\varphi} = 0$
C3	$u = N_{\varphi\theta} + \frac{M_{\varphi\theta}}{R} = w = \frac{\partial w}{\partial\varphi} = 0$
C4	$N_\varphi + \frac{M_\varphi}{R} = N_{\varphi\theta} + \frac{M_{\varphi\theta}}{R} = w = \frac{\partial w}{\partial\varphi} = 0$

7.2 Fundamental Equations of Thick Laminated Spherical Shells

Fundamental equations of thick laminated spherical shells will be derived in this section. Similarly, the equations that follow are a specialization of the general shear deformation shell theory (SDST).

7.2.1 Kinematic Relations

Based on the assumptions of SDST, the displacement field of a thick spherical shell can be expressed in terms of the middle surface displacements and rotation components as

$$
\begin{aligned}
U(\varphi, \theta, z) &= u(\varphi, \theta) + z\phi_\varphi \\
V(\varphi, \theta, z) &= v(\varphi, \theta) + z\phi_\theta \\
W(\varphi, \theta, z) &= w(\varphi, \theta)
\end{aligned}
\tag{7.13}
$$

where u, v and w are the middle surface shell displacements in the φ, θ and z directions, respectively, and ϕ_φ and ϕ_θ represent the rotations of the transverse normal respect to θ- and φ-axes. Substituting Eq. (7.1) into Eqs. (1.33) and (1.34), the normal and shear strains at any point of the thick spherical shell space can be defined in terms of the middle surface strains and curvature changes as

$$
\begin{aligned}
\varepsilon_\varphi &= \frac{1}{(1 + z/R)} \left(\varepsilon_\varphi^0 + z\chi_\varphi \right) \\
\varepsilon_\theta &= \frac{1}{(1 + z/R)} \left(\varepsilon_\theta^0 + z\chi_\theta \right) \\
\gamma_{\varphi\theta} &= \frac{1}{(1 + z/R)} \left(\gamma_{\varphi\theta}^0 + z\chi_{\varphi\theta} + \gamma_{\varphi\theta}^0 + z\chi_{\varphi\theta} \right) \\
\gamma_{\varphi z} &= \frac{\gamma_{\varphi z}^0}{(1 + z/R)} \\
\gamma_{\theta z} &= \frac{\gamma_{\theta z}^0}{(1 + z/R)}
\end{aligned}
\tag{7.14}
$$

where

$$\varepsilon_\varphi^0 = \frac{1}{R}\left(\frac{\partial u}{\partial \varphi} + w\right) \qquad \chi_\varphi = \frac{1}{R}\frac{\partial \phi_\varphi}{\partial \varphi}$$

$$\varepsilon_\theta^0 = \frac{1}{R}\left(\frac{u}{t} + \frac{\partial v}{s\partial \theta} + w\right) \qquad \chi_\theta = \frac{1}{R}\left(\frac{\partial \phi_\theta}{s\partial \theta} + \frac{\phi_\varphi}{t}\right)$$

$$\gamma_{\varphi\theta}^0 = \frac{1}{R}\frac{\partial v}{\partial \varphi} \qquad \chi_{\varphi\theta} = \frac{1}{R}\frac{\partial \phi_\theta}{\partial \varphi} \tag{7.15}$$

$$\gamma_{\theta\varphi}^0 = \frac{1}{R}\left(\frac{\partial u}{s\partial \theta} - \frac{v}{t}\right) \qquad \chi_{\theta\varphi} = \frac{1}{R}\left(\frac{\partial \phi_\varphi}{s\partial \theta} - \frac{\phi_\theta}{t}\right)$$

$$\gamma_{\varphi z}^0 = \frac{1}{R}\frac{\partial w}{\partial \varphi} - \frac{u}{R} + \phi_\varphi$$

$$\gamma_{\theta z}^0 = \frac{1}{Rs}\frac{\partial w}{\partial \theta} - \frac{v}{R} + \phi_\theta$$

7.2.2 Stress-Strain Relations and Stress Resultants

Substituting $\alpha = \varphi$ and $\beta = \theta$ into Eq. (1.35), the corresponding stress-strain relations in the kth layer of a thick spherical shell can be written as:

$$\begin{Bmatrix} \sigma_\varphi \\ \sigma_\theta \\ \tau_{\theta z} \\ \tau_{\varphi z} \\ \tau_{\varphi\theta} \end{Bmatrix}_k = \begin{bmatrix} Q_{11}^k & Q_{12}^k & 0 & 0 & Q_{16}^k \\ Q_{12}^k & Q_{22}^k & 0 & 0 & Q_{26}^k \\ 0 & 0 & Q_{44}^k & Q_{45}^k & 0 \\ 0 & 0 & Q_{45}^k & Q_{55}^k & 0 \\ Q_{16}^k & Q_{26}^k & 0 & 0 & Q_{66}^k \end{bmatrix} \begin{Bmatrix} \varepsilon_\varphi \\ \varepsilon_\theta \\ \gamma_{\theta z} \\ \gamma_{\varphi z} \\ \gamma_{\varphi\theta} \end{Bmatrix}_k \tag{7.16}$$

where $\tau_{\varphi z}$ and $\tau_{\theta z}$ are the corresponding shear stress components. The elastic stiffness coefficients $\overline{Q_{ij}^k}$ are given as in Eq. (1.39). Substituting Eq. (7.1) into (1.40), the force and moment resultants of a thick spherical shell can be obtained by following integration operation:

$$\begin{bmatrix} N_\varphi \\ N_{\varphi\theta} \\ Q_\varphi \end{bmatrix} = \int_{-h/2}^{h/2} \begin{bmatrix} \sigma_\varphi \\ \tau_{\varphi\theta} \\ \tau_{\varphi z} \end{bmatrix}\left(1 + \frac{z}{R}\right)dz, \qquad \begin{bmatrix} N_\theta \\ N_{\theta\varphi} \\ Q_\theta \end{bmatrix} = \int_{-h/2}^{h/2} \begin{bmatrix} \sigma_\theta \\ \tau_{\theta\varphi} \\ \tau_{\theta z} \end{bmatrix}\left(1 + \frac{z}{R}\right)dz$$

$$\begin{bmatrix} M_\varphi \\ M_{\varphi\theta} \end{bmatrix} = \int_{-h/2}^{h/2} \begin{bmatrix} \sigma_\varphi \\ \tau_{\varphi\theta} \end{bmatrix}\left(1 + \frac{z}{R}\right)zdz, \qquad \begin{bmatrix} M_\theta \\ M_{\theta\varphi} \end{bmatrix} = \int_{-h/2}^{h/2} \begin{bmatrix} \sigma_\theta \\ \tau_{\theta\varphi} \end{bmatrix}\left(1 + \frac{z}{R}\right)zdz \tag{7.17}$$

The deepness term $(1 + z/R)$ exist in both the numerator and denominator of the force and moment resultants, this will yields symmetric stress resultants ($N_{\varphi\theta} = N_{\theta\varphi}$

and $M_{\varphi\theta} = M_{\theta\varphi}$) and effectively reduce the equations to those of plates with the proper change of the coordinates. Performing the integration operation in Eq. (7.17) yields the force and moment resultants in terms of the middle surface strains and curvature changes as:

$$
\begin{bmatrix} N_{\varphi} \\ N_{\theta} \\ N_{\varphi\theta} \\ N_{\theta\varphi} \\ M_{\varphi} \\ M_{\theta} \\ M_{\varphi\theta} \\ M_{\theta\varphi} \end{bmatrix} = \begin{bmatrix} A_{11} & A_{12} & A_{16} & A_{16} & B_{11} & B_{12} & B_{16} & B_{16} \\ A_{12} & A_{22} & A_{26} & A_{26} & B_{12} & B_{22} & B_{26} & B_{26} \\ A_{16} & A_{26} & A_{66} & A_{66} & B_{16} & B_{26} & B_{66} & B_{66} \\ A_{16} & A_{26} & A_{66} & A_{66} & B_{16} & B_{26} & B_{66} & B_{66} \\ B_{11} & B_{12} & B_{16} & B_{16} & D_{11} & D_{12} & D_{16} & D_{16} \\ B_{12} & B_{22} & B_{26} & B_{26} & D_{12} & D_{22} & D_{26} & D_{26} \\ B_{16} & B_{26} & B_{66} & B_{66} & D_{16} & D_{26} & D_{66} & D_{66} \\ B_{16} & B_{26} & B_{66} & B_{66} & D_{16} & D_{26} & D_{66} & D_{66} \end{bmatrix} \begin{bmatrix} \varepsilon_{\varphi}^{0} \\ \varepsilon_{\theta}^{0} \\ \gamma_{\varphi\theta}^{0} \\ \gamma_{\theta\varphi}^{0} \\ \chi_{\varphi} \\ \chi_{\theta} \\ \chi_{\varphi\theta} \\ \chi_{\theta\varphi} \end{bmatrix} \tag{7.18}
$$

$$
\begin{bmatrix} Q_{\theta} \\ Q_{\varphi} \end{bmatrix} = \begin{bmatrix} A_{44} & A_{45} \\ A_{45} & A_{55} \end{bmatrix} \begin{bmatrix} \gamma_{\theta z}^{0} \\ \gamma_{\varphi z}^{0} \end{bmatrix}
$$

The stiffness coefficients A_{ij}, B_{ij} and D_{ij} are given as in Eq. (1.43). Note that when a spherical shell is symmetrically laminated with respect to its middle surface, the constants B_{ij} equal to zero, and, hence the equations are much simplified.

7.2.3 Energy Functions

The strain energy (U_s) of thick spherical shells during vibration can be defined in terms of the middle surface strains, curvature changes and corresponding stress resultants as

$$
U_s = \frac{1}{2} \int_{\varphi} \int_{\theta} \left\{ \begin{array}{l} N_{\varphi}\varepsilon_{\varphi}^{0} + N_{\theta}\varepsilon_{\theta}^{0} + N_{\varphi\theta}\gamma_{\varphi\theta}^{0} + N_{\theta\varphi}\gamma_{\theta\varphi}^{0} + M_{\varphi}\chi_{\varphi} \\ + M_{\theta}\chi_{\theta} + M_{\varphi\theta}\chi_{\varphi\theta} + M_{\theta\varphi}\chi_{\theta\varphi} + Q_{\varphi}\gamma_{\varphi z} + Q_{\theta}\gamma_{\theta z} \end{array} \right\} R^2 s d\theta d\varphi
$$

$$
\tag{7.19}
$$

Substituting Eqs. (7.17) and (7.20) into Eq. (7.21), the above equation can be expressed in terms of the middle surface displacements (u, v, w) and rotation components (ϕ_{φ}, ϕ_{θ}). The corresponding kinetic energy (T) of thick spherical shells during vibration can be written as:

$$
T = \frac{R^2}{2} \int\limits_{\varphi} \int\limits_{\theta} \int\limits_{-h/2}^{h/2} \rho_z \left\{ \begin{array}{l} \left(\dfrac{\partial u}{\partial t} + z \dfrac{\partial \phi_\varphi}{\partial t} \right)^2 \\[2mm] + \left(\dfrac{\partial v}{\partial t} + z \dfrac{\partial \phi_\theta}{\partial t} \right)^2 + \left(\dfrac{\partial w}{\partial t} \right)^2 \end{array} \right\} \left(1 + \frac{z}{R} \right)^2 s d\theta d\varphi dz
$$

$$
= \frac{R^2}{2} \int\limits_{\varphi} \int\limits_{\theta} \left\{ \begin{array}{l} \overline{I_0} \left(\dfrac{\partial u}{\partial t} \right)^2 + 2\overline{I_1} \dfrac{\partial u}{\partial t} \dfrac{\partial \phi_\varphi}{\partial t} + \overline{I_2} \left(\dfrac{\partial \phi_\varphi}{\partial t} \right)^2 + \overline{I_0} \left(\dfrac{\partial v}{\partial t} \right)^2 \\[3mm] + 2\overline{I_1} \dfrac{\partial v}{\partial t} \dfrac{\partial \phi_\theta}{\partial t} + \overline{I_2} \left(\dfrac{\partial \phi_\theta}{\partial t} \right)^2 + \overline{I_0} \left(\dfrac{\partial w}{\partial t} \right)^2 \end{array} \right\} s d\theta d\varphi \qquad (7.20)
$$

where the inertia terms are:

$$
\overline{I_0} = I_0 + \frac{2I_1}{R} + \frac{I_2}{R^2}
$$

$$
\overline{I_1} = I_1 + \frac{2I_2}{R} + \frac{I_3}{R^2}
$$

$$
\overline{I_2} = I_2 + \frac{2I_3}{R} + \frac{I_4}{R^2} \qquad (7.21)
$$

$$
[I_0, I_1, I_2, I_3, I_4] = \sum_{k=1}^{N} \int\limits_{z_k}^{z_{k+1}} \rho^k [1, z, z^2, z^3, z^4] dz
$$

in which ρ^k is the mass of the kth layer per unit middle surface area. Suppose the distributed external forces q_φ, q_θ and q_z are in the φ, θ and z directions, respectively, and m_φ and m_θ represent the external couples in the middle surface, thus, the work done by the external loads can be written as

$$
W_e = \int\limits_{\varphi} \int\limits_{\theta} \{ q_\varphi u + q_\theta v + q_z w + m_\varphi \phi_\varphi + m_\theta \phi_\theta \} R^2 s d\theta d\varphi. \qquad (7.22)
$$

Using the artificial spring boundary technique similar to that described earlier, the deformation strain energy (U_{sp}) of the boundary springs during vibration is:

$$
\begin{aligned}
U_{sp} = &\frac{1}{2} \int\limits_{\theta} \left\{ \begin{array}{l} s\left[k_{\varphi0}^u u^2 + k_{\varphi0}^v v^2 + k_{\varphi0}^w w^2 + K_{\varphi0}^\varphi \phi_\varphi^2 + K_{\varphi0}^\theta \phi_\theta^2 \right]\big|_{\varphi=\varphi_0} \\[2mm] + s\left[k_{\varphi1}^u u^2 + k_{\varphi1}^v v^2 + k_{\varphi1}^w w^2 + K_{\varphi1}^\varphi \phi_\varphi^2 + K_{\varphi1}^\theta \phi_\theta^2 \right]\big|_{\varphi=\varphi_1} \end{array} \right\} R d\theta \\[3mm]
&+ \frac{1}{2} \int\limits_{\varphi} \left\{ \begin{array}{l} \left[k_{\theta0}^u u^2 + k_{\theta0}^v v^2 + k_{\theta0}^w w^2 + K_{\theta0}^\varphi \phi_\varphi^2 + K_{\theta0}^\theta \phi_\theta^2 \right]\big|_{\theta=0} \\[2mm] + \left[k_{\theta1}^u u^2 + k_{\theta1}^v v^2 + k_{\theta1}^w w^2 + K_{\theta1}^\varphi \phi_\varphi^2 + K_{\theta1}^\theta \phi_\theta^2 \right]\big|_{\theta=\theta_0} \end{array} \right\} R d\varphi \qquad (7.23)
\end{aligned}
$$

where k^u_ψ, k^v_ψ, k^w_ψ, K^φ_ψ and $K^\theta_\psi (\psi = \varphi 0, \theta 0, \varphi 1$ and $\theta 1)$ represent the rigidities (per unit length) of the boundary springs at the boundaries $\varphi = \varphi_0$, $\theta = 0$, $\varphi = \varphi_1$ and $\theta = \theta_0$, respectively.

The energy functions presented here will be applied to search the approximate solutions of free vibration problems of spherical shells with general boundary conditions in the later sections.

7.2.4 Governing Equations and Boundary Conditions

Substituting Eq. (7.1) into Eq. (1.59), the governing equations of thick laminated spherical shells can be specialized from those of general thick shells. The resulting equations become

$$
\frac{\partial N_\varphi}{\partial \varphi} + \frac{N_\varphi}{t} - \frac{N_\theta}{t} + \frac{\partial N_{\theta\varphi}}{s\partial\theta} + Q_\varphi + Rq_\varphi = R\overline{I_0}\frac{\partial^2 u}{\partial t^2} + R\overline{I_1}\frac{\partial^2 \phi_\varphi}{\partial t^2}
$$

$$
\frac{\partial N_\theta}{s\partial\theta} + \frac{\partial N_{\varphi\theta}}{\partial\varphi} + \frac{N_{\varphi\theta}}{t} + \frac{N_{\theta\varphi}}{t} + Q_\theta + Rq_\theta = R\overline{I_0}\frac{\partial^2 v}{\partial t^2} + R\overline{I_1}\frac{\partial^2 \phi_\theta}{\partial t^2}
$$

$$
-(N_\varphi + N_\theta) + \frac{\partial Q_\varphi}{\partial\varphi} + \frac{Q_\varphi}{t} + \frac{\partial Q_\theta}{s\partial\theta} + Rq_z = R\overline{I_0}\frac{\partial^2 w}{\partial t^2} \qquad (7.24)
$$

$$
\frac{\partial M_\varphi}{\partial\varphi} + \frac{M_\varphi}{t} - \frac{M_\theta}{t} + \frac{\partial M_{\theta\varphi}}{s\partial\theta} - RQ_\varphi + Rm_\varphi = R\overline{I_1}\frac{\partial^2 u}{\partial t^2} + R\overline{I_2}\frac{\partial^2 \phi_\varphi}{\partial t^2}
$$

$$
\frac{\partial M_\theta}{s\partial\theta} + \frac{\partial M_{\varphi\theta}}{\partial\varphi} + \frac{M_{\varphi\theta}}{t} + \frac{M_{\theta\varphi}}{t} - RQ_\theta + Rm_\theta = R\overline{I_1}\frac{\partial^2 v}{\partial t^2} + R\overline{I_2}\frac{\partial^2 \phi_\theta}{\partial t^2}
$$

Substituting Eqs. (7.15) and (7.18) into Eq. (7.24), the governing equations of thick spherical shells can be written in terms of displacements as

$$
\left(\begin{bmatrix} L_{11} & L_{12} & L_{13} & L_{14} & L_{15} \\ L_{21} & L_{22} & L_{23} & L_{24} & L_{25} \\ L_{31} & L_{32} & L_{33} & L_{34} & L_{35} \\ L_{41} & L_{42} & L_{43} & L_{44} & L_{45} \\ L_{51} & L_{52} & L_{53} & L_{54} & L_{55} \end{bmatrix} - \omega^2 \begin{bmatrix} M_{11} & 0 & 0 & M_{14} & 0 \\ 0 & M_{22} & 0 & 0 & M_{25} \\ 0 & 0 & M_{33} & 0 & 0 \\ M_{41} & 0 & 0 & M_{44} & 0 \\ 0 & M_{52} & 0 & 0 & M_{55} \end{bmatrix}\right)\begin{bmatrix} u \\ v \\ w \\ \phi_\varphi \\ \phi_\theta \end{bmatrix} = R\begin{bmatrix} -p_\varphi \\ -p_\theta \\ -p_z \\ -m_\varphi \\ -m_\theta \end{bmatrix}
$$

$$(7.25)$$

The coefficients of the linear operator L_{ij} are

$$L_{11} = \frac{\partial}{\partial\varphi}\left(\frac{A_{11}}{R}\frac{\partial}{\partial\varphi} + \frac{A_{12}}{Rt} + \frac{A_{16}}{Rs}\frac{\partial}{\partial\theta}\right) + \frac{1}{t}\left(\frac{A_{11}}{R}\frac{\partial}{\partial\varphi} + \frac{A_{12}}{Rt} + \frac{A_{16}}{Rs}\frac{\partial}{\partial\theta}\right)$$

$$- \frac{1}{t}\left(\frac{A_{12}}{R}\frac{\partial}{\partial\varphi} + \frac{A_{22}}{Rt} + \frac{A_{26}}{Rs}\frac{\partial}{\partial\theta}\right) + \frac{\partial}{s\partial\theta}\left(\frac{A_{16}}{R}\frac{\partial}{\partial\varphi} + \frac{A_{26}}{Rt} + \frac{A_{66}}{Rs}\frac{\partial}{\partial\theta}\right) - \frac{A_{55}}{R}$$

$$L_{12} = \frac{\partial}{\partial\varphi}\left(\frac{A_{12}}{Rs}\frac{\partial}{\partial\theta} + \frac{A_{16}}{R}\frac{\partial}{\partial\varphi} - \frac{A_{16}}{Rt}\right) + \frac{1}{t}\left(\frac{A_{12}}{Rs}\frac{\partial}{\partial\theta} + \frac{A_{16}}{R}\frac{\partial}{\partial\varphi} - \frac{A_{16}}{Rt}\right)$$

$$- \frac{1}{t}\left(\frac{A_{22}}{Rs}\frac{\partial}{\partial\theta} + \frac{A_{26}}{R}\frac{\partial}{\partial\varphi} - \frac{A_{26}}{Rt}\right) + \frac{\partial}{s\partial\theta}\left(\frac{A_{26}}{Rs}\frac{\partial}{\partial\theta} + \frac{A_{66}}{R}\frac{\partial}{\partial\varphi} - \frac{A_{66}}{Rt}\right) - \frac{A_{45}}{R}$$

$$L_{13} = \frac{\partial}{\partial\varphi}\left(\frac{A_{11}}{R} + \frac{A_{12}}{R}\right) + \frac{1}{t}\left(\frac{A_{11}}{R} + \frac{A_{12}}{R}\right) - \frac{1}{t}\left(\frac{A_{12}}{R} + \frac{A_{22}}{R}\right)$$

$$+ \frac{\partial}{s\partial\theta}\left(\frac{A_{16}}{R} + \frac{A_{26}}{R}\right) + \left(\frac{A_{45}}{Rs}\frac{\partial}{\partial\theta} + \frac{A_{55}}{R}\frac{\partial}{\partial\varphi}\right)$$

$$L_{14} = \frac{\partial}{\partial\varphi}\left(\frac{B_{11}}{R}\frac{\partial}{\partial\varphi} + \frac{B_{12}}{Rt} + \frac{B_{16}}{Rs}\frac{\partial}{\partial\theta}\right) + \frac{1}{t}\left(\frac{B_{11}}{R}\frac{\partial}{\partial\varphi} + \frac{B_{12}}{Rt} + \frac{B_{16}}{Rs}\frac{\partial}{\partial\theta}\right)$$

$$- \frac{1}{t}\left(\frac{B_{12}}{R}\frac{\partial}{\partial\varphi} + \frac{B_{22}}{Rt} + \frac{B_{26}}{Rs}\frac{\partial}{\partial\theta}\right) + \frac{\partial}{s\partial\theta}\left(\frac{B_{16}}{R}\frac{\partial}{\partial\varphi} + \frac{B_{26}}{Rt} + \frac{B_{66}}{Rs}\frac{\partial}{\partial\theta}\right) + A_{55}$$

$$L_{15} = \frac{\partial}{\partial\varphi}\left(\frac{B_{12}}{Rs}\frac{\partial}{\partial\theta} + \frac{B_{16}}{R}\frac{\partial}{\partial\varphi} - \frac{B_{16}}{Rt}\right) + \frac{1}{t}\left(\frac{B_{12}}{Rs}\frac{\partial}{\partial\theta} + \frac{B_{16}}{R}\frac{\partial}{\partial\varphi} - \frac{B_{16}}{Rt}\right)$$

$$- \frac{1}{t}\left(\frac{B_{22}}{Rs}\frac{\partial}{\partial\theta} + \frac{B_{26}}{R}\frac{\partial}{\partial\varphi} - \frac{B_{26}}{Rt}\right) + \frac{\partial}{s\partial\theta}\left(\frac{B_{26}}{Rs}\frac{\partial}{\partial\theta} + \frac{B_{66}}{R}\frac{\partial}{\partial\varphi} - \frac{B_{66}}{Rt}\right) + A_{45}$$

$$L_{21} = \frac{\partial}{s\partial\theta}\left(\frac{A_{12}}{R}\frac{\partial}{\partial\varphi} + \frac{A_{22}}{Rt} + \frac{A_{26}}{Rs}\frac{\partial}{\partial\theta}\right) + \frac{\partial}{\partial\varphi}\left(\frac{A_{16}}{R}\frac{\partial}{\partial\varphi} + \frac{A_{26}}{Rt} + \frac{A_{66}}{Rs}\frac{\partial}{\partial\theta}\right)$$

$$+ \frac{2}{t}\left(\frac{A_{16}}{R}\frac{\partial}{\partial\varphi} + \frac{A_{26}}{Rt} + \frac{A_{66}}{Rs}\frac{\partial}{\partial\theta}\right) - \frac{A_{45}}{R}$$

$$L_{22} = \frac{\partial}{s\partial\theta}\left(\frac{A_{22}}{Rs}\frac{\partial}{\partial\theta} + \frac{A_{26}}{R}\frac{\partial}{\partial\varphi} - \frac{A_{26}}{Rt}\right) + \frac{\partial}{\partial\varphi}\left(\frac{A_{26}}{Rs}\frac{\partial}{\partial\theta} + \frac{A_{66}}{R}\frac{\partial}{\partial\varphi} - \frac{A_{66}}{Rt}\right)$$

$$+ \frac{2}{t}\left(\frac{A_{26}}{Rs}\frac{\partial}{\partial\theta} + \frac{A_{66}}{R}\frac{\partial}{\partial\varphi} - \frac{A_{66}}{Rt}\right) - \frac{A_{44}}{R}$$

$$L_{23} = \frac{\partial}{s\partial\theta}\left(\frac{A_{12}}{R} + \frac{A_{22}}{R}\right) + \frac{\partial}{\partial\varphi}\left(\frac{A_{16}}{R} + \frac{A_{26}}{R}\right) + \frac{2}{t}\left(\frac{A_{16}}{R} + \frac{A_{26}}{R}\right)$$

$$+ \left(\frac{A_{44}}{Rs}\frac{\partial}{\partial\theta} + \frac{A_{45}}{R}\frac{\partial}{\partial\varphi}\right)$$

$$L_{24} = \frac{\partial}{s\partial\theta}\left(\frac{B_{12}}{R}\frac{\partial}{\partial\varphi} + \frac{B_{22}}{Rt} + \frac{B_{26}}{Rs}\frac{\partial}{\partial\theta}\right) + \frac{\partial}{\partial\varphi}\left(\frac{B_{16}}{R}\frac{\partial}{\partial\varphi} + \frac{B_{26}}{Rt} + \frac{B_{66}}{Rs}\frac{\partial}{\partial\theta}\right)$$

$$+ \frac{2}{t}\left(\frac{B_{16}}{R}\frac{\partial}{\partial\varphi} + \frac{B_{26}}{Rt} + \frac{B_{66}}{Rs}\frac{\partial}{\partial\theta}\right) + A_{45}$$

$$L_{25} = \frac{\partial}{s\partial\theta}\left(\frac{B_{22}}{Rs}\frac{\partial}{\partial\theta} + \frac{B_{26}}{R}\frac{\partial}{\partial\varphi} - \frac{B_{26}}{Rt}\right) + \frac{\partial}{\partial\varphi}\left(\frac{B_{26}}{Rs}\frac{\partial}{\partial\theta} + \frac{B_{66}}{R}\frac{\partial}{\partial\varphi} - \frac{B_{66}}{Rt}\right)$$

$$+ \frac{2}{t}\left(\frac{B_{26}}{Rs}\frac{\partial}{\partial\theta} + \frac{B_{66}}{R}\frac{\partial}{\partial\varphi} - \frac{B_{66}}{Rt}\right) + A_{44}$$

$$L_{31} = -\left(\frac{A_{11}}{R}\frac{\partial}{\partial\varphi} + \frac{A_{12}}{Rt} + \frac{A_{16}}{Rs}\frac{\partial}{\partial\theta}\right) - \left(\frac{A_{12}}{R}\frac{\partial}{\partial\varphi} + \frac{A_{22}}{Rt} + \frac{A_{26}}{Rs}\frac{\partial}{\partial\theta}\right)$$

$$- \frac{A_{55}}{R}\frac{\partial}{\partial\varphi} + \frac{A_{55}}{Rt} - \frac{A_{45}}{Rs}\frac{\partial}{\partial\theta}$$

$$L_{32} = -\left(\frac{A_{12}}{Rs}\frac{\partial}{s\partial\theta} + \frac{A_{16}}{R}\frac{\partial}{\partial\varphi} - \frac{A_{16}}{Rt}\right) - \left(\frac{A_{22}}{Rs}\frac{\partial}{s\partial\theta} + \frac{A_{26}}{R}\frac{\partial}{\partial\varphi} - \frac{A_{26}}{Rt}\right)$$

$$- \frac{A_{45}}{R}\frac{\partial}{\partial\varphi} - \frac{A_{45}}{Rt} - \frac{A_{44}}{Rs}\frac{\partial}{\partial\theta}$$

$$L_{33} = -\left(\frac{A_{11}}{R} + \frac{A_{12}}{R}\right) - \left(\frac{A_{12}}{R} + \frac{A_{22}}{R}\right) + \frac{\partial}{\partial\varphi}\left(\frac{A_{45}}{Rs}\frac{\partial}{\partial\theta} + \frac{A_{55}}{R}\frac{\partial}{\partial\varphi}\right)$$

$$+ \frac{1}{t}\left(\frac{A_{45}}{Rs}\frac{\partial}{\partial\theta} + \frac{A_{55}}{R}\frac{\partial}{\partial\varphi}\right) + \frac{\partial}{s\partial\theta}\left(\frac{A_{44}}{Rs}\frac{\partial}{\partial\theta} + \frac{A_{45}}{R}\frac{\partial}{\partial\varphi}\right)$$

$$L_{34} = -\left(\frac{B_{11}}{R}\frac{\partial}{\partial\varphi} + \frac{B_{12}}{Rt} + \frac{B_{16}}{Rs}\frac{\partial}{\partial\theta}\right) - \left(\frac{B_{12}}{R}\frac{\partial}{\partial\varphi} + \frac{B_{22}}{Rt} + \frac{B_{26}}{Rs}\frac{\partial}{\partial\theta}\right)$$

$$+ A_{55}\frac{\partial}{\partial\varphi} + \frac{A_{55}}{t} + A_{45}\frac{\partial}{s\partial\theta}$$

$$L_{35} = -\left(\frac{B_{12}}{Rs}\frac{\partial}{\partial\theta} + \frac{B_{16}}{R}\frac{\partial}{\partial\varphi} - \frac{B_{16}}{Rt}\right) - \left(\frac{B_{22}}{Rs}\frac{\partial}{\partial\theta} + \frac{B_{26}}{R}\frac{\partial}{\partial\varphi} - \frac{B_{26}}{Rt}\right)$$

$$+ A_{45}\frac{\partial}{\partial\varphi} + \frac{A_{45}}{t} + A_{44}\frac{\partial}{s\partial\theta}$$

$$L_{41} = \frac{\partial}{\partial\varphi}\left(\frac{B_{11}}{R}\frac{\partial}{\partial\varphi} + \frac{B_{12}}{Rt} + \frac{B_{16}}{Rs}\frac{\partial}{\partial\theta}\right) + \frac{1}{t}\left(\frac{B_{11}}{R}\frac{\partial}{\partial\varphi} + \frac{B_{12}}{Rt} + \frac{B_{16}}{Rs}\frac{\partial}{\partial\theta}\right)$$

$$- \frac{1}{t}\left(\frac{B_{12}}{R}\frac{\partial}{\partial\varphi} + \frac{B_{22}}{Rt} + \frac{B_{26}}{Rs}\frac{\partial}{\partial\theta}\right) + \frac{\partial}{s\partial\theta}\left(\frac{B_{16}}{R}\frac{\partial}{\partial\varphi} + \frac{B_{26}}{Rt} + \frac{B_{66}}{Rs}\frac{\partial}{\partial\theta}\right) + A_{55}$$

$$L_{42} = \frac{\partial}{\partial\varphi}\left(\frac{B_{12}}{Rs}\frac{\partial}{s\partial\theta} + \frac{B_{16}}{R}\frac{\partial}{\partial\varphi} - \frac{B_{16}}{Rt}\right) + \frac{1}{t}\left(\frac{B_{12}}{Rs}\frac{\partial}{s\partial\theta} + \frac{B_{16}}{R}\frac{\partial}{\partial\varphi} - \frac{B_{16}}{Rt}\right)$$

$$- \frac{1}{t}\left(\frac{B_{22}}{Rs}\frac{\partial}{s\partial\theta} + \frac{B_{26}}{R}\frac{\partial}{\partial\varphi} - \frac{B_{26}}{Rt}\right) + \frac{\partial}{s\partial\theta}\left(\frac{B_{26}}{Rs}\frac{\partial}{s\partial\theta} + \frac{B_{66}}{R}\frac{\partial}{\partial\varphi} - \frac{B_{66}}{Rt}\right) + A_{45}$$

$$L_{43} = \frac{\partial}{\partial\varphi}\left(\frac{B_{11}}{R} + \frac{B_{12}}{R}\right) + \frac{1}{t}\left(\frac{B_{11}}{R} + \frac{B_{12}}{R}\right) - \frac{1}{t}\left(\frac{B_{12}}{R} + \frac{B_{22}}{R}\right)$$

$$+ \frac{\partial}{s\partial\theta}\left(\frac{B_{16}}{R} + \frac{B_{26}}{R}\right) - \left(\frac{A_{45}}{s}\frac{\partial}{\partial\theta} + A_{55}\frac{\partial}{\partial\varphi}\right)$$

$$L_{44} = \frac{\partial}{\partial\varphi}\left(\frac{D_{11}}{R}\frac{\partial}{\partial\varphi} + \frac{D_{12}}{Rt} + \frac{D_{16}}{Rs}\frac{\partial}{\partial\theta}\right) + \frac{1}{t}\left(\frac{D_{11}}{R}\frac{\partial}{\partial\varphi} + \frac{D_{12}}{Rt} + \frac{D_{16}}{Rs}\frac{\partial}{\partial\theta}\right)$$

$$- \frac{1}{t}\left(\frac{D_{12}}{R}\frac{\partial}{\partial\varphi} + \frac{D_{22}}{Rt} + \frac{D_{26}}{Rs}\frac{\partial}{\partial\theta}\right) + \frac{\partial}{s\partial\theta}\left(\frac{D_{16}}{R}\frac{\partial}{\partial\varphi} + \frac{D_{26}}{Rt} + \frac{D_{66}}{Rs}\frac{\partial}{\partial\theta}\right) - RA_{55}$$

$$L_{45} = \frac{\partial}{\partial \varphi}\left(\frac{D_{12}}{Rs}\frac{\partial}{\partial \theta} + \frac{D_{16}}{R}\frac{\partial}{\partial \varphi} - \frac{D_{16}}{Rt}\right) + \frac{1}{t}\left(\frac{D_{12}}{Rs}\frac{\partial}{\partial \theta} + \frac{D_{16}}{R}\frac{\partial}{\partial \varphi} - \frac{D_{16}}{Rt}\right)$$
$$- \frac{1}{t}\left(\frac{D_{22}}{Rs}\frac{\partial}{\partial \theta} + \frac{D_{26}}{R}\frac{\partial}{\partial \varphi} - \frac{D_{26}}{Rt}\right) + \frac{\partial}{s\partial \theta}\left(\frac{D_{26}}{Rs}\frac{\partial}{\partial \theta} + \frac{D_{66}}{R}\frac{\partial}{\partial \varphi} - \frac{D_{66}}{Rt}\right) - RA_{45}$$

$$L_{51} = \frac{\partial}{s\partial \theta}\left(\frac{B_{12}}{R}\frac{\partial}{\partial \varphi} + \frac{B_{22}}{Rt} + \frac{B_{26}}{Rs}\frac{\partial}{\partial \theta}\right) + \frac{\partial}{\partial \varphi}\left(\frac{B_{16}}{R}\frac{\partial}{\partial \varphi} + \frac{B_{26}}{Rt} + \frac{B_{66}}{Rs}\frac{\partial}{\partial \theta}\right)$$
$$+ \frac{2}{t}\left(\frac{B_{16}}{R}\frac{\partial}{\partial \varphi} + \frac{B_{26}}{Rt} + \frac{B_{66}}{Rs}\frac{\partial}{\partial \theta}\right) + A_{45}$$

$$L_{52} = \frac{\partial}{s\partial \theta}\left(\frac{B_{22}}{Rs}\frac{\partial}{s\partial \theta} + \frac{B_{26}}{R}\frac{\partial}{\partial \varphi} - \frac{B_{26}}{Rt}\right) + \frac{\partial}{\partial \varphi}\left(\frac{B_{26}}{Rs}\frac{\partial}{s\partial \theta} + \frac{B_{66}}{R}\frac{\partial}{\partial \varphi} - \frac{B_{66}}{Rt}\right)$$
$$+ \frac{2}{t}\left(\frac{B_{26}}{Rs}\frac{\partial}{s\partial \theta} + \frac{B_{66}}{R}\frac{\partial}{\partial \varphi} - \frac{B_{66}}{Rt}\right) + A_{44}$$

$$L_{53} = \frac{\partial}{s\partial \theta}\left(\frac{B_{12}}{R} + \frac{B_{22}}{R}\right) + \frac{\partial}{\partial \varphi}\left(\frac{B_{16}}{R} + \frac{B_{26}}{R}\right) + \frac{2}{t}\left(\frac{B_{16}}{R} + \frac{B_{26}}{R}\right) \qquad (7.26)$$
$$- \left(\frac{A_{44}}{s}\frac{\partial}{\partial \theta} + A_{45}\frac{\partial}{\partial \varphi}\right)$$

$$L_{54} = \frac{\partial}{s\partial \theta}\left(\frac{D_{12}}{R}\frac{\partial}{\partial \varphi} + \frac{D_{22}}{Rt} + \frac{D_{26}}{Rs}\frac{\partial}{\partial \theta}\right) + \frac{\partial}{\partial \varphi}\left(\frac{D_{16}}{R}\frac{\partial}{\partial \varphi} + \frac{D_{26}}{Rt} + \frac{D_{66}}{Rs}\frac{\partial}{\partial \theta}\right)$$
$$+ \frac{2}{t}\left(\frac{D_{16}}{R}\frac{\partial}{\partial \varphi} + \frac{D_{26}}{Rt} + \frac{D_{66}}{Rs}\frac{\partial}{\partial \theta}\right) - RA_{45}$$

$$L_{55} = \frac{\partial}{s\partial \theta}\left(\frac{D_{22}}{Rs}\frac{\partial}{\partial \theta} + \frac{D_{26}}{R}\frac{\partial}{\partial \varphi} - \frac{D_{26}}{Rt}\right) + \frac{\partial}{\partial \varphi}\left(\frac{D_{26}}{Rs}\frac{\partial}{\partial \theta} + \frac{D_{66}}{R}\frac{\partial}{\partial \varphi} - \frac{D_{66}}{Rt}\right)$$
$$+ \frac{2}{t}\left(\frac{D_{26}}{Rs}\frac{\partial}{\partial \theta} + \frac{D_{66}}{R}\frac{\partial}{\partial \varphi} - \frac{D_{66}}{Rt}\right) - RA_{44}$$

$$M_{11} = M_{22} = M_{33} = -R\overline{I_0}$$
$$M_{14} = M_{41} = M_{15} = M_{51} = -R\overline{I_1}$$
$$M_{44} = M_{55} = -R\overline{I_2}$$

The above equations show the level of complexity in solving governing equations of a general laminated spherical shell. According to Eqs. (1.60), (1.61) and (7.1), the general boundary conditions of thick laminated spherical shells are

$$\varphi = \varphi_0 : \begin{cases} N_\varphi - k_{\varphi 0}^u u = 0 \\ N_{\varphi \theta} - k_{\varphi 0}^v v = 0 \\ Q_\varphi - k_{\varphi 0}^w w = 0 \\ M_\varphi - K_{\varphi 0}^\varphi \phi_\varphi = 0 \\ M_{\varphi \theta} - K_{\varphi 0}^\theta \phi_\theta = 0 \end{cases} \qquad \varphi = \varphi_1 : \begin{cases} N_\varphi + k_{\varphi 1}^u u = 0 \\ N_{\varphi \theta} + k_{\varphi 1}^v v = 0 \\ Q_\varphi + k_{\varphi 1}^w w = 0 \\ M_\varphi + K_{\varphi 1}^\varphi \phi_\varphi = 0 \\ M_{\varphi \theta} + K_{\varphi 1}^\theta \phi_\theta = 0 \end{cases}$$
$$\qquad\qquad\qquad\qquad\qquad\qquad\qquad\qquad\qquad\qquad\qquad (7.27)$$
$$\theta = 0 : \begin{cases} N_{\theta \varphi} - k_{\theta 0}^u u = 0 \\ N_\theta - k_{\theta 0}^v v = 0 \\ Q_\theta - k_{\theta 0}^w w = 0 \\ M_{\theta \varphi} - K_{\theta 0}^\varphi \phi_\varphi = 0 \\ M_\theta - K_{\theta 0}^\theta \phi_\theta = 0 \end{cases} \qquad \theta = \theta_0 : \begin{cases} N_{\theta \varphi} + k_{\theta 1}^u u = 0 \\ N_\theta + k_{\theta 1}^v v = 0 \\ Q_\theta + k_{\theta 1}^w w = 0 \\ M_{\theta \varphi} + K_{\theta 1}^\varphi \phi_\varphi = 0 \\ M_\theta + K_{\theta 1}^\theta \phi_\theta = 0 \end{cases}$$

Similarly, thick spherical shells can have up to 24 possible classical boundary conditions at each boundary (see Table 7.2) which leads a high number of combinations of boundary conditions, particular for the open ones.

The classical boundary conditions given in Table 7.2 can be readily realized by applying the artificial spring boundary technique. In this chapter we mainly consider four typical boundary conditions which are frequently encountered in practices, i.e., F, SD, S and C. Taking edge $\varphi = \varphi_0$ for example, the corresponding spring rigidities for the four classical boundary conditions are given as follows:

$$
\begin{aligned}
\text{F:} \quad & k_{\varphi 0}^u = k_{\varphi 0}^v = k_{\varphi 0}^w = K_{\varphi 0}^\varphi = K_{\varphi 0}^\theta = 0 \\
\text{SD:} \quad & k_{\varphi 0}^v = k_{\varphi 0}^w = K_{\varphi 0}^\theta = 10^7 D, \ k_{\varphi 0}^u = K_{\varphi 0}^\varphi = 0 \\
\text{S:} \quad & k_{\varphi 0}^u = k_{\varphi 0}^v = k_{\varphi 0}^w = K_{\varphi 0}^\theta = 10^7 D, \ K_{\varphi 0}^\varphi = 0 \\
\text{C:} \quad & k_{\varphi 0}^u = k_{\varphi 0}^v = k_{\varphi 0}^w = K_{\varphi 0}^\varphi = K_{\varphi 0}^\theta = 10^7 D
\end{aligned}
\tag{7.28}
$$

where $D = E_1 h^3 / 12(1 - \mu_{12}\mu_{21})$ is the flexural stiffness.

Table 7.2 Possible classical boundary conditions for thick spherical shells at each boundary of $\varphi = $ constant

Boundary type	Conditions
Free boundary conditions	
F	$N_\varphi = N_{\varphi\theta} = Q_\varphi = M_\varphi = M_{\varphi\theta} = 0$
F2	$u = N_{\varphi\theta} = Q_\varphi = M_\varphi = M_{\varphi\theta} = 0$
F3	$N_\varphi = v = Q_\varphi = M_\varphi = M_{\varphi\theta} = 0$
F4	$u = v = Q_\varphi = M_\varphi = M_{\varphi\theta} = 0$
F5	$N_\varphi = N_{\varphi\theta} = Q_\varphi = M_\varphi = \phi_\theta = 0$
F6	$u = N_{\varphi\theta} = Q_\varphi = M_\varphi = \phi_\theta = 0$
F7	$N_\varphi = v = Q_\varphi = M_\varphi = \phi_\theta = 0$
F8	$u = v = Q_\varphi = M_\varphi = \phi_\theta = 0$
Simply supported boundary conditions	
S	$u = v = w = M_\varphi = \phi_\theta = 0$
SD	$N_\varphi = v = w = M_\varphi = \phi_\theta = 0$
S3	$u = N_{\varphi\theta} = w = M_\varphi = \phi_\theta = 0$
S4	$N_\varphi = N_{\varphi\theta} = w = M_\varphi = \phi_\theta = 0$
S5	$u = v = w = M_\varphi = M_{\varphi\theta} = 0$
S6	$N_\varphi = v = w = M_\varphi = M_{\varphi\theta} = 0$
S7	$u = N_{\varphi\theta} = w = M_\varphi = M_{\varphi\theta} = 0$
S8	$N_\varphi = N_{\varphi\theta} = w = M_\varphi = M_{\varphi\theta} = 0$
Clamped boundary conditions	
C	$u = v = w = \phi_\varphi = \phi_\theta = 0$
C2	$N_\varphi = v = w = \phi_\varphi = \phi_\theta = 0$
C3	$u = N_{\varphi\theta} = w = \phi_\varphi = \phi_\theta = 0$
C4	$N_\varphi = N_{\varphi\theta} = w = \phi_\varphi = \phi_\theta = 0$
C5	$u = v = w = \phi_\varphi = M_{\varphi\theta} = 0$
C6	$N_\varphi = v = w = \phi_\varphi = M_{\varphi\theta} = 0$
C7	$u = N_{\varphi\theta} = w = \phi_\varphi = M_{\varphi\theta} = 0$
C8	$N_\varphi = N_{\varphi\theta} = w = \phi_\varphi = M_{\varphi\theta} = 0$

7.3 Vibration of Laminated Closed Spherical Shells

This section contains vibration results of laminated closed spherical shells including natural frequencies and mode shapes. The open ones will then be treated in the later section. Various classical and elastic boundary conditions, lamination schemes and different geometry parameters will be considered in the subsequent analysis.

For a closed laminated spherical shell, there exists two boundaries, i.e., $\varphi = \varphi_0$ and $\varphi = \varphi_1$. In a similar treatment as previous chapters, a two-letter string is employed to denote the boundary conditions of a closed spherical shell, such as F-C identifies the shell with completely free and clamped boundary conditions at the edges $\varphi = \varphi_0$ and $\varphi = \varphi_1$, respectively. Unless otherwise stated, the non-dimensional frequency parameter $\Omega = \omega R \sqrt{\rho/E_2}$ is used in the presentation and closed laminated spherical shells under consideration are assumed to be composed of composite layers having following material properties: $E_2 = 10$ GPa, $E_1/E_2 =$ open, $\mu_{12} = 0.25$, $G_{12} = G_{13} = 0.6E_2$, $G_{23} = 0.5\ E_2$, $\rho = 1{,}500$ kg/m^3.

Considering the circumferential symmetry of closed spherical shells, each displacement and rotation component of a closed spherical shell is expanded as a 1-D modified Fourier series of the following form through Fourier decomposition of the circumferential wave motion:

$$u(\varphi,\theta) = \sum_{m=0}^{M}\sum_{n=0}^{N} A_{mn}\cos\lambda_m\varphi\cos n\theta + \sum_{l=1}^{2}\sum_{n=0}^{N} a_{ln}P_l(\varphi)\cos n\theta$$

$$v(\varphi,\theta) = \sum_{m=0}^{M}\sum_{n=0}^{N} B_{mn}\cos\lambda_m\varphi\sin n\theta + \sum_{l=1}^{2}\sum_{n=0}^{N} b_{ln}P_l(\varphi)\sin n\theta$$

$$w(\varphi,\theta) = \sum_{m=0}^{M}\sum_{n=0}^{N} C_{mn}\cos\lambda_m\varphi\cos n\theta + \sum_{l=1}^{2}\sum_{n=0}^{N} c_{ln}P_l(\varphi)\cos n\theta \qquad (7.29)$$

$$\phi_\varphi(\varphi,\theta) = \sum_{m=0}^{M}\sum_{n=0}^{N} D_{mn}\cos\lambda_m\varphi\cos n\theta + \sum_{l=1}^{2}\sum_{n=0}^{N} d_{ln}P_l(\varphi)\cos n\theta$$

$$\phi_\theta(\varphi,\theta) = \sum_{m=0}^{M}\sum_{n=0}^{N} E_{mn}\cos\lambda_m\varphi\sin n\theta + \sum_{l=1}^{2}\sum_{n=0}^{N} e_{ln}P_l(\varphi)\sin n\theta$$

where $\lambda_m = m\pi/\Delta\varphi$, in which $\Delta\varphi$ is the included angle in the meridional direction, defined as $\Delta\varphi = \phi_1 - \phi_0$. n represents the circumferential wave number of the corresponding mode. n is non-negative integer. Interchanging of $\sin n\theta$ and $\cos n\theta$ in Eq. (7.29), another set of free vibration modes (anti-symmetric modes) can be obtained. It should be noted that the modified Fourier series presented in Eq. (7.29) are complete series defined over the range $[0, \Delta\phi]$. Therefore, a linear transformation for coordinate φ from $\varphi \in [\varphi_0, \varphi_1]$ to $\overline{\varphi} \in [0, \Delta\varphi]$ need to be introduced for the practical programming and computing, i.e., $\varphi = \overline{\varphi} + \varphi_0$. $P_l(\varphi)$ are two auxiliary functions introduced in each displacement expression to remove all the discontinuities potentially associated with the first-order derivatives at the boundaries. These auxiliary functions are in the same form as those of Eq. (5.39).

7.3.1 Convergence Studies and Result Verification

Table 7.3 shows the convergence behavior of the lowest six natural frequencies (Hz) of a single-layered composite ([0°]) spherical shell with different boundary conditions and meridional included angles. Two different boundary conditions (i.e. F-F, C-C) and two meridional included angles (i.e. $\varphi_1 = \pi/3$ and $5\pi/6$) are performed to determine the optimal number of truncation terms required for satisfactory solutions. The geometric constants and material properties of the shell are: $\phi_0 = \pi/6$, $R = 1$ m, $E_2 = 10$ GPa, $E_1/E_2 = 15$. Six truncation schemes (i.e. $M = 5$–10, $N = 10$) are performed for each case. From Table 7.3, it can be seen that the natural

Table 7.3 Convergence of frequencies (Hz) for a single-layered composite ([0°]) spherical shell

$\Delta\varphi$	B. C.	M	Mode number					
			1	2	3	4	5	6
$\varphi_0 = \pi/6$ $\varphi_1 = \pi/3$	F-F	5	64.748	174.50	202.49	319.84	406.81	450.91
		6	64.747	174.50	202.48	319.84	406.81	450.91
		7	64.747	174.50	202.48	319.84	406.81	450.90
		8	64.747	174.49	202.47	319.84	406.81	450.90
		9	64.747	174.49	202.47	319.84	406.81	450.90
		10	64.747	174.49	202.47	319.84	406.81	450.90
	C-C	5	1,957.0	2,054.4	2,091.7	2,124.3	2,141.3	2,163.6
		6	1,957.0	2,054.4	2,091.7	2,124.3	2,141.3	2,163.6
		7	1,957.0	2,054.4	2,091.7	2,124.3	2,141.3	2,163.6
		8	1,957.0	2,054.4	2,091.7	2,124.3	2,141.3	2,163.6
		9	1,957.0	2,054.4	2,091.7	2,124.3	2,141.3	2,163.6
		10	1,957.0	2,054.4	2,091.7	2,124.3	2,141.3	2,163.6
$\varphi_0 = \pi/6$ $\varphi_1 = 5\pi/6$	F-F	5	139.83	148.61	334.48	349.64	366.51	412.93
		6	139.83	148.21	333.81	349.63	366.47	412.93
		7	139.69	148.21	333.81	349.24	366.47	412.93
		8	139.69	148.02	333.56	349.24	366.45	412.93
		9	139.62	148.02	333.56	349.08	366.45	412.93
		10	139.62	147.93	333.45	349.08	366.44	412.93
	C-C	5	470.77	502.73	505.70	593.78	706.29	734.59
		6	470.76	502.72	505.69	593.77	706.29	734.55
		7	470.75	502.72	505.68	593.77	706.28	734.55
		8	470.75	502.72	505.68	593.77	706.28	734.55
		9	470.75	502.72	505.68	593.77	706.28	734.55
		10	470.75	502.72	505.68	593.77	706.28	734.55

frequencies of the shell converge monotonically and rapidly as the truncation number M increases. The maximum differences between the '10 × 10' and '5 × 10' form results for the F-F and C-C boundary conditions are less than 0.46 and 0.01 %, respectively. Furthermore, Table 7.3 shows that the C-C solutions converge faster than those of F-F boundary conditions and solutions of spherical shells with smaller meridional included angle converge faster. Unless otherwise stated, the truncated number of the displacement expressions will be uniformly selected as $M = 15$ in all the following examples.

Table 7.4 shows the comparison of the first longitudinal mode frequency parameters Ω for a three-layered, cross-ply [0°/90°/0°] spherical shell with three different boundary conditions to prove the validity of the present method for vibration analysis of composite laminated spherical shells. The first ten circumferential wave number, i.e., $n = 0$–9 are considered in the comparison. The geometric and material constants of the spherical shell are: $E_2 = 10.6$ GPa, $E_1 = 138$ GPa, $\mu_{12} = 0.28$, $G_{12} = 6$ GPa, $G_{13} = G_{23} = 3.9$ GPa, $\rho = 1{,}500$ kg/m^3, $R = 1$ m, $h = 0.05$ m, $\varphi_0 = 60°$, $\varphi_1 = 90°$. The results provided by Qu et al. (2013a) based on a FSDT formulation are selected as the benchmark solutions. It can be seen that the accuracy of the present method compares well with the referential data. The discrepancies are very small and do not exceed 0.051 % for the worst case although different solution approaches were used in the literature.

7.3.2 Closed Laminated Spherical Shells with General Boundary Conditions

As the first case, a four-layered, cross-ply [0°/90°/0°/90°] spherical shell is considered. The start and end meridional angles of the shell are $\varphi_0 = 15°$, 30° or 45° and $\varphi_1 = 90°$, other structure quantities used in the calculation are: $E_1/E_2 = 15$,

Table 7.4 Comparison of the frequency parameters Ω for a three-layered, cross-ply [0°/90°/0°] spherical shell with different restraints ($R = 1$ m, $h = 0.05$ m, $\varphi_0 = 60°$, $\varphi_1 = 90°$, $m = 1$)

n	FSDT (Qu et al. 2013a)			Present		
	SD–SD	S-S	C-C	SD–SD	S-S	C-C
0	0.5030	3.8955	4.3368	0.5030	3.8940	4.3355
1	0.9123	3.5823	3.9707	0.9117	3.5810	3.9696
2	1.6095	3.3768	3.8241	1.6083	3.3754	3.8229
3	2.1245	3.2947	3.7732	2.1234	3.2933	3.7720
4	2.1239	3.2719	3.7616	2.1228	3.2705	3.7604
5	2.1640	3.2846	3.7736	2.1629	3.2832	3.7724
6	2.2506	3.3258	3.8062	2.2495	3.3244	3.8050
7	2.3790	3.3949	3.8607	2.3779	3.3936	3.8595
8	2.5469	3.4938	3.9401	2.5457	3.4925	3.9390
9	2.7523	3.6247	4.0477	2.7512	3.6235	4.0466

$R = 1$ m and $h = 0.1$ m. In Table 7.5, the lowest five frequency parameters Ω of the spherical shell subjected to as many as six possible boundary conditions are presented. The table shows that the frequency parameters of the shell decrease with initial meridional angle increases. Figure 7.2 shows some selected mode shapes and corresponding frequency parameters $\Omega_{n,m}$ for the spherical shell with initial meridional angle $\varphi_0 = 30°$ and F-C boundary condition. From the figure, it is obvious that the frequency parameters of circumferential wave number $n = 2$ are smaller than those of $n = 1$ and 3.

Table 7.6 shows the lowest five frequency parameters Ω for a two-layered, cross-ply [0°/90°] spherical shell with different boundary conditions and thickness-to-radius ratios. Six different classical boundary combinations (i.e. F-F, F-S, F-C, S-S, S-C and C-C) and four kinds of thickness-to-radius ratios ($h/R = 0.01$, 0.02, 0.05 and 0.1) are included in the table. The geometric and material constants of the spherical shell are: $R = 1$ m, $\varphi_0 = 60°$, $\varphi_1 = 120°$, $E_1/E_2 = 15$. It can be seen from the table that the frequency parameters of the shell increase in general as the thickness-radius ratio increases. The second observation that needs to be made here is that boundary conditions have a conspicuous effect on the vibration frequencies of spherical shells. Increase of the restraint stiffness always results in increments of the frequency parameters.

By appropriately selecting each lamina of the composite laminated spherical shells, desired strength and stiffness parameters can be achieved. In Fig. 7.3, the first fifteen dimensionless frequencies Ω of a two-layered, [0°/90°] cross-ply

Table 7.5 Frequency parameters Ω for a four-layered, cross-ply [0°/90°/0°/90°] spherical shell with different boundary conditions and initial meridional angles ($\varphi_1 = 90°$, $R = 1$ m, $h = 0.1$ m, $E_1/E_2 = 15$)

φ_0	Mode	Boundary conditions					
		F-F	F-S	F-C	S-S	S-C	C-C
15°	1	0.2414	0.8844	0.9318	1.9310	2.0209	2.0283
	2	0.6192	1.0955	1.1491	2.0032	2.1837	2.1841
	3	1.0981	1.8795	1.9782	2.1941	2.3925	2.3943
	4	1.1571	1.9253	1.9962	2.3735	2.4103	2.5464
	5	1.4791	2.0133	2.1605	2.3781	2.4352	2.5779
30°	1	0.2323	0.7254	0.8008	2.2576	2.3487	2.4341
	2	0.6120	0.9782	1.0378	2.3134	2.4555	2.5689
	3	0.6143	1.3630	1.4261	2.4216	2.5983	2.6928
	4	1.0961	2.0498	2.1984	2.6493	2.8630	2.9204
	5	1.4375	2.0975	2.2187	2.7832	2.9672	3.1576
45°	1	0.2198	0.7417	0.8861	2.5362	2.7911	3.0317
	2	0.3763	0.9916	1.1069	2.5800	2.8360	3.0773
	3	0.5991	1.1813	1.2721	2.6548	2.8538	3.1500
	4	0.9870	1.6005	1.6928	2.7641	3.0383	3.3090
	5	1.0827	2.3083	2.4053	3.0887	3.3560	3.5825

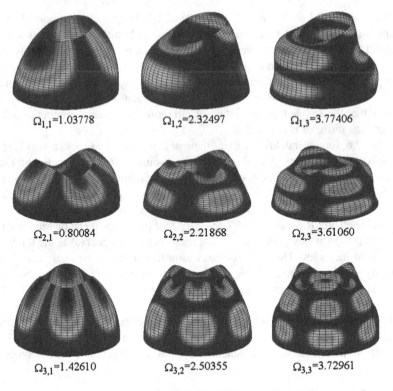

$\Omega_{1,1}=1.03778$ $\Omega_{1,2}=2.32497$ $\Omega_{1,3}=3.77406$

$\Omega_{2,1}=0.80084$ $\Omega_{2,2}=2.21868$ $\Omega_{2,3}=3.61060$

$\Omega_{3,1}=1.42610$ $\Omega_{3,2}=2.50355$ $\Omega_{3,3}=3.72961$

Fig. 7.2 Mode shapes for a F-C supported [0°/90°/0°/90°] spherical shell ($\Omega_{n,m}$, $n = 1$–3, $m = 1$–3)

spherical shells with F-F, S-S and C-C boundary conditions and various anisotropic degrees are presented, respectively, for the sake of investigating the influence of the material properties. In each case, four types of anisotropic degrees, i.e., $\eta = E_1/E_2 = 5$, 15, 25 and 35 are considered. The other geometric and material properties of the spherical shell are: $R = 1$ m, $h = 0.1$ m. $\varphi_0 = 45°$ and $\varphi_1 = 90°$. From the figure, we can see that the influence of the anisotropic degree on the frequency parameters varies with boundary conditions. The second observation is that the effects of anisotropic degree (η) are much higher for the S-S and C-C supported shells than it is for the F-F one.

Furthermore, Table 7.7 shows the lowest five frequency parameters Ω of slightly thick spherical shells with different lamination schemes for various boundary conditions. Four lamination schemes, i.e., [0°], [90°], [0°/90°] and [0°/90°/0°] are studied for each case. The geometric and material constants used in the study are: $\varphi_0 = 60°$, $\varphi_1 = 120°$, $R = 1$ m, $h/R = 0.05$, $E_1/E_2 = 15$. Compare the results in Table 7.7, the first observation is that when the spherical shells with F-F, F-S and F-C boundary conditions, the frequency parameters for spherical shells with fibers oriented in the meridional direction (i.e., [0°]) are lower than those of fibers oriented in the circumferential direction (i.e., [90°]). This is not the case of S-S, S-C and C-C

Table 7.6 Frequency parameters Ω of a two-layered, cross-ply $[0°/90°]$ spherical shell with different sets of boundary conditions and thickness-to-radius ratios ($\varphi_0 = 60°$, $\varphi_1 = 120°$, $R = 1$ m, $E_1/E_2 = 15$)

h/R	Mode	Boundary conditions					
		F-F	F-S	F-C	S-S	S-C	C-C
0.01	1	0.0237	0.1333	0.1359	1.5433	1.5577	1.5744
	2	0.0297	0.1480	0.1486	1.5550	1.5644	1.5789
	3	0.0703	0.2270	0.2271	1.5768	1.5902	1.6098
	4	0.0824	0.2903	0.2934	1.5800	1.5956	1.6216
	5	0.1382	0.3242	0.3242	1.5975	1.6138	1.6246
0.02	1	0.0457	0.1813	0.1858	1.5771	1.6160	1.6655
	2	0.0570	0.2697	0.2704	1.5978	1.6336	1.6845
	3	0.1337	0.2975	0.3049	1.6336	1.6714	1.7166
	4	0.1560	0.4179	0.4179	1.6632	1.6930	1.7439
	5	0.2585	0.5909	0.5909	1.7570	1.7788	1.8298
0.05	1	0.1053	0.3232	0.3430	1.8336	1.9347	2.1167
	2	0.1306	0.3355	0.3437	1.8484	1.9399	2.1235
	3	0.2979	0.5778	0.5789	1.9201	2.0213	2.2147
	4	0.3460	0.6819	0.6963	2.0134	2.0731	2.2369
	5	0.5593	0.8863	0.8863	2.0850	2.1800	2.2698
0.10	1	0.1881	0.3759	0.4158	2.0157	2.1568	2.3446
	2	0.2347	0.5564	0.5723	2.0438	2.1690	2.4592
	3	0.5102	0.6902	0.7218	2.0548	2.1738	2.5138
	4	0.6057	0.9704	0.9772	2.1662	2.2747	2.6023
	5	0.9257	1.4611	1.4660	2.3662	2.4709	2.7598

boundary conditions. The $[0°]$ lamination give the higher frequency parameter than those of $[90°]$ lamination. The second observation is that the two-layered, unsymmetrical lamination scheme $[0°/90°]$ produces lower frequency parameter than the three-layered, symmetrical lamination scheme $[0°/90°/0°]$. This is the case for all boundary conditions under consideration except the F-S type.

Influence of the elastic restraint stiffness on the frequency parameters of laminated spherical shells is investigated in subsequent example. For simplicity and convenience in the analysis, five non-dimensional spring parameters Γ_λ ($\lambda = u, v, w, \varphi, \theta$), which are defined as the ratios of the corresponding spring stiffness to the flexural stiffness D are introduced here, i.e., $\Gamma_u = k_u/D$, $\Gamma_v = k_v/D$, $\Gamma_w = k_w/D$, $\Gamma_\varphi = K_\varphi/D$ and $\Gamma_\theta = K_\theta/D$. Also, a frequency parameter $\Delta\Omega$ which is defined as the difference of the frequency parameters Ω to those of the elastic restraint parameters Γ_λ equal to 10^{-2} are used in the investigation, i.e., $\Delta\Omega = \Omega(\Gamma_\lambda) - \Omega(\Gamma_\lambda = 10^{-2})$. In Figs. 7.4 and 7.5, variation of the 1st, 3rd and 5th mode frequency parameters $\Delta\Omega$ versus the elastic restraint parameters Γ_λ for a two-layered, $[0°/90°]$ spherical shell ($\varphi_0 = 45°$, $\varphi_1 = 90°$, $R = 1$ m, $h/R = 0.1$, $E_1/E_2 = 15$) with various elastic restraints.

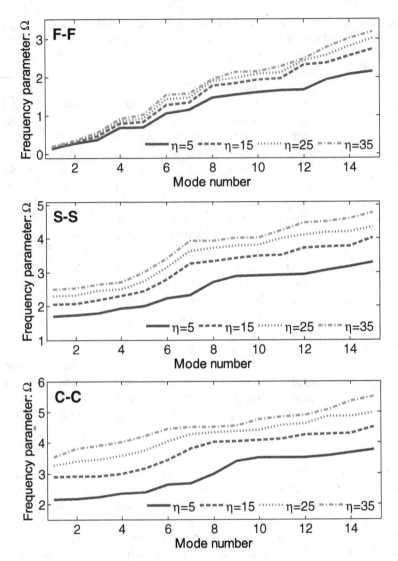

Fig. 7.3 The variation of frequency parameters Ω versus mode number for a two-layered, cross-ply [0°/90°] spherical shell with various material properties ($\eta = E_1/E_2$)

are presented. In Fig. 7.4, the spherical shell is assumed to be clamped at edge $\varphi = \varphi_0$ whilst the other edge is elastically restrained by only one kind of spring components with various stiffness (denoted by C-F^e). It is clearly that in a certain range, the frequency parameters $\Delta\Omega$ increase rapidly as the elastic restraint parameter Γ_λ increasing. And beyond this range, there is little variation in the frequency parameters. In Fig. 7.5, the edge $\varphi = \varphi_1$ of the shell is supported by all the

Table 7.7 Frequency parameters Ω of laminated spherical shells with different sets of boundary conditions and various lamination schemes ($\varphi_0 = 60°$, $\varphi_1 = 120°$, $R = 1$ m, $h/R = 0.05$, $E_1/E_2 = 15$)

Lamination schemes	Mode	Boundary conditions					
		F-F	F-S	F-C	S-S	S-C	C-C
[0°]	1	0.0924	0.2682	0.3071	1.9167	2.1857	2.3187
	2	0.1112	0.2787	0.3651	1.9334	2.1992	2.4786
	3	0.2458	0.4459	0.4509	1.9832	2.2410	2.4901
	4	0.2675	0.5803	0.6497	2.0642	2.3096	2.5247
	5	0.4351	0.6552	0.6552	2.1734	2.3189	2.5822
[90°]	1	0.1804	0.3155	0.3321	1.2350	1.2521	1.2717
	2	0.1937	0.5417	0.5453	1.2997	1.3166	1.3352
	3	0.5140	0.6289	0.6454	1.3638	1.3777	1.3938
	4	0.5662	0.9878	0.9888	1.6262	1.6363	1.6472
	5	0.9615	1.3481	1.3635	1.6604	1.6701	1.6813
[0°/90°]	1	0.1053	0.3232	0.3430	1.8336	1.9347	2.1167
	2	0.1306	0.3355	0.3437	1.8484	1.9399	2.1235
	3	0.2979	0.5778	0.5791	1.9201	2.0213	2.2147
	4	0.3460	0.6819	0.6962	2.0134	2.0731	2.2369
	5	0.5593	0.8863	0.8862	2.0850	2.1800	2.2698
[0°/90°/0°]	1	0.1201	0.3090	0.3503	2.2789	2.4427	2.4509
	2	0.1253	0.3212	0.3870	2.3116	2.4838	2.6692
	3	0.3104	0.5330	0.5361	2.3369	2.5121	2.7288
	4	0.3135	0.6760	0.7330	2.4024	2.5331	2.7357
	5	0.5178	0.7775	0.7774	2.4411	2.5919	2.7528

five spring components in which four groups of them with infinite stiffness ($10^7 D$) and the rest one is assigned at changed stiffness (C^e). The spherical shell is assumed to be clamped at edge $\varphi = \varphi_0$ as usual (denoted by C-C^e). The similar observations as those of Fig. 7.4 can be seen in this figure. This study shows the active ranges of the elastic restraint parameters on the vibration characteristic of laminated shells vary with mode sequences, lamination schemes and spring components. In this case, they can be defined as Γ_u: 10^1 to 10^4, Γ_v: 10^0 to 10^3, Γ_w: 10^{-1} to 10^3, Γ_φ: 10^{-1} to 10^2 and Γ_θ: 10^{-1} to 10^2, respectively.

7.4 Vibration of Laminated Open Spherical Shells

An open laminated spherical shell can be obtained by rotating the generator in desired circumferential angle (less than one full revolution) about the axis. It can be produced by cutting a segment of the closed ones as well. For an open spherical shell, the assumption of whole periodic wave numbers in the circumferential

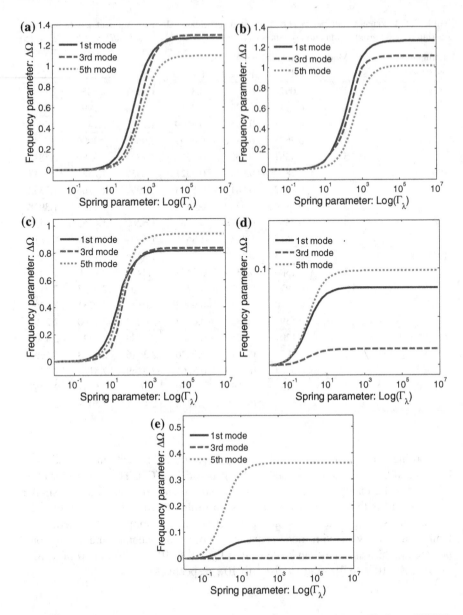

Fig. 7.4 Variation of frequency parameters $\Delta\Omega$ versus spring parameters Γ_λ for a [0°/90°] laminated spherical shell with C-F^e boundary conditions: **a** Γ_u; **b** Γ_v; **c** Γ_w; **d** Γ_φ; **e** Γ_θ

direction is inappropriate, and thus, a set of complete two-dimensional analysis is required and resort must be made to a full two-dimensional solution scheme. In this section, we consider free vibration of laminated deep open spherical shells. As was done previously for the open cylindrical and conical shells, each displacement and

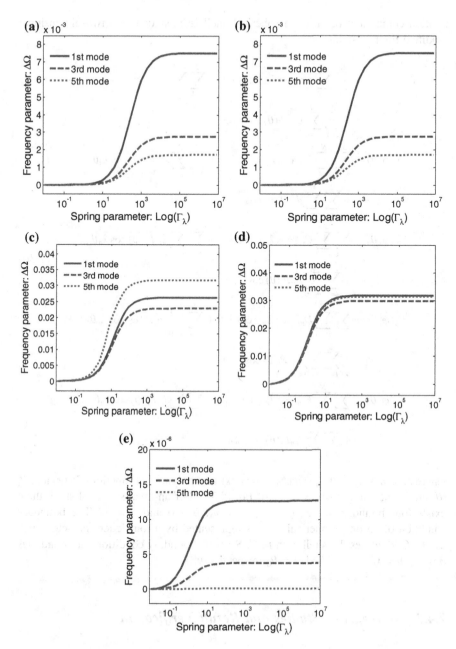

Figs 7.5 Variation of frequency parameters $\Delta\Omega$ versus spring parameters Γ_λ for a [0°/90°] laminated spherical shell with C-C^e boundary conditions: **a** Γ_u; **b** Γ_v; **c** Γ_w; **d** Γ_φ; **e** Γ_θ

rotation component of an open spherical shell is expanded as a two-dimensional modified Fourier series as:

$$u(\varphi, \theta) = \sum_{m=0}^{M} \sum_{n=0}^{N} A_{mn} \cos \lambda_m \varphi \cos \lambda_n \theta + \sum_{l=1}^{2} \sum_{n=0}^{N} a_{ln} P_l(\varphi) \cos \lambda_n \theta$$
$$+ \sum_{l=1}^{2} \sum_{m=0}^{M} b_{lm} P_l(\theta) \cos \lambda_m \varphi$$

$$v(\varphi, \theta) = \sum_{m=0}^{M} \sum_{n=0}^{N} B_{mn} \cos \lambda_m \varphi \cos \lambda_n \theta + \sum_{l=1}^{2} \sum_{n=0}^{N} c_{ln} P_l(\varphi) \cos \lambda_n \theta$$
$$+ \sum_{l=1}^{2} \sum_{m=0}^{M} d_{lm} P_l(\theta) \cos \lambda_m \varphi$$

$$w(\varphi, \theta) = \sum_{m=0}^{M} \sum_{n=0}^{N} C_{mn} \cos \lambda_m \varphi \cos \lambda_n \theta + \sum_{l=1}^{2} \sum_{n=0}^{N} e_{ln} P_l(\varphi) \cos \lambda_n \theta$$
$$+ \sum_{l=1}^{2} \sum_{m=0}^{M} f_{lm} P_l(\theta) \cos \lambda_m \varphi$$

$$\phi_\varphi(\varphi, \theta) = \sum_{m=0}^{M} \sum_{n=0}^{N} D_{mn} \cos \lambda_m \varphi \cos \lambda_n \theta + \sum_{l=1}^{2} \sum_{n=0}^{N} g_{ln} P_l(\varphi) \cos \lambda_n \theta$$
$$+ \sum_{l=1}^{2} \sum_{m=0}^{M} h_{lm} P_l(\theta) \cos \lambda_m \varphi$$

$$\phi_\theta(\varphi, \theta) = \sum_{m=0}^{M} \sum_{n=0}^{N} E_{mn} \cos \lambda_m \varphi \cos \lambda_n \theta + \sum_{l=1}^{2} \sum_{n=0}^{N} i_{ln} P_l(\varphi) \cos \lambda_n \theta$$
$$+ \sum_{l=1}^{2} \sum_{m=0}^{M} j_{lm} P_l(\theta) \cos \lambda_m \varphi$$

(7.30)

where $\lambda_m = m\pi/\Delta\varphi$ and $\lambda_n = n\pi/\theta_0$. The auxiliary polynomial functions $P_l(x)$ and $P_l(\theta)$ are in the same form as those of Eq. (5.39). For an open spherical shell, there exists four boundaries, i.e., $\varphi = \varphi_0$, $\varphi = \varphi_1$, $\theta = 0$ and $\theta = \theta_0$. The boundary condition of an open spherical shell is represented by a four-letter character, such as FCSC identifies the shell with F, C, S and C boundary conditions at boundaries $\varphi = \varphi_0$, $\theta = 0$, $\varphi = \varphi_1$ and $\theta = \theta_0$, respectively.

7.4.1 Convergence Studies and Result Verification

Several numerical examples are given to confirm the convergence, reliability and accuracy of the present method for open laminated spherical shells..

Table 7.8 shows the convergence and comparison of the lowest six frequency parameters $\Omega = \omega L_\varphi^2/h \sqrt{\rho/E_1}$ (where L_φ is the length of the shell in the meridional direction, namely, $L_\varphi = R\Delta\varphi$) of a four-layered, [30°/−30°/−30°/30°] graphite/epoxy

Table 7.8 Convergence and comparison of frequency parameters $\Omega = \omega L_\varphi^2/h\sqrt{\rho/E_1}$ of a [30°/−30°/−30°/30°] laminated graphite/epoxy open spherical shell ($R = 2$ m, $h = 0.01$ m, $\varphi_0 = \pi/2 - 1/4$, $\varphi_1 = \pi/2 + 1/4$, $\theta_0 = 1/2$)

B.C.	$M \times N$	Mode number					
		1	2	3	4	5	6
FFFF	14 × 14	2.3388	3.3944	6.0043	6.8884	8.5215	11.330
	15 × 15	2.3379	3.3937	6.0013	6.8850	8.5160	11.328
	16 × 16	2.3369	3.3929	6.0000	6.8832	8.5143	11.320
	17 × 17	2.3364	3.3925	5.9977	6.8805	8.5098	11.319
	18 × 18	2.3357	3.3919	5.9967	6.8791	8.5086	11.312
	Qatu (2004)	2.283	3.323	5.871	6.781	8.934	–
SSSS	14 × 14	27.519	27.574	35.023	37.004	38.510	38.977
	15 × 15	27.516	27.571	35.018	37.000	38.508	38.972
	16 × 16	27.511	27.569	35.015	36.997	38.506	38.971
	17 × 17	27.509	27.567	35.012	36.995	38.505	38.969
	18 × 18	27.506	27.565	35.010	36.993	38.504	38.968
CCCC	14 × 14	28.491	30.958	37.153	37.912	39.323	40.941
	15 × 15	28.487	30.950	37.148	37.909	39.320	40.931
	16 × 16	28.485	30.944	37.140	37.901	39.318	40.923
	17 × 17	28.482	30.940	37.137	37.899	39.316	40.918
	18 × 18	28.481	30.936	37.133	37.894	39.314	40.914

open spherical shell with FFFF, SSSS and CCCC boundary conditions. The geometry and material properties of the layers of the shell are: $R = 2$ m, $h = 0.01$ m, $\varphi_0 = \pi/2 - 1/4$, $\varphi_1 = \pi/2 + 1/4$, $\theta_0 = 1/2$, $E_1 = 138$ GPa, $E_2 = 8.96$ GPa, $\mu_{12} = 0.3$, $G_{12} = G_{13} = 7.1$ GPa, $G_{23} = 3.9$ GPa, $\rho = 1,500$ kg/m^3. Five different truncation forms (i.e. $M \times N = 14 \times 14$, 15×15, 16×16, 17×17 and 18×18) are considered. It is apparent that as the truncated numbers increase the convergence is achieved. Furthermore, in Table 7.8, a comparison between the current results and those reported by Qatu (2004) (by using the classical shallow theory) is performed for the shell with FFFF boundary condition. The symbols "–" are missing data that were not considered by Qatu (2004). The discrepancy between the two results is acceptable. The small discrepancy in the results may be attributed to different shell theories and solution procedures are used in the referential work. In the following studies, the numerical results will be obtained by using truncation form of $M \times N = 18 \times 18$.

Since there are no suitable comparison results in the literature, two illustrative examples are presented for the laminated shallow spherical panels. As the first case, Table 7.9 shows the first six frequency parameters $\Omega = \omega L_\varphi^2/h\sqrt{\rho/E_2}$ of a four-layered, [45°/−45°/45°/−45°] open spherical shell with different classical restraints (i.e., FFFF, FSFS, FCFC, SSSS, SCSC and CCCC), together with the modified Fourier series solutions obtained by Ye et al. (2013). The shell parameters used in the comparison are: $E_1/E_2 = 15$, $\mu_{12} = 0.23$, $G_{12} = G_{13} = G_{23} = 0.5E_2$, $R = 10$ m,

Table 7.9 Comparison of frequency parameters $\Omega = \omega L_\varphi^2 / h \sqrt{\rho / E_2}$ of a $[45°/-45°/45°/-45°]$ open spherical shell with various boundary conditions ($R = 10$ m, $h = 0.01$ m, $\varphi_0 = \pi/2 - 1/20$, $\varphi_1 = \pi/2 + 1/20$, $\theta_0 = 1/5$)

Theory	Mode	Boundary conditions					
		FFFF	FSFS	FCFC	SSSS	SCSC	CCCC
CST (Ye et al. 2013)	1	2.3732	7.6886	10.119	31.799	32.310	35.898
	2	5.8131	10.895	11.429	34.178	34.599	36.700
	3	8.2261	13.397	13.425	34.947	35.113	40.645
	4	12.141	16.703	17.729	37.594	38.487	42.597
	5	14.152	17.830	20.388	42.428	44.898	48.469
	6	19.369	22.328	24.034	43.625	45.858	52.539
Present FSDT	1	2.3722	7.6509	10.062	31.717	32.218	35.750
	2	5.7876	10.866	11.406	34.068	34.473	36.547
	3	8.1743	13.378	13.402	34.821	34.979	40.442
	4	12.081	16.642	17.631	37.433	38.324	42.368
	5	14.077	17.762	20.274	42.269	44.735	48.191
	6	19.247	22.200	23.933	43.481	45.660	52.236

$h = 0.01$ m, $\varphi_0 = \pi/2 - 1/20$, $\varphi_1 = \pi/2 + 1/20$, $\theta_0 = 1/5$. Our results are in a closed agreement with the existing ones. In order to further verify the present formulation, Table 7.10 shows the comparison of the first nine non-dimensional frequencies $\Omega = \omega L_\varphi^2 / h \sqrt{\rho / E_1}$ of certain three-layered shallow spherical shells subjected to SDSDSDSD boundary condition, with results provided by Fazzolari and Carrera (2013) by using the Carrera unified formulation and shallow shell theory. The shell parameters used in the comparison are: $E_1 = 60.7$ GPa, $E_2 = 24.8$ GPa, $\mu_{12} = 0.23$, $G_{12} = G_{13} = G_{23} = 12$ GPa, $R = 2$ m, $h = 0.05$ m, $\varphi_0 = \pi/2 - 1/4$, $\varphi_1 = \pi/2 + 1/4$, $\theta_0 = 1/2$. Four sets of lamination schemes, i.e., $[0°]$, $[15°/-15°/15°]$, $[30°/-30°/30°]$ and $[45°/-45°/45°]$ are performed in the comparison. It is clearly evident that the

Table 7.10 Comparison of frequency parameters $\Omega = \omega L_\varphi^2 / h \sqrt{\rho / E_1}$ of certain three-layered open spherical shells with SDSDSDSD boundary conditions ($R = 2$ m, $h = 0.05$ m, $\varphi_0 = \pi/2 - 1/4$, $\varphi_1 = \pi/2 + 1/4$, $\theta_0 = 1/2$)

Layout	Theory	Mode number				
		1	2	3	4	5
$[0°]$	Fazzolari and Carrera (2013)	8.1600	12.455	14.027	18.335	20.121
	Present	8.1430	12.374	13.956	18.131	20.053
$[15°/-15°/15°]$	Fazzolari and Carrera (2013)	8.4908	12.750	14.008	18.562	20.570
	Present	8.4676	12.639	13.912	18.319	20.453
$[30°/-30°/30°]$	Fazzolari and Carrera (2013)	9.1409	13.116	14.027	18.993	21.614
	Present	9.0932	12.958	13.855	18.674	21.408
$[45°/-45°/45°]$	Fazzolari and Carrera (2013)	9.4813	13.164	14.077	19.201	22.670
	Present	9.3961	12.997	13.855	18.845	22.431

present solutions match well with the reference data. The differences between the two results are attributed to different shell theories are used in the literature. It has been proved that the shallow shell theories will give inaccurate results when applied to deep open shells Qatu (2004).

7.4.2 Laminated Open Spherical Shells with General Boundary Conditions

Some further vibration results for laminated deep open spherical shells with various combinations of boundary conditions and shell parameters are presented. In the following examples, unless otherwise stated, the material constants of the layers of spherical shells under consideration are: $E_2 = 10$ GPa, $E_1/E_2 =$ open, $\mu_{12} = 0.25$, $G_{12} = G_{13} = 0.5E_2$, $G_{23} = 0.2E_2$, $\rho = 1,500$ kg/m^3 and the non-dimensional frequency parameter $\Omega = \omega R\sqrt{\rho/E_2}$ is used in the subsequent calculations.

Table 7.11 shows the first five frequency parameters Ω of a three-layered, cross-ply [0°/90°/0°] open spherical shell with various boundary conditions and

Table 7.11 Frequency parameters Ω of a three-layered, cross-ply [0°/90°/0°] open spherical shell with various boundary conditions and circumferential included angles ($E_1/E_2 = 15$, $R = 1$ m, $h/R = 0.1$, $\varphi_0 = 30°$, $\varphi_1 = 90°$)

θ_0	Mode	Boundary conditions					
		FSFS	FCFC	SFSF	CFCF	SSSS	CCCC
45°	1	2.0987	2.2193	1.8423	1.8462	3.4330	3.7494
	2	2.2218	2.7411	2.3813	2.6423	3.6881	3.9026
	3	2.4927	2.7499	2.5450	2.7554	3.7096	4.2086
	4	3.2649	3.5364	2.8010	2.8835	4.3550	4.8337
	5	3.5147	3.9621	2.8930	2.9655	5.2045	5.4755
90°	1	0.7994	0.9795	1.9122	1.9144	2.7830	2.9932
	2	1.1343	1.2014	2.4170	2.6431	3.0436	3.2609
	3	1.4850	1.5991	2.4395	2.6824	3.1293	3.3746
	4	1.6120	1.8785	2.6830	2.8074	3.2461	3.3923
	5	2.2076	2.2646	2.8318	2.9126	3.5980	3.7936
135°	1	0.4444	0.5272	1.9351	1.9366	2.5935	2.7310
	2	0.5904	0.6673	2.4213	2.5769	2.7831	2.9717
	3	0.9489	1.0065	2.4296	2.6700	2.9482	3.1588
	4	1.0845	1.1871	2.5292	2.7106	3.0528	3.2204
	5	1.2694	1.3760	2.7486	2.8692	3.0662	3.2307
180°	1	0.2672	0.3142	1.9467	1.9479	2.4576	2.5506
	2	0.3349	0.3881	2.3793	2.4489	2.6797	2.8439
	3	0.6236	0.6863	2.4303	2.6625	2.7862	2.9722
	4	0.6871	0.7683	2.4460	2.6858	2.9373	3.1266
	5	0.9963	1.0681	2.6539	2.7960	2.9846	3.1406

circumferential included angles. The shell parameters used are: $E_1/E_2 = 15$, $R = 1$ m, $h/R = 0.1$, $\varphi_0 = 30°$, $\varphi_1 = 90°$. Six different boundary conditions (i.e., FSFS, FCFC, SFSF, CFCF, SSSS and CCCC) are considered for each circumferential included angle. It is obvious from the table that the increasing of the circumferential included angle will result in a decrease in the frequency parameter. Meanwhile, we can see that an open spherical shell with greater restraining rigidity will have higher vibration frequencies. For any given frequency parameters, the corresponding mode shapes of the open spherical shell can be readily determined by Eq. (7.30) after solving the standard matrix eigenproblem. For instance, the first three mode shapes for the spherical shell with CCCC boundary condition and circumferential included angle of $\theta_0 = 45°$, $90°$ and $135°$ are plotted in Fig. 7.6.

Table 7.12 shows the lowest five frequency parameters Ω for a two-layered, [0°/90°] open spherical shell with different boundary conditions and thickness-to-radius ratios. Four kinds of thickness-to-radius ratios ($h/R = 0.01, 0.02, 0.05$ and 0.1) are included in the table. The geometric and material constants of the spherical shell are: $R = 1$ m, $\varphi_0 = 30°$, $\varphi_1 = 120°$, $\theta_0 = 135°$, $E_1/E_2 = 15$. It can be seen from

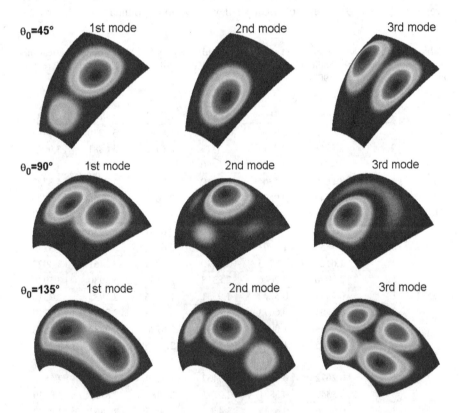

Fig. 7.6 Mode shapes for a three-layered, cross-ply [0°/90°/0°] open spherical shell with CCCC boundary conditions

Table 7.12 Frequency parameters Ω of a two-layered, [0°/90°] open spherical shell with various boundary conditions and thickness-to-radius ratios ($E_1/E_2 = 15$, $R = 1$ m, $\varphi_0 = 30°$, $\varphi_1 = 120°$, $\theta_0 = 135°$)

h/R	Mode	Boundary conditions					
		FSFS	FCFC	SFSF	CFCF	SSSS	CCCC
0.01	1	0.2170	0.2385	0.3064	0.3220	1.3909	1.4042
	2	0.2311	0.2497	0.3084	0.3316	1.3956	1.4096
	3	0.4366	0.4740	0.3241	0.3349	1.4012	1.4101
	4	0.4396	0.4764	0.3501	0.3661	1.4051	1.4141
	5	0.5461	0.5660	0.6141	0.6430	1.4484	1.4692
0.02	1	0.3261	0.3670	0.4443	0.4685	1.4315	1.4542
	2	0.3353	0.3696	0.4516	0.4712	1.4406	1.4646
	3	0.6198	0.6567	0.4744	0.5460	1.4588	1.4968
	4	0.6499	0.7466	0.5139	0.5736	1.4621	1.4996
	5	0.7735	0.8258	0.8558	0.9847	1.5877	1.6112
0.05	1	0.4795	0.5336	0.6014	0.7401	1.5142	1.5887
	2	0.5296	0.6386	0.6132	0.7515	1.5270	1.6471
	3	0.9001	0.9595	0.9518	1.0171	1.7359	1.7908
	4	0.9123	0.9649	0.9944	1.1131	1.7555	1.8162
	5	1.0897	1.1762	1.1657	1.2290	1.8956	2.0089
0.10	1	0.5744	0.6831	0.8821	1.0671	1.6844	1.8280
	2	0.7306	0.8728	0.8972	1.1196	1.7455	1.9653
	3	1.0258	1.1040	1.1061	1.1499	1.8224	2.0112
	4	1.0906	1.1740	1.5524	1.6027	2.0816	2.1895
	5	1.4615	1.6460	1.6362	1.6600	2.2685	2.4354

the table that the frequency parameters of the shell increase in general as the thickness-to-radius ratio increases. This is true due to the fact that the stiffness of the shell increases with the thickness increases. The second observation is that boundary conditions have a conspicuous effect on the vibration frequencies of the shell. Increasing the restraint stiffness always results in increments of the frequency parameters.

Table 7.13 lists the first five frequency parameters Ω of a laminated open spherical shell with various boundary conditions and initial meridional angles (φ_0). The shell with two-layered, cross-ply [0°/90°] lamination scheme is considered in the investigation. The shell is assumed to be thin ($R = 1$ m, $h/R = 0.05$) and with end meridional angle $\varphi_1 = 150°$, circumferential included angle $\theta_0 = 135°$. The composite layers of the shell are the same material as those in Table 7.12. Results show that the CCCC open spherical shell has the highest, whereas the SFSF one has the lowest frequency parameters. When the shell is subjected to FSFS and FCFC boundary conditions, the increase of the initial meridional angle φ_0 will result in a decrease of the fundamental frequency parameters. And for SFSF, CFCF, SSSS and CCCC boundary conditions, open spherical shells with higher initial meridional

Table 7.13 Frequency parameters Ω of a two-layered, $[0°/90°]$ open spherical shell with various boundary conditions and initial meridional angles ($E_1/E_2 = 15$, $R = 1$ m, $h/R = 0.05$, $\varphi_1 = 150°$, $\theta_0 = 135°$)

φ_0	Mode	Boundary conditions					
		FSFS	FCFC	SFSF	CFCF	SSSS	CCCC
30°	1	0.7913	0.8606	0.3562	0.4241	1.3948	1.4372
	2	0.8675	0.9443	0.4263	0.4738	1.4165	1.4871
	3	0.9428	0.9772	0.4988	0.5397	1.4629	1.4919
	4	1.1628	1.3136	0.5800	0.6621	1.5091	1.5722
	5	1.1711	1.3531	0.7213	0.7663	1.7329	1.7520
60°	1	0.4795	0.5336	0.6014	0.7401	1.5142	1.5887
	2	0.5296	0.6386	0.6132	0.7515	1.5270	1.6471
	3	0.9001	0.9595	0.9518	1.0171	1.7359	1.7908
	4	0.9123	0.9649	0.9944	1.1131	1.7555	1.8162
	5	1.0897	1.1762	1.1657	1.2290	1.8956	2.0089
90°	1	0.3593	0.4350	1.0909	1.4384	1.8532	2.1094
	2	0.4142	0.5162	1.0978	1.4497	1.8793	2.1130
	3	0.8098	0.9343	1.7959	1.8772	2.1059	2.2805
	4	0.8458	0.9411	1.8827	1.9611	2.2063	2.3482
	5	1.1502	1.1620	1.9047	2.0362	2.2108	2.4266
120°	1	0.3350	0.4949	1.8017	3.0062	2.2359	3.2276
	2	0.4561	0.5823	1.8037	3.0096	2.2756	3.2373
	3	0.8565	1.0304	2.1716	3.1549	2.6541	3.5375
	4	1.0307	1.2126	2.3206	3.2555	2.8179	3.7615
	5	1.4168	1.4915	2.4204	3.2888	3.4268	4.1770

angle yield higher frequency parameters. Furthermore, it is interesting to find that the frequency parameters of the shell with S and C boundary conditions in the $\varphi = $ constant edges are lower than those of $\theta = $ constant edges when $\varphi_0 = 30°$.

At the end, the influence of circumferential included angle θ_0 on the frequency parameters of open laminated spherical shells is investigated. A four-layered $[0°/90°/0°/90°]$ open spherical shell with mean radius $R = 1$ m, end meridional angle $\varphi_1 = 120°$, thickness-to-radius ratio $h/R = 0.1$ is investigated. The layers of the shell are composed of composite material having orthotropy ratio $E_1/E_2 = 15$. Three different initial meridional angles, i.e., $\varphi_0 = 30°$, $60°$ and $90°$ are performed in the investigation. Figure 7.7 shows the variation of the lowest three frequency parameters Ω of the shell with SSSS boundary condition against the circumferential included angle θ_0. As observed from the figure, the frequency parameter traces of the shell decline when the circumferential included angle θ_0 is varied from $30°$ to $330°$ by a step of $10°$. It is attributed to the stiffness of the shell decreases with circumferential included angle θ_0 increases. Furthermore, it is obvious that the effect of the circumferential included angle θ_0 is much higher for open spherical

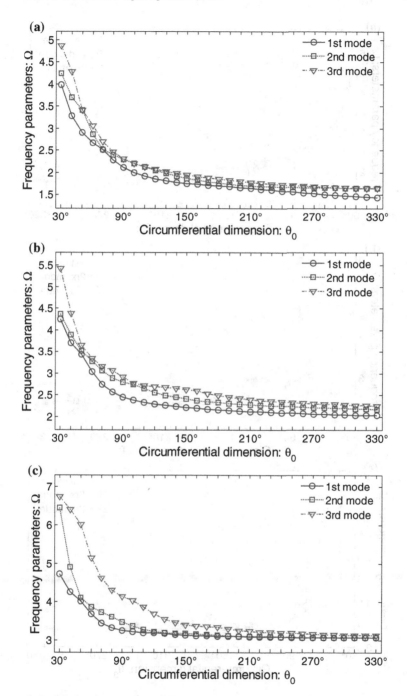

Fig. 7.7 Variation of frequency parameters Ω versus θ_0 for a four-layered [0°/90°/0°/90°] open spherical shell with SSSS boundary conditions: **a** $\varphi_0 = 30°$; **b** $\varphi_0 = 60°$; **c** $\varphi_0 = 90°$

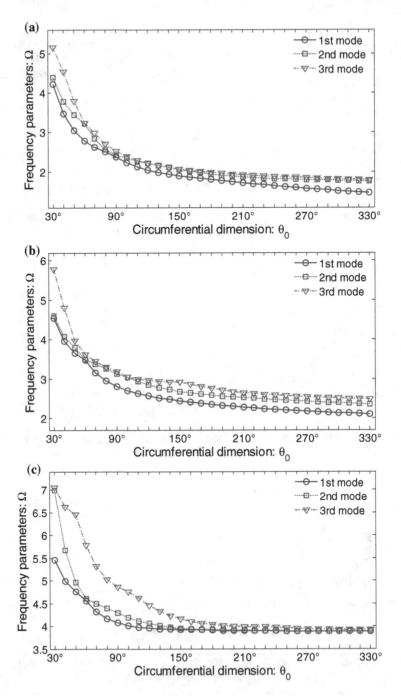

Fig. 7.8 Variation of frequency parameters Ω versus θ_0 for a four-layered $[0°/90°/0°/90°]$ open spherical shell with CCCC boundary conditions: **a** $\varphi_0 = 30°$; **b** $\varphi_0 = 60°$; **c** $\varphi_0 = 90°$

with lower circumferential included angle. Change θ_0 from 30° to 150° can result in frequency parameters that are more than twice lower. However, increase θ_0 from 150° to 330° yields less than 20 % decrements of the frequency parameters. Figure 7.8 shows the similar studies for the open spherical shell with CCCC boundary conditions. The similar observations can be seen in this figure.

Chapter 8
Shallow Shells

Shallow shells are open shells that have small curvatures (i.e. large radii of curvatures compared with other shell parameters such as length and width). Vlasov
(1951) and Leissa (1973) describe a shallow shell as follows:

> Consider a shell outlined in part by some surface and which is a thin-walled spatial
> structure with a comparatively small rise above the plane covered by this structure. We call
> such shells shallow. If, for example, a building which has a rectangular floor plan is covered
> by a shell with a rise of not more than 1/5 of the smallest side of the rectangle lying in the
> plane of the supporting points of the structure, then we class such a spatial structure in the
> category of shallow shells.

These shells are sometimes referred to as curved plates. The general shell theory
given in Chap. 1 can be readily applied to shallow shells. However, the general shell
equations are rare complicated because the bending of a general shell is coupling with
its stretching. When a shell is shallow as previous described, certain additional
assumptions can be made to reduce the complexity in general shell equations considerably. The resulting set of equations is referred to as shallow shell theory. The
development of the shallow shell theory is principally credited to, Leissa (1973),
Leissa et al. (1984) and Qatu (2004), etc. The classical shallow shell theories (CSST)
and shear deformation shallow shell theory (SDSST) are obtained by making following additional assumptions to the general shell theories (Qatu 2004):

1. The radii of curvature are very large compared to the inplane displacements (i.e.,
   the curvature changes caused by the tangential displacement components u and
   v are small in a shallow shell, in comparison with changes caused by the normal
   component w). Also, the transverse shear forces are much smaller than the term
   $R_i \partial N_i / \partial i$:

$$\frac{u_i}{R_i} \ll 1 \qquad \frac{Q_i}{R_i} \ll \frac{\partial N_i}{\partial i} \qquad (8.1)$$

where u_i is either of the inplane displacement components u and v; Q_i is either of
the shear forces Q_α and Q_β; N_i is N_α, N_β or $N_{\alpha\beta}$; and R_i is R_α, R_β or $R_{\alpha\beta}$; The term
∂i indicates derivative with respect to either α or β;

© Science Press, Beijing and Springer-Verlag Berlin Heidelberg 2015
G. Jin et al., *Structural Vibration*, DOI 10.1007/978-3-662-46364-2_8

2. The deepness term $(1 + z/R_i)$ is close to 1; where R_i can be R_α, R_β or $R_{\alpha\beta}$;
3. The shell is shallow enough to be represented by the plane coordinate systems. For the case of rectangular orthotropy, this leads to constant Lamé parameters.

It should be stressed that the shallow shell theories should be used for maximum span to minimum radius ratio of 1/2 or less.

With the progress of composite materials, shallow shells constructed by composite laminas are extensively used in many fields of modern engineering practices requiring high strength-weight and stiffness-weight ratios such as aircraft structures, space vehicles. A complete understanding of the bucking, bending, vibration and other characteristics of these shells is of particular importance. Laminated shallow shells can be formed as rectangular, triangular, trapezoidal, circular or any other planforms and various types of curvatures such as singly-curved (e.g., cylindrical), double-curved (e.g., spherical, hyperbolic paraboloidal) or other complex shapes such as turbomachinery blades. In the context of this chapter, we consider laminated shallow shells formed in rectangular planform with rectangular orthotropy, in which the fibers in each layer to the planform being straight.

In recent decades, a huge amount of research efforts have been devoted to the vibration analysis of laminated shallow shells. So far, some of the static and dynamic behaviors of these shells with classical boundary conditions had being presented precisely. Some research papers and articles oriented to such contributions may be found in following enumeration. Fazzolari and Carrera (2013) developed a hierarchical trigonometric Ritz formulation for free vibration and dynamic response analysis of doubly-curved anisotropic laminated shallow and deep shells. Reddy and Asce (1984) presented the exact solutions of the equations and fundamental frequencies for simply supported, doubly-curved, cross-ply laminated shells. Khdeir and Reddy (1997) predicted free and force vibration of cross-ply laminated composite shallow arches by a generalized modal approach. Qatu (1995a) studied natural vibration of completely free laminated composite triangular and trapezoidal shallow shells. Soldatos and Shu (1999) used the five-degrees-of-freedom shallow shell theory in the stress analysis of cross-ply laminated plates and shallow shell panels having a rectangular plan-form. Some other contributors in this subject are Dogan and Arslan (2009), Ghavanloo and Fazelzadeh (2013), Kurpa et al. (2010), Leissa and Chang (1996), Librescu et al. (1989a, b), Qatu (1995a, b, 1996, 2011), Qatu and Leissa (1991), Singh and Kumar (1996). More detailed and systematic summarizations can be seen in the excellent monographs by Leissa (1973), Qatu (2002a, b, 2004), Qatu et al. (2010), and Reddy (2003).

In this chapter, we consider vibration of thin and moderately thick laminated shallow shells with general boundary conditions. Fundamental equations of thin and thick shallow shells are presented in the first and second sections, respectively. Then, numerous vibration results of thin and thick laminated shallow shells with different boundary conditions, lamination schemes and geometry parameters are given in the third section by using the SDSST and the modified Fourier series. The results are obtained by applying the weak form solution procedure.

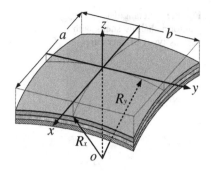

Fig. 8.1 Schematic diagram of laminated shallow shells with rectangular planform

Consider a laminated shallow shell in rectangular planform as shown in Fig. 8.1, the length, width and thickness of the shell are represented by a, b and h, respectively. The shell is shallow enough so that is can be represented by the orthogonal Cartesian coordinate system (x, y and z). The laminated shallow shell is characterized by its middle surface, which can be defined by (Qatu 2004):

$$z = -\frac{1}{2}\left(\frac{x^2}{R_x} + \frac{xy}{R_{xy}} + \frac{y^2}{R_y}\right) \tag{8.2}$$

where R_x and R_y represent the radii of curvature in the x, y directions as depicted in Fig. 8.1. R_{xy} is the corresponding radius of twist. In the present chapter, we focus on the cases when R_x, R_y and R_{xy} are constants. In addition, the x and y coordinates are conveniently oriented to be parallel to boundaries so that $R_{xy} = \infty$. The displacements of the shell in the x, y and z directions are denoted by u, v and w, respectively.

In Fig. 8.2, shallow shells constructed as various types of curvatures are plotted. It can be flat (i.e. $R_x = R_y = R_{xy} = \infty$), spherical (e.g., $R_x/R_y > 0$, $R_{xy} = \infty$), circular cylindrical (e.g., $R_y = R_{xy} = \infty$) and hyperbolic paraboloidal (e.g., $R_x/R_y < 0$, $R_{xy} = \infty$).

Considering the shallow shell in Fig. 8.1 and its Cartesian coordinate system, the coordinates, characteristics of the Lamé parameters and radius of curvatures are:

$$\alpha = x \quad \beta = y \quad A = B = 1 \quad R_\alpha = R_x \quad R_\beta = R_y \tag{8.3}$$

The equations of shallow shells are a reduction of the general shell equations given in Chap. 1 by substituting Eq. (8.3) into such shell equations.

8.1 Fundamental Equations of Thin Laminated Shallow Shells

Fundamental equations of thin laminated shallow shells are given here by substituting Eq. (8.3) into the general thin shell equations developed in Sect. 1.2. Similarly, the equations are given for the general dynamic analysis.

Fig. 8.2 Rectangular
planform shallow shells with
various types of curvatures

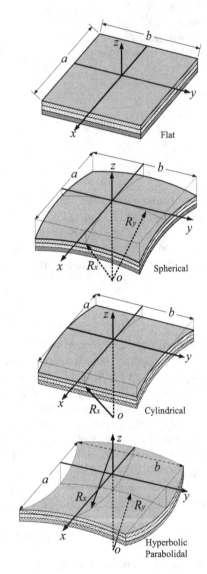

Flat

Spherical

Cylindrical

Hyperbolic
Parabolidal

8.1.1 Kinematic Relations

Substituting Eq. (8.3) into Eq. (1.7), the middle surface strains and curvature
changes of thin laminated shallow shells can be written in terms of middle surface
displacements. Taking Eq. (8.1) into consideration, they are given as:

$$\varepsilon_x^0 = \frac{\partial u}{\partial x} + \frac{w}{R_x}, \quad \chi_x = -\frac{\partial^2 w}{\partial x^2}$$

$$\varepsilon_y^0 = \frac{\partial v}{\partial y} + \frac{w}{R_y}, \quad \chi_y = -\frac{\partial^2 w}{\partial y^2} \tag{8.4}$$

$$\gamma_{xy}^0 = \frac{\partial u}{\partial y} + \frac{\partial v}{\partial x}, \quad \chi_{xy} = -2\frac{\partial^2 w}{\partial x \partial y}$$

where ε_x^0, ε_y^0 and γ_{xy}^0 indicate the strains in middle surface; χ_x, χ_y and χ_{xy} are the curvature changes. Thus, the liner strains in the kth layer space of a laminated shallow shell can be defined as:

$$\varepsilon_x = \varepsilon_x^0 + z\chi_x$$

$$\varepsilon_y = \varepsilon_y^0 + z\chi_y \tag{8.5}$$

$$\gamma_{xy} = \gamma_{xy}^0 + z\chi_{xy}$$

where $Z_k < z < Z_{k+1}$. Z_k and Z_{k+1} denote the distances from the top surface and the bottom surface of the layer to the referenced middle surface, respectively.

8.1.2 Stress-Strain Relations and Stress Resultants

The stress-strain relations, force and moment resultants for thin laminated shallow shells formed in rectangular planform are the same as those derived earlier for thin laminated rectangular plates, see Eqs. (4.4) and (4.6).

8.1.3 Energy Functions

The strain energy (U_s) of thin laminated shallow shells during vibration can be defined in terms of middle surface strains, curvature changes and stress resultants as:

$$U_s = \frac{1}{2} \int_0^a \int_0^b \left\{ \begin{array}{l} N_x\varepsilon_x^0 + N_y\varepsilon_y^0 + N_{xy}\gamma_{xy}^0 \\ + M_x\chi_x + M_y\chi_y + M_{xy}\chi_{xy} \end{array} \right\} dxdy \tag{8.6}$$

And the kinetic energy (T) of thin laminated shallow shells during vibration is:

$$T = \frac{1}{2} \int_0^a \int_0^b I_0 \left\{ \left(\frac{\partial u}{\partial t}\right)^2 + \left(\frac{\partial v}{\partial t}\right)^2 + \left(\frac{\partial w}{\partial t}\right)^2 \right\} dxdy \tag{8.7}$$

where the inertia term I_0 is given in Eq. (1.19). Suppose q_x, q_y and q_z are the external loads in the x, y and z directions, respectively. Thus, the external work can be expressed as:

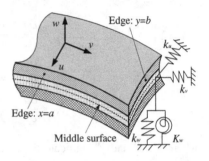

$$W_e = \int\limits_0^a \int\limits_0^b \{q_x u + q_y v + q_z w\} dx dy \qquad (8.8)$$

The artificial spring boundary technique is adopted here to realize the general boundary conditions of a thin shallow shell. Under the current framework, symbols k_ψ^u, k_ψ^v, k_ψ^w and K_ψ^w ($\psi = x0$, $y0$, $x1$ and $y1$) are used to indicate the stiffness of the boundary spring components at the boundaries $x = 0$, $y = 0$, $x = a$ and $y = b$, respectively, see Fig. 8.3. Thus, the deformation strain energy about the boundary springs (U_{sp}) can be defined as:

$$U_{sp} = \frac{1}{2} \int\limits_0^b \left\{ \begin{array}{l} \left[k_{x0}^u u^2 + k_{x0}^v v^2 + k_{x0}^w w^2 + K_{x0}^w (\partial w / \partial x) \right]|_{x=0} \\ + \left[k_{x1}^u u^2 + k_{x1}^v v^2 + k_{x1}^w w^2 + K_{x1}^w (\partial w / \partial x) \right]|_{x=a} \end{array} \right\} dy$$

$$+ \frac{1}{2} \int\limits_0^a \left\{ \begin{array}{l} \left[k_{y0}^u u^2 + k_{y0}^v v^2 + k_{y0}^w w^2 + K_{y0}^w (\partial w / \partial y) \right]|_{y=0} \\ + \left[k_{y1}^u u^2 + k_{y1}^v v^2 + k_{y1}^w w^2 + K_{y1}^w (\partial w / \partial y) \right]|_{y=b} \end{array} \right\} dx \qquad (8.9)$$

8.1.4 Governing Equations and Boundary Conditions

Substituting Eq. (8.3) into Eq. (1.28) and then simplifying the expressions and taking Eq. (8.1) into consideration, the reduced governing equations of thin laminated shallow shells are:

$$\frac{\partial N_x}{\partial x} + \frac{\partial N_{xy}}{\partial y} + q_x = I_0 \frac{\partial^2 u}{\partial t^2}$$

$$\frac{\partial N_{xy}}{\partial x} + \frac{\partial N_y}{\partial y} + q_y = I_0 \frac{\partial^2 v}{\partial t^2} \qquad (8.10)$$

$$-\left(\frac{N_x}{R_x} + \frac{N_y}{R_y} \right) + \frac{\partial^2 M_x}{\partial x^2} + 2 \frac{\partial^2 M_{xy}}{\partial x \partial y} + \frac{\partial^2 M_y}{\partial y^2} + q_z = I_0 \frac{\partial^2 w}{\partial t^2}$$

Substituting Eq. (8.4) and the stress resultants equations Eq. (4.6) into Eq. (8.10), the governing equations of thin laminated shallow shells can be writing in matrix form as:

$$
\left(
\begin{bmatrix}
L_{11} & L_{12} & L_{13} \\
L_{21} & L_{22} & L_{23} \\
L_{31} & L_{32} & L_{33}
\end{bmatrix}
- \omega^2
\begin{bmatrix}
-I_0 & 0 & 0 \\
0 & -I_0 & 0 \\
0 & 0 & -I_0
\end{bmatrix}
\right)
\begin{bmatrix}
u \\ v \\ w
\end{bmatrix}
=
\begin{bmatrix}
-p_x \\ -p_y \\ -p_z
\end{bmatrix}
\tag{8.11}
$$

The coefficients of the linear operator L_{ij} are written as

$$
L_{11} = \frac{\partial}{\partial x}\left(A_{11}\frac{\partial}{\partial x} + A_{16}\frac{\partial}{\partial y}\right) + \frac{\partial}{\partial y}\left(A_{16}\frac{\partial}{\partial x} + A_{66}\frac{\partial}{\partial y}\right)
$$

$$
L_{12} = L_{21} = \frac{\partial}{\partial x}\left(A_{12}\frac{\partial}{\partial y} + A_{16}\frac{\partial}{\partial x}\right) + \frac{\partial}{\partial y}\left(A_{26}\frac{\partial}{\partial y} + A_{66}\frac{\partial}{\partial x}\right)
$$

$$
L_{13} = \frac{\partial}{\partial x}\left(\frac{A_{11}}{R_x} + \frac{A_{12}}{R_y} - B_{11}\frac{\partial^2}{\partial x^2} - B_{12}\frac{\partial^2}{\partial y^2} - 2B_{16}\frac{\partial^2}{\partial x\partial y}\right)
$$
$$
+ \frac{\partial}{\partial y}\left(\frac{A_{16}}{R_x} + \frac{A_{26}}{R_y} - B_{16}\frac{\partial^2}{\partial x^2} - B_{26}\frac{\partial^2}{\partial y^2} - 2B_{66}\frac{\partial^2}{\partial x\partial y}\right)
$$

$$
L_{22} = \frac{\partial}{\partial x}\left(A_{26}\frac{\partial}{\partial y} + A_{66}\frac{\partial}{\partial x}\right) + \frac{\partial}{\partial y}\left(A_{22}\frac{\partial}{\partial y} + A_{26}\frac{\partial}{\partial x}\right)
$$

$$
L_{23} = \frac{\partial}{\partial x}\left(\frac{A_{16}}{R_x} + \frac{A_{26}}{R_y} - B_{16}\frac{\partial^2}{\partial x^2} - B_{26}\frac{\partial^2}{\partial y^2} - 2B_{66}\frac{\partial^2}{\partial x\partial y}\right)
$$
$$
+ \frac{\partial}{\partial y}\left(\frac{A_{12}}{R_x} + \frac{A_{22}}{R_y} - B_{12}\frac{\partial^2}{\partial x^2} - B_{22}\frac{\partial^2}{\partial y^2} - 2B_{26}\frac{\partial^2}{\partial x\partial y}\right)
$$

$$
L_{31} = -\frac{1}{R_x}\left(A_{11}\frac{\partial}{\partial x} + A_{16}\frac{\partial}{\partial y}\right) - \frac{1}{R_y}\left(A_{12}\frac{\partial}{\partial x} + A_{26}\frac{\partial}{\partial y}\right)
$$
$$
+ \frac{\partial^2}{\partial x^2}\left(B_{11}\frac{\partial}{\partial x} + B_{16}\frac{\partial}{\partial y}\right) + 2\frac{\partial^2}{\partial x\partial y}\left(B_{16}\frac{\partial}{\partial x} + B_{66}\frac{\partial}{\partial y}\right)
$$
$$
+ \frac{\partial^2}{\partial y^2}\left(B_{12}\frac{\partial}{\partial x} + B_{26}\frac{\partial}{\partial y}\right)
$$

$$
L_{32} = -\frac{1}{R_x}\left(A_{12}\frac{\partial}{\partial y} + A_{16}\frac{\partial}{\partial x}\right) - \frac{1}{R_y}\left(A_{22}\frac{\partial}{\partial y} + A_{26}\frac{\partial}{\partial x}\right)
$$
$$
+ \frac{\partial^2}{\partial x^2}\left(B_{12}\frac{\partial}{\partial y} + B_{16}\frac{\partial}{\partial x}\right) + 2\frac{\partial^2}{\partial x\partial y}\left(B_{26}\frac{\partial}{\partial y} + B_{66}\frac{\partial}{\partial x}\right)
$$
$$
+ \frac{\partial^2}{\partial y^2}\left(B_{22}\frac{\partial}{\partial y} + B_{26}\frac{\partial}{\partial x}\right)
$$

$$
L_{33} = -\frac{1}{R_x}\left(\frac{A_{11}}{R_x} + \frac{A_{12}}{R_y} - B_{11}\frac{\partial^2}{\partial x^2} - B_{12}\frac{\partial^2}{\partial y^2} - 2B_{16}\frac{\partial^2}{\partial x\partial y}\right)
$$
$$
- \frac{1}{R_y}\left(\frac{A_{12}}{R_x} + \frac{A_{22}}{R_y} - B_{12}\frac{\partial^2}{\partial x^2} - B_{22}\frac{\partial^2}{\partial y^2} - 2B_{26}\frac{\partial^2}{\partial x\partial y}\right)
$$
$$
+ \frac{\partial^2}{\partial x^2}\left(\frac{B_{11}}{R_x} + \frac{B_{12}}{R_y} - D_{11}\frac{\partial^2}{\partial x^2} - D_{12}\frac{\partial^2}{\partial y^2} - 2D_{16}\frac{\partial^2}{\partial x\partial y}\right)
$$
$$
+ 2\frac{\partial^2}{\partial x\partial y}\left(\frac{B_{16}}{R_x} + \frac{B_{26}}{R_y} - D_{16}\frac{\partial^2}{\partial x^2} - D_{26}\frac{\partial^2}{\partial y^2} - 2D_{66}\frac{\partial^2}{\partial x\partial y}\right)
$$
$$
+ \frac{\partial^2}{\partial y^2}\left(\frac{B_{12}}{R_x} + \frac{B_{22}}{R_y} - D_{12}\frac{\partial^2}{\partial x^2} - D_{22}\frac{\partial^2}{\partial y^2} - 2D_{26}\frac{\partial^2}{\partial x\partial y}\right)
$$

$$
\tag{8.12}
$$

And according to Eqs. (1.29) and (1.30), the boundary conditions of thin shallow shells are:

$$
x = 0 : \begin{cases} N_x + \frac{M_x}{R_x} - k_{x0}^u u = 0 \\ N_{xy} + \frac{M_{xy}}{R_y} - k_{x0}^v v = 0 \\ Q_x + \frac{\partial M_{xy}}{\partial y} - k_{x0}^w w = 0 \\ -M_x - K_{x0}^w \frac{\partial w}{\partial x} = 0 \end{cases}
\qquad
x = a : \begin{cases} N_x + \frac{M_x}{R_x} + k_{x1}^u u = 0 \\ N_{xy} + \frac{M_{xy}}{R_y} + k_{x1}^v v = 0 \\ Q_x + \frac{\partial M_{xy}}{\partial y} + k_{x1}^w w = 0 \\ -M_x + K_{x1}^w \frac{\partial w}{\partial x} = 0 \end{cases}
$$

$$
y = 0 : \begin{cases} N_{xy} + \frac{M_{xy}}{R_x} - k_{y0}^u u = 0 \\ N_y + \frac{M_y}{R_y} - k_{y0}^v v = 0 \\ Q_y + \frac{\partial M_{xy}}{\partial x} - k_{y0}^w w = 0 \\ -M_y - K_{y0}^w \frac{\partial w}{\partial y} = 0 \end{cases}
\qquad
y = 0 : \begin{cases} N_{xy} + \frac{M_{xy}}{R_x} + k_{y1}^u u = 0 \\ N_y + \frac{M_y}{R_y} + k_{y1}^v v = 0 \\ Q_y + \frac{\partial M_{xy}}{\partial x} + k_{y1}^w w = 0 \\ -M_y + K_{y1}^w \frac{\partial w}{\partial y} = 0 \end{cases}
\qquad (8.13)
$$

In each boundary of thin laminated shallow shells, there exists 12 possible classical boundary conditions. Taking boundaries x = constant for example, the possible classical boundary conditions are given in Table 8.1, similar boundary conditions can be obtained for boundaries y = constant.

Table 8.1 Possible classical boundary conditions for thin shallow shells at each boundary of x = constant

Boundary type	Conditions
Free boundary conditions	
F	$N_x + \frac{M_x}{R_x} = N_{xy} + \frac{M_{xy}}{R_y} = Q_x + \frac{\partial M_{xy}}{\partial y} = M_x = 0$
F2	$u = N_{xy} + \frac{M_{xy}}{R_y} = Q_x + \frac{\partial M_{xy}}{\partial y} = M_x = 0$
F3	$N_x + \frac{M_x}{R_x} = v = Q_x + \frac{\partial M_{xy}}{\partial y} = M_x = 0$
F4	$u = v = Q_x + \frac{\partial M_{xy}}{\partial y} = M_x = 0$
Simply supported boundary conditions	
S	$u = v = w = M_x = 0$
SD	$N_x + \frac{M_x}{R_x} = v = w = M_x = 0$
S3	$u = N_{xy} + \frac{M_{xy}}{R_y} = w = M_x = 0$
S4	$N_x + \frac{M_x}{R_x} = N_{xy} + \frac{M_{xy}}{R_y} = w = M_x = 0$
Clamped boundary conditions	
C	$u = v = w = \frac{\partial w}{\partial x} = 0$
C2	$N_x + \frac{M_x}{R_x} = v = w = \frac{\partial w}{\partial x} = 0$
C3	$u = N_{xy} + \frac{M_{xy}}{R_y} = w = \frac{\partial w}{\partial x} = 0$
C4	$N_x + \frac{M_x}{R_x} = N_{xy} + \frac{M_{xy}}{R_y} = w = \frac{\partial w}{\partial x} = 0$

8.2 Fundamental Equations of Thick Laminated Shallow Shells

Like general thin shell theory, the classical shallow shell theory (CSST) is applicable where the thickness is smaller than 1/20 of the smallest of the wave length and/or radii of curvature (Qatu 2004). For thick shallow shells, the Kirchhoff hypothesis should be relaxed and the shear deformation and rotary inertia should be included in the formulation. Fundamental equations of thick laminated shallow shells will be derived in this section. The following equations are derived from the general shear deformation shell theory (SDST) described in Sect. 1.3 by imposing Eq. (8.3) into those of general shells.

8.2.1 Kinematic Relations

As was done for general shells and plates, we assume that normals to the undeformed middle surface remain straight but do not normal to the deformed middle surface and the shell inplane displacements are expanded in terms of shell thickness of first order expansion. Thus, the displacement field of a thick shallow shell can be expressed as

$$
\begin{aligned}
U(x,y,z) &= u(x,y) + z\phi_x \\
V(x,y,z) &= v(x,y) + z\phi_y \\
W(x,y,z) &= w(x,y)
\end{aligned}
\tag{8.14}
$$

where u, v and w are the middle surface displacements of the shallow shell in the x, y and z directions, respectively, and ϕ_x and ϕ_y represent the rotations of transverse normal respect to y- and x-axes. Substituting $\alpha = x$, $\beta = y$ into Eq. (1.33) and deleting the z/R_α and z/R_β terms, the normal and shear strains at any point in the shallow shell then can be written in terms of middle surface strains and curvature changes as:

$$
\begin{aligned}
\varepsilon_x &= \varepsilon_x^0 + z\chi_x, \quad \gamma_{xz} = \gamma_{xz}^0 \\
\varepsilon_y &= \varepsilon_y^0 + z\chi_y, \quad \gamma_{yz} = \gamma_{yz}^0 \\
\gamma_{xy} &= \gamma_{xy}^0 + z\chi_{xy}
\end{aligned}
\tag{8.15}
$$

Substituting Eq. (8.3) into Eq. (1.34) and taking Eq. (8.1) into consideration, the middle surface strains and curvature changes are:

$$
\begin{aligned}
\varepsilon_x^0 &= \frac{\partial u}{\partial x} + \frac{w}{R_x}, & \chi_x &= \frac{\partial \phi_x}{\partial x} \\
\varepsilon_y^0 &= \frac{\partial v}{\partial y} + \frac{w}{R_y}, & \chi_y &= \frac{\partial \phi_y}{\partial y} \\
\gamma_{xy}^0 &= \frac{\partial v}{\partial x} + \frac{\partial u}{\partial y}, & \chi_{xy} &= \frac{\partial \phi_y}{\partial x} + \frac{\partial \phi_x}{\partial y} \\
\gamma_{xz}^0 &= \frac{\partial w}{\partial x} + \phi_x, & \gamma_{yz}^0 &= \frac{\partial w}{\partial y} + \phi_y
\end{aligned}
\tag{8.16}
$$

The above equations constitute the fundamental strain-displacement relations of a thick shallow shell formed in rectangular planform.

8.2.2 Stress-Strain Relations and Stress Resultants

The stress-strain relations and stress resultants for thick laminated shallow shells formed in rectangular planform are the same as those derived earlier for thick laminated rectangular plates, see Eqs. (4.19), (4.21) and (4.22).

8.2.3 Energy Functions

The strain energy (U_s) of thick shallow shells during vibration can be defined as

$$
U_s = \frac{1}{2} \int_0^a \int_0^b \left\{ \begin{array}{l} N_x \varepsilon_x^0 + N_y \varepsilon_y^0 + N_{xy} \varepsilon_{xy}^0 + N_{yx} \varepsilon_{yx}^0 + M_x \chi_x \\ + M_y \chi_y + M_{xy} \chi_{xy} + M_{yx} \chi_{yx} + Q_y \gamma_{yz} + Q_x \gamma_{xz} \end{array} \right\} dy dx \tag{8.17}
$$

And the corresponding kinetic energy function (T) is:

$$
T = \frac{1}{2} \int_0^a \int_0^b \left\{ \begin{array}{l} I_0 \left(\frac{\partial u}{\partial t}\right)^2 + 2I_1 \frac{\partial u}{\partial t}\frac{\partial \phi_x}{\partial t} + I_2 \left(\frac{\partial \phi_x}{\partial t}\right)^2 + I_0 \left(\frac{\partial v}{\partial t}\right)^2 \\ + 2I_1 \frac{\partial v}{\partial t}\frac{\partial \phi_y}{\partial t} + I_2 \left(\frac{\partial \phi_y}{\partial t}\right)^2 + I_0 \left(\frac{\partial w}{\partial t}\right)^2 \end{array} \right\} dy dx \tag{8.18}
$$

The inertia terms I_i ($i = 0, 1, 2$) are written as in Eq. (1.52). Assuming the distributed external forces q_x, q_y and q_z are in the x, y and z directions, respectively and m_x and m_y represent the external couples in the middle surface, thus, the work done by the external forces and moments is

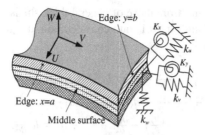

Fig. 8.4 Boundary conditions of thick laminated shallow shells

$$W_e = \int_0^a \int_0^b \{q_x u + q_y v + q_z w + m_x \phi_x + m_y \phi_y\} dy dx \qquad (8.19)$$

As was done for thin laminated shallow shells, the artificial spring boundary technique is adopted here to realize the general boundary conditions of thick laminated shallow shells, in which three groups of linear springs (k_u, k_v, k_w) and two groups of rotational springs (K_x, K_y) are distributed uniformly along each shell boundary artificially, see Fig. 8.4. Therefore, the deformation strain energy (U_{sp}) of the boundary springs during vibration can be written as:

$$U_{sp} = \frac{1}{2} \int_0^b \left\{ \begin{array}{l} \left[k_{x0}^u u^2 + k_{x0}^v v^2 + k_{x0}^w w^2 + K_{x0}^x \phi_x^2 + K_{x0}^y \phi_y^2\right]|_{x=0} \\ + \left[k_{x1}^u u^2 + k_{x1}^v v^2 + k_{x1}^w w^2 + K_{x1}^x \phi_x^2 + K_{x1}^y \phi_y^2\right]|_{x=a} \end{array} \right\} dy$$

$$+ \frac{1}{2} \int_0^a \left\{ \begin{array}{l} \left[k_{y0}^u u^2 + k_{y0}^v v^2 + k_{y0}^w w^2 + K_{y0}^x \phi_x^2 + K_{y0}^y \phi_y^2\right]|_{y=0} \\ + \left[k_{y1}^u u^2 + k_{y1}^v v^2 + k_{y1}^w w^2 + K_{y0}^x \phi_x^2 + K_{y0}^y \phi_y^2\right]|_{y=b} \end{array} \right\} dx \qquad (8.20)$$

where k_ψ^u, k_ψ^v, k_ψ^w, K_ψ^x and K_ψ^y ($\psi = x0$, $y0$, $x1$ and $y1$) represent the rigidities (per unit length) of the boundary springs at the boundaries $x = 0$, $y = 0$, $x = a$ and $y = b$, respectively. In such case, the stiffness of the boundary springs can take any value from zero to infinity to better model many real-world boundary conditions including all the classical boundaries and the uniform elastic ones. For instance, the clamped restraint condition is essentially obtained by setting the spring stiffness substantially larger than the bending rigidity of the involved shallow shell ($10^7 \times D$).

8.2.4 Governing Equations and Boundary Conditions

The governing equations and boundary conditions of thick laminated shallow shells can be derived by the Hamilton's principle in the same manner as described in

Sect. 1.2.4 or specializing from those of general thick shells. Substituting Eq. (8.3) into Eq. (1.59) and taking Eq. (8.1) into consideration, the governing equations for thick shallow shells are

$$
\begin{aligned}
&\frac{\partial N_x}{\partial x} + \frac{\partial N_{xy}}{\partial y} + q_x = I_0 \frac{\partial^2 u}{\partial t^2} + I_1 \frac{\partial^2 \phi_x}{\partial t^2} \\
&\frac{\partial N_{xy}}{\partial x} + \frac{\partial N_y}{\partial y} + q_y = I_0 \frac{\partial^2 v}{\partial t^2} + I_1 \frac{\partial^2 \phi_y}{\partial t^2} \\
&-\left(\frac{N_x}{R_x} + \frac{N_y}{R_y}\right) + \frac{\partial Q_x}{\partial x} + \frac{\partial Q_y}{\partial y} + q_z = I_0 \frac{\partial^2 w}{\partial t^2} \\
&\frac{\partial M_x}{\partial x} + \frac{\partial M_{xy}}{\partial y} - Q_x + m_x = I_1 \frac{\partial^2 u}{\partial t^2} + I_2 \frac{\partial^2 \phi_x}{\partial t^2} \\
&\frac{\partial M_{xy}}{\partial x} + \frac{\partial M_y}{\partial y} - Q_y + m_y = I_1 \frac{\partial^2 v}{\partial t^2} + I_2 \frac{\partial^2 \phi_y}{\partial t^2}
\end{aligned}
\tag{8.21}
$$

Furthermore, the above equations can be written in terms of displacements and rotation components as

$$
\begin{aligned}
&\left(\begin{bmatrix}
L_{11} & L_{12} & L_{13} & L_{14} & L_{15} \\
L_{21} & L_{22} & L_{23} & L_{24} & L_{25} \\
L_{31} & L_{32} & L_{33} & L_{34} & L_{35} \\
L_{41} & L_{42} & L_{43} & L_{44} & L_{45} \\
L_{51} & L_{52} & L_{53} & L_{54} & L_{55}
\end{bmatrix} - \omega^2 \begin{bmatrix}
M_{11} & 0 & 0 & M_{14} & 0 \\
0 & M_{22} & 0 & 0 & M_{25} \\
0 & 0 & M_{33} & 0 & 0 \\
M_{41} & 0 & 0 & M_{44} & 0 \\
0 & M_{52} & 0 & 0 & M_{55}
\end{bmatrix}\right)\begin{bmatrix}
u \\ v \\ w \\ \phi_x \\ \phi_y
\end{bmatrix} \\
&= \begin{bmatrix}
-p_x \\ -p_y \\ -p_z \\ -m_x \\ -m_y
\end{bmatrix}
\end{aligned}
\tag{8.22}
$$

The coefficients of the linear operator L_{ij} are listed below

$$
\begin{aligned}
L_{11} &= \frac{\partial}{\partial x}\left(A_{11}\frac{\partial}{\partial x} + A_{16}\frac{\partial}{\partial y}\right) + \frac{\partial}{\partial y}\left(A_{16}\frac{\partial}{\partial x} + A_{66}\frac{\partial}{\partial y}\right) \\
L_{12} &= \frac{\partial}{\partial x}\left(A_{12}\frac{\partial}{\partial y} + A_{16}\frac{\partial}{\partial x}\right) + \frac{\partial}{\partial y}\left(A_{26}\frac{\partial}{\partial y} + A_{66}\frac{\partial}{\partial x}\right) \\
L_{13} &= \frac{\partial}{\partial x}\left(\frac{A_{11}}{R_x} + \frac{A_{12}}{R_y}\right) + \frac{\partial}{\partial y}\left(\frac{A_{16}}{R_x} + \frac{A_{26}}{R_y}\right) \\
L_{14} &= \frac{\partial}{\partial x}\left(B_{11}\frac{\partial}{\partial x} + B_{16}\frac{\partial}{\partial y}\right) + \frac{\partial}{\partial y}\left(B_{16}\frac{\partial}{\partial x} + B_{66}\frac{\partial}{\partial y}\right) \\
L_{15} &= \frac{\partial}{\partial x}\left(B_{12}\frac{\partial}{\partial y} + B_{16}\frac{\partial}{\partial x}\right) + \frac{\partial}{\partial y}\left(B_{26}\frac{\partial}{\partial y} + B_{66}\frac{\partial}{\partial x}\right) \\
L_{21} &= \frac{\partial}{\partial x}\left(A_{16}\frac{\partial}{\partial x} + A_{66}\frac{\partial}{\partial y}\right) + \frac{\partial}{\partial y}\left(A_{12}\frac{\partial}{\partial x} + A_{26}\frac{\partial}{\partial y}\right)
\end{aligned}
\tag{8.23}
$$

$$L_{22} = \frac{\partial}{\partial x}\left(A_{26}\frac{\partial}{\partial y} + A_{66}\frac{\partial}{\partial x}\right) + \frac{\partial}{\partial y}\left(A_{22}\frac{\partial}{\partial y} + A_{26}\frac{\partial}{\partial x}\right)$$

$$L_{23} = \frac{\partial}{\partial x}\left(\frac{A_{16}}{R_x} + \frac{A_{26}}{R_y}\right) + \frac{\partial}{\partial y}\left(\frac{A_{12}}{R_x} + \frac{A_{22}}{R_y}\right)$$

$$L_{24} = \frac{\partial}{\partial x}\left(B_{16}\frac{\partial}{\partial x} + B_{66}\frac{\partial}{\partial y}\right) + \frac{\partial}{\partial y}\left(B_{12}\frac{\partial}{\partial x} + B_{26}\frac{\partial}{\partial y}\right)$$

$$L_{25} = \frac{\partial}{\partial x}\left(B_{26}\frac{\partial}{\partial y} + B_{66}\frac{\partial}{\partial x}\right) + \frac{\partial}{\partial y}\left(B_{22}\frac{\partial}{\partial y} + B_{26}\frac{\partial}{\partial x}\right)$$

$$L_{31} = -\frac{1}{R_x}\left(A_{11}\frac{\partial}{\partial x} + A_{16}\frac{\partial}{\partial y}\right) - \frac{1}{R_y}\left(A_{12}\frac{\partial}{\partial x} + A_{26}\frac{\partial}{\partial y}\right)$$

$$L_{32} = -\frac{1}{R_x}\left(A_{12}\frac{\partial}{\partial y} + A_{16}\frac{\partial}{\partial x}\right) - \frac{1}{R_y}\left(A_{22}\frac{\partial}{\partial y} + A_{26}\frac{\partial}{\partial x}\right)$$

$$L_{33} = -\frac{1}{R_x}\left(\frac{A_{11}}{R_x} + \frac{A_{12}}{R_y}\right) - \frac{1}{R_y}\left(\frac{A_{11}}{R_x} + \frac{A_{12}}{R_y}\right) + \frac{\partial}{\partial x}\left(A_{45}\frac{\partial}{\partial y} + A_{55}\frac{\partial}{\partial x}\right)$$
$$+ \frac{\partial}{\partial y}\left(A_{44}\frac{\partial}{\partial y} + A_{45}\frac{\partial}{\partial x}\right)$$

$$L_{34} = -\frac{1}{R_x}\left(B_{11}\frac{\partial}{\partial x} + B_{16}\frac{\partial}{\partial y}\right) - \frac{1}{R_y}\left(B_{12}\frac{\partial}{\partial x} + B_{26}\frac{\partial}{\partial y}\right) + A_{55}\frac{\partial}{\partial x} + A_{45}\frac{\partial}{\partial y}$$

$$L_{35} = -\frac{1}{R_x}\left(B_{12}\frac{\partial}{\partial y} + B_{16}\frac{\partial}{\partial x}\right) - \frac{1}{R_y}\left(B_{22}\frac{\partial}{\partial y} + B_{26}\frac{\partial}{\partial x}\right) + A_{45}\frac{\partial}{\partial x} + A_{44}\frac{\partial}{\partial y}$$

$$L_{41} = \frac{\partial}{\partial x}\left(B_{11}\frac{\partial}{\partial x} + B_{16}\frac{\partial}{\partial y}\right) + \frac{\partial}{\partial y}\left(B_{16}\frac{\partial}{\partial x} + B_{66}\frac{\partial}{\partial y}\right)$$

$$L_{42} = \frac{\partial}{\partial x}\left(B_{12}\frac{\partial}{\partial y} + B_{16}\frac{\partial}{\partial x}\right) + \frac{\partial}{\partial y}\left(B_{26}\frac{\partial}{\partial y} + B_{66}\frac{\partial}{\partial x}\right)$$

$$L_{43} = \frac{\partial}{\partial x}\left(\frac{B_{11}}{R_x} + \frac{B_{12}}{R_y}\right) + \frac{\partial}{\partial y}\left(\frac{B_{16}}{R_x} + \frac{B_{26}}{R_y}\right) - \left(A_{45}\frac{\partial}{\partial y} + A_{55}\frac{\partial}{\partial x}\right)$$

$$L_{44} = \frac{\partial}{\partial x}\left(D_{11}\frac{\partial}{\partial x} + D_{16}\frac{\partial}{\partial y}\right) + \frac{\partial}{\partial y}\left(D_{16}\frac{\partial}{\partial x} + D_{66}\frac{\partial}{\partial y}\right) - A_{55}$$

$$L_{45} = \frac{\partial}{\partial x}\left(D_{12}\frac{\partial}{\partial y} + D_{16}\frac{\partial}{\partial x}\right) + \frac{\partial}{\partial y}\left(D_{26}\frac{\partial}{\partial y} + D_{66}\frac{\partial}{\partial x}\right) - A_{45}$$

$$L_{51} = \frac{\partial}{\partial x}\left(B_{16}\frac{\partial}{\partial x} + B_{66}\frac{\partial}{\partial y}\right) + \frac{\partial}{\partial y}\left(B_{12}\frac{\partial}{\partial x} + B_{26}\frac{\partial}{\partial y}\right)$$

$$L_{52} = \frac{\partial}{\partial x}\left(B_{26}\frac{\partial}{\partial y} + B_{66}\frac{\partial}{\partial x}\right) + \frac{\partial}{\partial y}\left(B_{22}\frac{\partial}{\partial y} + B_{26}\frac{\partial}{\partial x}\right)$$

$$L_{53} = \frac{\partial}{\partial x}\left(\frac{B_{16}}{R_x} + \frac{B_{26}}{R_y}\right) + \frac{\partial}{\partial y}\left(\frac{B_{12}}{R_x} + \frac{B_{22}}{R_y}\right) - \left(A_{44}\frac{\partial}{\partial y} + A_{45}\frac{\partial}{\partial x}\right)$$

$$L_{54} = \frac{\partial}{\partial x}\left(D_{16}\frac{\partial}{\partial x} + D_{66}\frac{\partial}{\partial y}\right) + \frac{\partial}{\partial y}\left(D_{12}\frac{\partial}{\partial x} + D_{26}\frac{\partial}{\partial y}\right) - A_{45}$$

$$L_{55} = \frac{\partial}{\partial x}\left(D_{26}\frac{\partial}{\partial y} + D_{66}\frac{\partial}{\partial x}\right) + \frac{\partial}{\partial y}\left(D_{22}\frac{\partial}{\partial y} + D_{26}\frac{\partial}{\partial x}\right) - A_{44}$$

$$M_{11} = M_{22} = M_{33} = -I_0$$

$$M_{14} = M_{41} = M_{15} = M_{51} = -I_1$$

$$M_{44} = M_{55} = -I_2$$

And according to Eqs. (1.60), (1.61) and (8.3), the general boundary conditions of thick shallow shells are

$$x = 0: \begin{cases} N_x - k_{x0}^u u = 0 \\ N_{xy} - k_{x0}^v v = 0 \\ Q_x - k_{x0}^w w = 0 \\ M_x - K_{x0}^x \phi_x = 0 \\ M_{xy} - K_{x0}^y \phi_y = 0 \end{cases} \quad x = a: \begin{cases} N_x + k_{x1}^u u = 0 \\ N_{xy} + k_{x1}^v v = 0 \\ Q_x + k_{x1}^w w = 0 \\ M_x + K_{x1}^x \phi_x = 0 \\ M_{xy} + K_{x1}^y \phi_y = 0 \end{cases}$$

$$y = 0: \begin{cases} N_{yx} - k_{y0}^u u = 0 \\ N_y - k_{y0}^v v = 0 \\ Q_y - k_{y0}^w w = 0 \\ M_{yx} - K_{y0}^x \phi_x = 0 \\ M_y - K_{y0}^y \phi_y = 0 \end{cases} \quad y = 0: \begin{cases} N_{yx} + k_{y1}^u u = 0 \\ N_y + k_{y1}^v v = 0 \\ Q_y + k_{y1}^w w = 0 \\ M_{yx} + K_{y1}^x \phi_x = 0 \\ M_y + K_{y1}^y \phi_y = 0 \end{cases} \quad (8.24)$$

For thick shallow shells, there exists 24 possible classical boundary conditions in each boundary. The classification of the classical boundary conditions shown in Table 4.2 for thick rectangular plates is applicable for thick shallow shells in rectangular planform.

The classical boundary conditions of laminated shallow shells can be readily realized by applying the artificial spring boundary technique. In this chapter, we mainly consider the F, SD, S and C boundary conditions. Taking edge $x = 0$ for example, the corresponding spring rigidities for the four classical boundary conditions are given as in Eq. (4.30).

8.3 Vibration of Laminated Shallow Shells

Free vibration of laminated shallow shells formed on rectangular planforms with general boundary conditions, arbitrary lamination schemes and various types of curvatures will be considered, including shallow cylindrical, spherical and hyperbolic paraboloidal shells, see Fig. 8.2. Solutions in the framework of shear deformation shallow shell theory (SDSST) will be presented.

From Eq. (8.23), it is obvious that each displacements/rotation component of thick laminated shallow shells is required to have up to the second derivatives. Therefore, regardless of boundary conditions, each displacements/rotation component of a laminated shallow shell is expressed as the proposed modified Fourier series expansion in which several auxiliary functions are introduced to ensure and accelerate the convergence of the series expansion:

$$
\begin{aligned}
u(x,y) &= \sum_{m=0}^{M}\sum_{n=0}^{N} A_{mn}\cos\lambda_m x\cos\lambda_n y + \sum_{l=1}^{2}\sum_{n=0}^{N} a_{ln}P_l(x)\cos\lambda_n y \\
&\quad + \sum_{l=1}^{2}\sum_{m=0}^{M} b_{lm}P_l(y)\cos\lambda_m x \\
v(x,y) &= \sum_{m=0}^{M}\sum_{n=0}^{N} B_{mn}\cos\lambda_m x\cos\lambda_n y + \sum_{l=1}^{2}\sum_{n=0}^{N} c_{ln}P_l(x)\cos\lambda_n y \\
&\quad + \sum_{l=1}^{2}\sum_{m=0}^{M} d_{lm}P_l(y)\cos\lambda_m x \\
w(x,y) &= \sum_{m=0}^{M}\sum_{n=0}^{N} C_{mn}\cos\lambda_m x\cos\lambda_n y + \sum_{l=1}^{2}\sum_{n=0}^{N} e_{ln}P_l(x)\cos\lambda_n y \\
&\quad + \sum_{l=1}^{2}\sum_{m=0}^{M} f_{lm}P_l(y)\cos\lambda_m x \\
\phi_x(x,y) &= \sum_{m=0}^{M}\sum_{n=0}^{N} D_{mn}\cos\lambda_m x\cos\lambda_n y + \sum_{l=1}^{2}\sum_{n=0}^{N} g_{ln}P_l(x)\cos\lambda_n y \\
&\quad + \sum_{l=1}^{2}\sum_{m=0}^{M} h_{lm}P_l(y)\cos\lambda_m x \\
\phi_y(x,y) &= \sum_{m=0}^{M}\sum_{n=0}^{N} E_{mn}\cos\lambda_m x\cos\lambda_n y + \sum_{l=1}^{2}\sum_{n=0}^{N} i_{ln}P_l(x)\cos\lambda_n y \\
&\quad + \sum_{l=1}^{2}\sum_{m=0}^{M} j_{lm}P_l(y)\cos\lambda_m x
\end{aligned}
\tag{8.25}
$$

where $\lambda_m = m\pi/a$ and $\lambda_n = n\pi/b$. The auxiliary functions $P_l(x)$ and $P_l(y)$ are given in Eqs. (4.32) and (4.33).

Like rectangular plates, each shallow shell formed on rectangular planforms has four boundaries. Thus, for the sake of simplicity, a four-letter string is employed to represent the boundary condition of a shallow shell, such as FCSSD identifies the shell having F, C, S and SD boundary conditions at boundaries $x = 0$, $y = 0$, $x = a$, $y = b$, respectively. Furthermore, for all numerical examples, unless otherwise stated, the layers of the considered laminated shallow shells are of equal thickness and made of the same composite material: $E_2 = 10$ GPa, $E_1/E_2 =$ open, $\mu_{12} = 0.25$,

$G_{12} = 0.6E_2$, $G_{13} = 0.6E_2$, $G_{23} = 0.5E_2$ and $\rho = 1{,}450$ kg/m^3. In addition, the natural frequencies of the considered shells will be expressed by the non-dimensional parameters as $\Omega = \omega a^2 \sqrt{\rho/E_2 h^2}$.

8.3.1 Convergence Studies and Result Verification

Table 8.2 shows the first six frequency parameters $\Omega = \omega a^2 \sqrt{\rho/E_1 h^2}$ for completely free, [30°/−30°/−30°/30°] laminated graphite/epoxy shallow cylindrical, spherical and hyperbolic paraboloidal shells with different truncation schemes (i.e. $M = N = 11, 12, 14, 15$), respectively, together with results presented by Qatu (2004).

Table 8.2 Convergence of frequency parameters $\Omega = \omega a^2 \sqrt{\rho/E_1 h^2}$ for [30°/−30°/−30°/30°] laminated graphite/epoxy shallow cylindrical, spherical and hyperbolic paraboloidal shells

R_y/R_x	B.C.	$M \times N$	Mode number				
			1	2	3	4	5
0	FFFF	11 × 11	1.912	2.975	4.889	6.697	8.226
		12 × 12	1.912	2.975	4.887	6.695	8.223
		14 × 14	1.912	2.974	4.885	6.693	8.219
		15 × 15	1.911	2.974	4.884	6.693	8.216
	CCCC	11 × 11	15.38	15.91	19.71	21.86	24.21
		12 × 12	15.38	15.91	19.70	21.85	24.20
		14 × 14	15.38	15.90	19.69	21.85	24.19
		15 × 15	15.38	15.89	19.69	21.85	24.18
1	FFFF	11 × 11	2.282	3.320	5.847	6.746	8.341
		12 × 12	2.280	3.318	5.843	6.742	8.333
		14 × 14	2.278	3.316	5.838	6.736	8.325
		15 × 15	2.277	3.315	5.836	6.734	8.324
		Qatu (2004)	2.283	3.323	5.871	6.781	8.394
	CCCC	11 × 11	28.54	31.05	37.24	37.94	39.44
		12 × 12	28.54	31.03	37.23	37.93	39.43
		14 × 14	28.53	31.01	37.21	37.91	39.43
		15 × 15	28.53	31.00	37.20	37.90	39.42
−1	FFFF	11 × 11	2.111	4.873	4.875	8.771	9.060
		12 × 12	2.110	4.870	4.874	8.767	9.055
		14 × 14	2.109	4.866	4.872	8.762	9.048
		15 × 15	2.108	4.865	4.871	8.761	9.046
		Qatu (2004)	2.115	4.882	4.887	8.809	9.106
	CCCC	11 × 11	22.05	24.51	27.85	27.87	32.86
		12 × 12	22.04	24.51	27.84	27.86	32.83
		14 × 14	22.04	24.49	27.83	27.84	32.81
		15 × 15	22.04	24.49	27.82	27.83	32.81

The FFFF and CCCC boundary conditions are performed in the study. The geometry and material constants of the layers of the considered shells are given as: $b/a = 1$, $h/a = 0.01$, $R_y/a = 2$, $E_1 = 138$ GPa, $E_2 = 8.96$ GPa $\mu_{12} = 0.3$, $G_{12} = G_{13} = 7.1$ GPa, $G_{23} = 3.9$ GPa. It is obvious that the modified Fourier series solution converges quickly. The maximum difference between the 11×11-term solutions and those of 15×15-term is less than 0.24 %. Based on this analysis, the truncated number of the displacement expressions will be uniformly selected as $M = N = 15$ in all subsequent calculations. Furthermore, by comparing with results published by Qatu (2004), we can find a well agreement between the two results. The differences between these two results are attributed a different shallow theory was used by Qatu (2004).

Table 8.3 shows the comparison of the fundamental parameters Ω for cross-ply cylindrical, spherical and hyperbolic paraboloidal shallow shells with SDSDSDSD boundary conditions and different curvature ratios (a/R). The shells are composed of graphite/epoxy ($E_1/E_2 = 15$, $\mu_{12} = 0.25$, $G_{12} = G_{13} = G_{23} = 0.5E_2$) and having geometry properties: $a/b = 1$, $h/a = 0.1$. The symmetric lamination $[0°/90°/90°/0°]$ and two asymmetric laminations $[0°/90°]$ and $[90°/0°]$ are used in the comparison. "–" represents results that were not considered in the referential work. Curvature ratios are varied from 0 (flat plate) to 0.5 (limit of shallow shell theory). The table shows that the present solutions match very well with those reported by Qatu (2004). Except the case of shallow cylindrical shell with $[90°/0°]$ and curvature ratio $a/R = 0.5$, the maximum differences between the two results are less than 1.12, 0.012 and 0.012 % for the shallow cylindrical, spherical and hyperbolic parabo-loidal shells, respectively.

Table 8.4 shows the comparison of the fundamental parameters Ω for a three-layered, $[0°/90°/0°]$ shallow spherical shell with different boundary conditions and

Table 8.3 Comparison of the fundamental frequency parameters Ω for cross-ply cylindrical, spherical and hyperbolic paraboloidal shallow shells with SDSDSDSD boundary conditions and different curvature ratios

R_y/R_x	a/R_y	SDSST (Qatu 2004)			Present		
		$[0°/90°]$	$[90°/0°]$	$[0°/90°/90°/0°]$	$[0°/90°]$	$[90°/0°]$	$[0°/90°/90°/0°]$
0	0	8.1196	8.1196	10.972	8.1205	8.1205	10.972
	0.1	8.1259	8.1448	10.987	8.1268	8.1602	10.987
	0.2	8.1774	8.1538	11.032	8.1783	8.2454	11.032
	0.5	8.5784	8.0831	11.334	8.5790	8.7529	11.335
+1	0	–	8.1196	10.972	8.1205	8.1205	10.972
	0.1	–	8.2190	11.043	8.2199	8.2199	11.043
	0.2	–	8.5084	11.252	8.5092	8.5092	11.253
	0.5	–	10.249	12.572	10.249	10.249	12.571
−1	0	8.1196	8.1196	10.972	8.1205	8.1205	10.972
	0.1	8.0785	8.1448	10.961	8.0794	8.1458	10.961
	0.2	8.0223	8.1538	10.927	8.0233	8.1548	10.928
	0.5	7.7739	8.0831	10.703	7.7748	8.0841	10.704

Table 8.4 Comparison of the fundamental frequency parameters Ω for a three-layered, [0°/90°/0°] shallow spherical shell with different boundary conditions and curvature ratios

Theory	a/R_y	Boundary conditions					
		FSFS	FSSS	FSCS	SSSS	SSCS	CSCS
Present	0	3.788	4.320	6.144	12.163	14.248	16.383
	0.05	3.795	4.325	6.146	12.179	14.265	16.487
	0.20	3.898	4.404	6.164	12.417	14.514	17.966
SDSST (Librescu et al. 1989a)	0	3.788	4.320	6.144	12.163	14.248	16.383
	0.05	3.794	4.325	6.146	12.178	14.264	16.487
	0.20	3.891	4.397	6.163	12.394	14.499	17.959
Difference (%)	0	0.01	0.01	0.01	0.00	0.00	0.00
	0.05	0.02	0.00	0.01	0.01	0.01	0.00
	0.20	0.17	0.16	0.01	0.19	0.11	0.04

curvature ratios (a/R). The shell having SD boundary conditions at the edges $y = 0$, $y = b$ and the other two edges having arbitrary boundary conditions. The shell is formed on square planform and composed of composite layers having following material and geometry parameters: $a/b = 1$, $h/a = 0.1$, $E_1/E_2 = 25$, $\mu_{12} = 0.25$, $G_{12} = G_{13} = 0.5E_2$, $G_{23} = 0.2E_2$. The state space solutions provided by Librescu et al. (1989a) are selected as the benchmark solutions. Comparing the two results, we can find that the current solutions match very well with the referential data. The maximum difference between the two results is very small and less than 0.17 %.

8.3.2 Laminated Shallow Shells with General Boundary Conditions

Table 8.5 shows the first four non-dimension frequency parameters Ω of a three-layered, [0°/90°/0°] shallow cylindrical shell with different boundary conditions and thickness-length ratios. Four different thickness-length ratios i.e. $h/a = 0.01$, 0.05, 0.1 and 0.15, corresponding to thin to thick shallow shells are performed in the analysis. The shell parameters used are: $a/b = 1$, $R_x = \infty$, $a/R_y = 0.1$, $E_1/E_2 = 15$. It can be seen from the table that the augmentation of the thickness-length ratio leads to the decrease of the frequency parameters. Tables 8.6 and 8.7 show the similar results for shallow spherical ($a/R_x = a/R_y = 0.1$) and hyperbolic paraboloidal ($a/R_x = 0.1$, $a/R_y = -0.1$) shells, respectively. The similar observation can be seen in the tables as well. In addition, comparing the three tables, it can be found that the frequency parameters of the hyperbolic paraboloidal shell are higher than the cylindrical and spherical ones. And the results of the shallow cylindrical shell are the smallest. Furthermore, from Table 8.5, we can see that frequency parameters of

Table 8.5 Frequency parameters Ω for a three-layered, $[0°/90°/0°]$ shallow cylindrical shell with different boundary conditions and thickness-length ratios ($a/b = 1$, $R_x = \infty$, $a/R_y = 0.1$, $E_1/E_2 = 15$)

h/a	Mode	Mode number					
		FCCC	FFCC	FFFC	CFFF	CCFF	CCCF
0.01	1	21.441	5.3519	1.2520	4.7214	5.3519	25.358
	2	23.308	11.164	3.1822	5.1009	11.164	28.704
	3	33.542	23.718	7.8374	10.943	23.718	36.292
	4	38.015	25.838	11.515	23.927	25.838	53.380
0.05	1	10.038	4.4519	1.2491	3.8493	4.4519	21.643
	2	22.125	9.8673	3.1181	4.8754	9.8673	23.978
	3	24.983	22.408	7.7257	10.495	22.408	32.230
	4	33.952	22.695	11.156	21.836	22.695	48.200
0.10	1	8.8956	4.2056	1.2411	3.6585	4.2056	16.423
	2	19.795	9.2023	3.0083	4.5636	9.2023	18.571
	3	20.306	18.167	7.4109	9.8304	18.167	26.348
	4	28.154	20.440	10.139	10.921	20.440	34.952
0.15	1	8.0767	3.9166	1.2282	3.4237	3.9166	12.584
	2	16.565	8.4317	2.8750	4.1729	8.4317	14.695
	3	17.212	14.606	6.9028	7.2807	14.606	21.834
	4	23.254	17.471	6.9649	9.0797	17.471	25.490

cylindrical shell with curved edge cantilevered (CFFF) are higher than those of straight edge cantilevered (FFFC).

Influence of the fiber orientations on the modal frequencies of shallow shells is also investigated. In Fig. 8.5, variations of the lowest three frequency parameters Ω of certain single-layered ($[\vartheta]$) shallow shells with various fiber orientations are given. Three types of curvature considered in the study are: cylindrical ($R_x = \infty$, $a/R_y = 0.1$), spherical ($a/R_x = a/R_y = 0.1$) and hyperbolic paraboloidal ($a/R_x = 0.1$, $a/R_y = -0.1$). In each case, the fiber orientation ϑ is varies from 0° to 180° by an increment of 10°. The shallow shells under consideration are assumed to be with SDSDSDSD boundary conditions. The material constants and geometry parameters of the laminas of the shallow shells are: $b/a = 1$, $h/a = 0.05$, $E_1/E_2 = 15$. Many interesting characteristics can be observed from the figure. The first observation is that all the subfigures are symmetrical about central line (i.e., $\vartheta = 90°$). The second observation is that the fundamental frequency parameter traces climb up and then decline, and may reach their crests around $\vartheta = 45°$ (when ϑ is increased from 0° to 90°). The similar characteristics can be observed in the subfigure of the second mode. However, for the third mode, the maximum frequency parameters may occur around $\vartheta = 30°$ or $\vartheta = 60°$. In addition, it is obvious that frequency parameters of the shallow spherical shell are larger than the shallow cylindrical and hyperbolic paraboloidal ones. Furthermore, since the length-to-radius ratio used in each case in this investigation is small ($a/R = 0.1$), thus, the differences between these results are

Table 8.6 Frequency parameters Ω for a three-layered, [0°/90°/0°] shallow spherical shell with different boundary conditions and thickness-length ratios ($a/b = 1$, $a/R_x = a/R_y = 0.1$, $E_1/E_2 = 15$)

h/a	Mode	Mode number					
		FCCC	FFCC	FFFC	CFFF	CCFF	CCCF
0.01	1	22.125	6.6469	2.4195	4.4589	6.6469	36.435
	2	25.223	12.090	3.2764	5.0746	12.090	39.743
	3	35.576	25.341	10.435	13.354	25.341	45.528
	4	42.189	26.814	11.922	25.096	26.814	60.365
0.05	1	10.101	4.5350	1.3341	3.8447	4.5350	22.313
	2	22.225	9.9327	3.1223	4.8711	9.9327	24.631
	3	25.075	22.393	7.8779	10.657	22.393	32.715
	4	34.138	22.832	11.170	21.583	22.832	48.537
0.10	1	8.9124	4.2279	1.2627	3.6560	4.2279	16.657
	2	19.821	9.2191	3.0095	4.5592	9.2191	18.790
	3	20.325	18.145	7.4477	9.8768	18.145	26.497
	4	28.208	20.485	10.123	10.912	20.485	34.936
0.15	1	8.0840	3.9271	1.2375	3.4215	3.9271	12.723
	2	16.570	8.4387	2.8757	4.1680	8.4387	14.818
	3	17.223	14.587	6.9003	7.2788	14.587	21.912
	4	23.280	17.503	6.9802	9.1028	17.503	25.477

small as well. The maximum differences between the results of cylindrical and spherical shallow shells, cylindrical and hyperbolic paraboloidal shallow shells, spherical and hyperbolic paraboloidal shallow shells are less than 1.14, 0.61, 6.24 %, respectively.

Figure 8.6 performs the similar study for the shallow shells with CCCC boundary conditions. The different observation is that the variation tendencies of the fundamental modes decline firstly and then climb up. The minimum frequency parameters may occur around $\vartheta = 45°$. For the second mode, the frequency parameters increase when ϑ increases from 0° to 40° and decrease when ϑ increases from 50° to 90°. The variation tendencies of the third mode are the same as those in Fig. 8.5. The figure also shows that the differences between the three results are small due to the fact that a low length-to-radius ratio is used in this investigation.

Figure 8.7 shows the similar study for the shallow shells with SDSDSDSD boundary conditions and a higher length-to-radius ratio, i.e., $a/R = 0.2$ (limit of the shallow theory). The material and other geometry properties are the same as those used in Figs. 8.6 and 8.7. The observations from this figure are the same as those of Fig. 8.5. However, the maximum differences between the results of cylindrical and spherical shallow shells, cylindrical and hyperbolic paraboloidal shallow shells, spherical and hyperbolic paraboloidal shallow shells can be as many as 5.29, 6.50, 22.9 %, respectively. It is attributed to the effects of curvatures on the frequency parameters of shallow shells are more significant when the length-to-radius ratio is

Table 8.7 Frequency parameters Ω for a three-layered, [0°/90°/0°] shallow hyperbolic paraboloidal shell with different boundary conditions and thickness-length ratios ($a/b = 1$, $a/R_x = 0.1$, $a/R_y = -0.1$, $E_1/E_2 = 15$)

h/a	Mode	Mode number					
		FCCC	FFCC	FFFC	CFFF	CCFF	CCCF
0.01	1	22.736	4.935666	2.4375	4.3788	4.9357	36.722
	2	26.242	14.70847	3.2710	5.0823	14.708	38.502
	3	34.129	26.91862	10.876	15.157	26.919	44.979
	4	39.603	27.18928	11.896	25.575	27.189	59.766
0.05	1	10.143	4.429413	1.3368	3.8471	4.4294	22.327
	2	22.268	10.09015	3.1212	4.8706	10.090	24.564
	3	25.002	22.46707	7.8884	10.750	22.467	32.691
	4	34.013	22.82693	11.175	21.644	22.827	48.511
0.10	1	8.9240	4.19818	1.2634	3.6569	4.1982	16.662
	2	19.834	9.264119	3.0085	4.5580	9.2641	18.769
	3	20.302	18.163	7.4502	9.9013	18.163	26.492
	4	28.166	20.48119	10.159	10.917	20.481	34.944
0.15	1	8.0894	3.912118	1.2379	3.4220	3.9121	12.726
	2	16.558	8.461952	2.8748	4.1665	8.4620	14.807
	3	17.230	14.59844	6.9054	7.2823	14.598	21.911
	4	23.255	17.46116	6.9813	9.1142	17.461	25.483

higher. The figure also shows that the variation tendency of each frequency parameter of the cylindrical shallow shell is the same as the corresponding ones of the spherical and hyperbolic paraboloidal shells. Comparing Figs. 8.5, 8.6 and 8.7, it can be found that the influence of fiber orientations on the modal frequencies of shallow shells vary with mode sequence and boundary conditions.

Effects of the fiber orientations on the frequency parameters and mode shapes of laminated shallow shells are further reported. In Tables 8.8, 8.9 and 8.10, the lowest four frequency parameters Ω of certain three-layered, [0°/ϑ/0°] shallow shells with various boundary conditions and fiber orientations are presented. The aspect ratio is chosen to be $b/a = 1$. The thickness-to-width ratio $h/b = 0.1$ is used in the calculation. As usual, the shallow cylindrical ($R_x = \infty$, $a/R_y = 0.1$), spherical ($a/R_x = a/R_y = 0.1$) and hyperbolic paraboloidal ($a/R_x = 0.1$, $a/R_y = -0.1$) shells are considered in the study. The fiber direction angle ϑ is varied from 0° to 90° with an increment of 30°. The layers of the shells are made of composite materials with orthotropy ratio $E_1/E_2 = 15$. From Table 8.8, it can be noticed that the results of the shell with FFCC boundary conditions equal to those of CCFF boundary conditions. The similar observation can be seen from Tables 8.9 and 8.10 as well.

Figures 8.8, 8.9 and 8.10 give the first three contour plots of the mode shapes and frequency parameters with various lamination angles for CFFF, single-layered [ϑ] shallow cylindrical, spherical and hyperbolic paraboloidal shells, respectively. The aspect ratio $a/b = 1$, thickness-to-length ratio $h/a = 0.05$ and orthotropy ratio

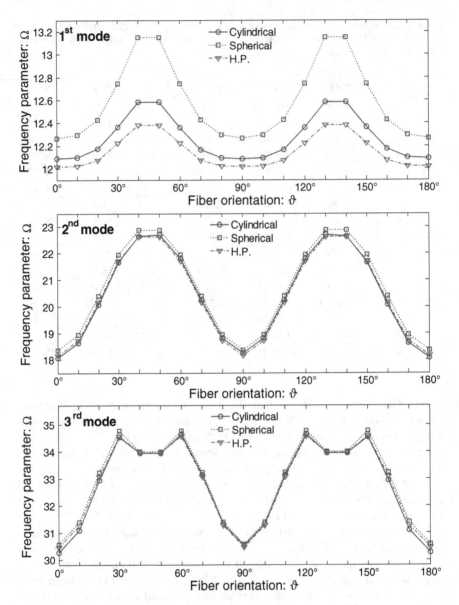

Fig. 8.5 The lowest three frequency parameters Ω of single-layered ([9]) shallow shells with SDSDSDSD boundary condition and various fiber orientations ($b/a = 1$, $h/a = 0.05$, $a/R = 0.1$)

$E_1/E_2 = 25$ are used in the presentation. Mode shapes and frequency parameters are shown for $\vartheta = 0$, $30°$, $60°$ and $90°$. From the figures, we can find that when $\vartheta = 0$, the mode shapes of the shallow cylindrical, spherical and hyperbolic paraboloidal shells

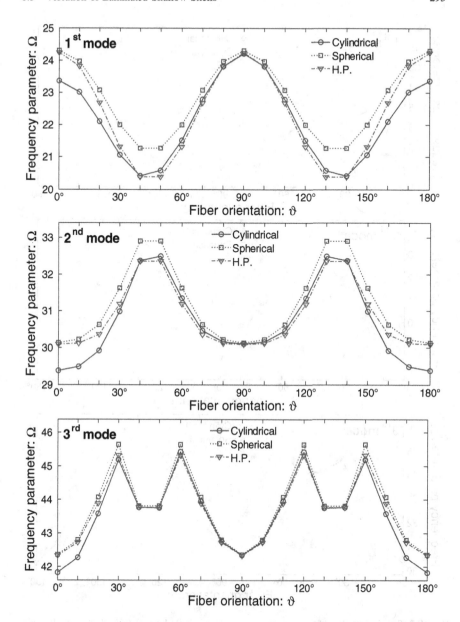

Fig. 8.6 The lowest three frequency parameters Ω of single-layered ([9]) shallow shells with CCCC boundary condition and different fiber orientations ($b/a = 1$, $h/a = 0.05$, $a/R = 0.1$)

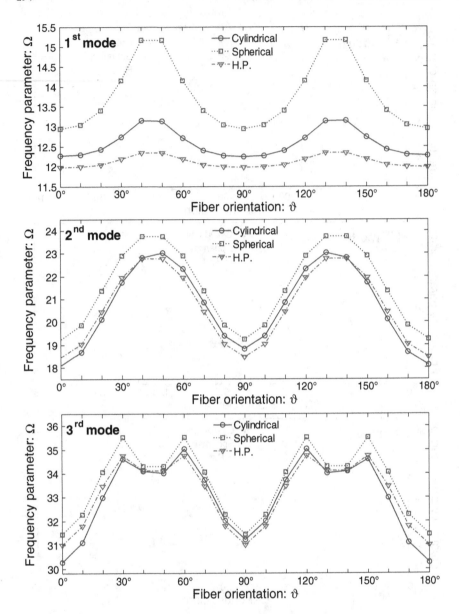

Fig. 8.7 The lowest three frequency parameters Ω of single-layered ([9]) shallow shells with SDSDSDSD boundary condition and different fiber orientations ($b/a = 1$, $h/a = 0.05$, $a/R = 0.2$)

Table 8.8 Frequency parameters Ω for a three-layered, $[0°/\vartheta/0°]$ shallow cylindrical shell with different boundary conditions and fiber orientations ($a/b = 1$, $h/a = 0.1$, $R_x = \infty$, $a/R_y = 0.1$, $E_1/E_2 = 15$)

ϑ	Mode	Boundary conditions						
		FFFC	FFCC	FCCC	CCCC	CCCF	CCFF	CFFF
0°	1	1.0086	4.1935	7.6932	18.084	16.790	4.1935	3.7272
	2	2.8658	8.2303	16.985	23.818	18.445	8.2303	4.6176
	3	6.0857	17.494	20.188	34.495	24.372	17.494	8.8185
	4	7.1446	18.539	26.655	36.670	28.413	18.539	11.145
30°	1	1.0255	4.2544	7.8266	18.081	16.690	4.2544	3.6999
	2	3.0243	8.4954	17.287	24.067	18.506	8.4954	4.6980
	3	6.1868	17.845	20.260	34.990	24.685	17.845	9.0384
	4	8.2781	18.522	27.064	36.540	35.586	18.522	14.682
60°	1	1.1380	4.2553	8.4006	18.180	16.509	4.2553	3.6655
	2	3.0852	9.0720	18.670	24.977	18.569	9.0720	4.6604
	3	6.8488	18.270	20.334	36.237	25.658	18.270	9.5119
	4	9.8179	19.325	27.760	37.001	35.157	19.325	13.338
90°	1	1.2411	4.2056	8.8956	18.288	16.423	4.2056	3.6585
	2	3.0083	9.2023	19.795	25.634	18.571	9.2023	4.5636
	3	7.4109	18.167	20.306	36.071	26.348	18.167	9.8304
	4	10.139	20.440	28.154	38.556	34.952	20.440	10.921

Table 8.9 Frequency parameters Ω for a three-layered, $[0°/\vartheta/0°]$ shallow spherical shell with different boundary conditions and fiber orientations ($a/b = 1$, $h/a = 0.1$, $a/R_x = a/R_y = 0.1$, $E_1/E_2 = 15$)

ϑ	Mode	Boundary conditions						
		FFFC	FFCC	FCCC	CCCC	CCCF	CCFF	CFFF
0°	1	1.0154	4.2034	7.7061	18.405	17.124	4.2034	3.7246
	2	2.8665	8.2380	17.010	24.060	18.756	8.2380	4.6131
	3	6.1081	17.534	20.201	34.662	24.601	17.534	8.8466
	4	7.1418	18.517	26.706	36.665	28.418	18.517	11.135
30°	1	1.0337	4.2659	7.8541	18.388	16.976	4.2659	3.6970
	2	3.0252	8.5066	17.329	24.284	18.783	8.5066	4.6937
	3	6.2174	17.892	20.297	35.138	24.887	17.892	9.0737
	4	8.2752	18.510	27.144	36.544	35.579	18.510	14.641
60°	1	1.1525	4.2700	8.4395	18.439	16.748	4.2700	3.6627
	2	3.0868	9.0865	18.706	25.148	18.803	9.0865	4.6564
	3	6.8821	18.248	20.374	36.246	25.821	18.248	9.5526
	4	9.8140	19.381	27.828	37.117	35.147	19.381	13.313
90°	1	1.2627	4.2279	8.9124	18.510	16.657	4.2279	3.6560
	2	3.0095	9.2191	19.821	25.791	18.790	9.2191	4.5592
	3	7.4477	18.145	20.325	36.068	26.497	18.145	9.8768
	4	10.123	20.485	28.208	38.661	34.936	20.485	10.912

Table 8.10 Frequency parameters Ω for a three-layered, [0°/ϑ/0°] shallow hyperbolic parabo-loidal shell with different boundary conditions and fiber orientations ($a/b = 1$, $h/a = 0.1$, $a/R_x = 0.1$, $a/R_y = -0.1$, $E_1/E_2 = 15$)

ϑ	Mode	Boundary conditions						
		FFFC	FFCC	FCCC	CCCC	CCCF	CCFF	CFFF
0°	1	1.0157	4.1873	7.7187	18.386	17.124	4.1873	3.7256
	2	2.8653	8.2775	17.026	24.050	18.738	8.2775	4.6122
	3	6.1094	17.538	20.183	34.655	24.596	17.538	8.8732
	4	7.1476	18.505	26.670	36.658	28.411	18.505	11.140
30°	1	1.0341	4.2485	7.8638	18.312	16.953	4.2485	3.6985
	2	3.0238	8.5641	17.344	24.249	18.735	8.5641	4.6926
	3	6.2200	17.904	20.263	35.119	24.861	17.904	9.1066
	4	8.2838	18.501	27.099	36.529	35.567	18.501	14.653
60°	1	1.1531	4.2492	8.4234	18.363	16.743	4.2492	3.6641
	2	3.0852	9.1477	18.718	25.127	18.763	9.1477	4.6551
	3	6.8868	18.258	20.324	36.219	25.805	18.258	9.5801
	4	9.8311	19.378	27.781	37.102	35.140	19.378	13.324
90°	1	1.2634	4.1982	8.9240	18.492	16.662	4.1982	3.6569
	2	3.0085	9.2641	19.834	25.782	18.769	9.2641	4.5580
	3	7.4502	18.163	20.302	36.060	26.492	18.163	9.9013
	4	10.159	20.481	28.166	38.655	34.944	20.481	10.917

are symmetrical with respect to the longitudinal center line. It is attributed to the boundary conditions, geometry and material properties of the shallow shells under consideration are symmetrical with respect to the longitudinal center line as well. Furthermore, it is obviously that the maximum frequency parameters for the first, second and third modes in each case occur at $\vartheta = 0$, $\vartheta = 0$ and $\vartheta = 30°$, respectively. The figures also show that that the node lines (i.e., lines with zero displacements) of the mode shapes vary with respect to the fiber orientation. In addition, since the length-to-radius ratio used in this investigation is small ($a/R = 0.1$), thus, the differences between the frequency parameters are small as well.

Tables 8.11, 8.12 and 8.13 present the first four non-dimension frequency parameters Ω of certain two-layered, angle-ply [45°/−45°] shallow cylindrical, spherical and hyperbolic paraboloidal shells with different boundary conditions and length-to-radius ratios, respectively. Four different length-to-radius ratios i.e. $a/R = 0.05, 0.1, 0.15$ and 0.2, corresponding to considerably to slightly shallow shells are performed in the analysis. The shells are formed in rectangular planform with following geometry and material properties: $a/b = 1/2$, $h/a = 0.1$, $E_1/E_2 = 15$, $R_x = \infty$ for shallow cylindrical shells, $R_y = R_x$ for the spherical shells and $R_y = -R_x$ for the hyperbolic paraboloidal ones. Seven different boundary conditions, i.e.,

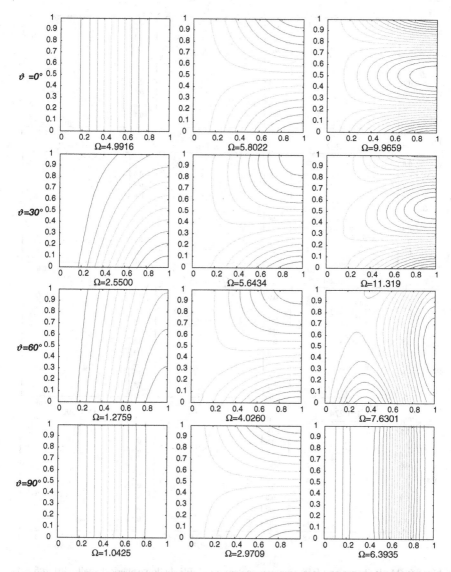

Fig. 8.8 Mode shapes and frequency parameters for CFFF shallow cylindrical shells ([9], $a/b = 1$, $h/a = 0.05$, $R_x = \infty$, $a/R_y = 0.1$, $E_1/E_2 = 25$)

FFFC, FFCC, FCCC, CCCC, CCCF, CCFF and CFFF are performed in the calculation. It can be seen from the table that the augmentation of the length-to-radius ratio leads to the increases of the frequency parameters. Table 8.11 also shows that the results of shallow cylindrical shells cantilevered in the curved edge (CFFF) are

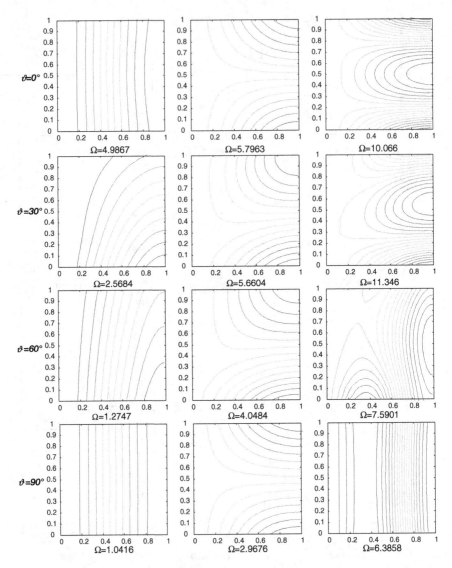

Fig. 8.9 Mode shapes and frequency parameters for CFFF shallow spherical shells ([9], $a/b = 1$, $h/a = 0.05$, $a/R_x = a/R_y = 0.1$, $E_1/E_2 = 25$)

higher than those of straight edge cantilevered (FFFC). From Table 8.12, it can be noticed that the results for the spherical shells with FFCC boundary conditions equal to those of CCFF boundary conditions. The similar observation can be seen from Tables 8.10 and 8.13 as well.

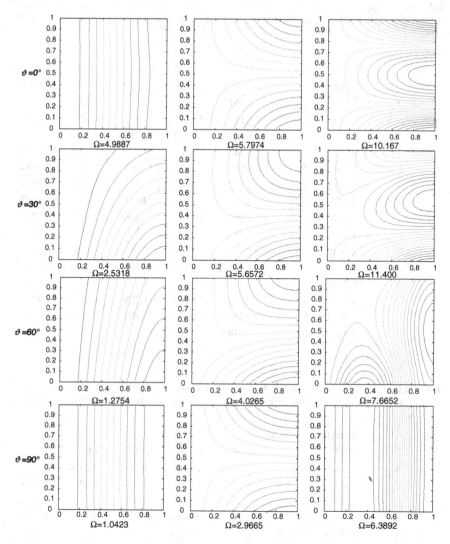

Fig. 8.10 Mode shapes and frequency parameters for CFFF hyperbolic paraboloidal shallow shells ([9], $a/b = 1$, $h/a = 0.05$, $a/R_x = 0.1$, $a/R_y = -0.1$, $E_1/E_2 = 25$)

As the last case, parameter studies are carried out in Figs. 8.11 and 8.12 to further investigate the effects of length-to-radius ratios a/R_x and a/R_y on the frequency parameters Ω of laminated shallow shells. Figure 8.11 depicts the lowest three frequency parameters Ω versus length-to-radius ratios a/R_x and a/R_y for a [0°/90°] layered shallow shell with aspect ratio of $a/b = 5$ and CCCC boundary conditions. The other shell parameters used in the study are: $h/b = 0.1$, $E_1/E_2 = 15$. It is

Table 8.11 Frequency parameters Ω for two-layered, angle-ply [45°/−45°] shallow cylindrical shells with different boundary conditions and length-to-radius ratios ($a/b = 1/2$, $h/a = 0.1$, $R_x = \infty$, $E_1/E_2 = 15$)

a/R_y	Mode	Boundary conditions						
		FFFC	FFCC	FCCC	CCCC	CCCF	CCFF	CFFF
0.05	1	0.3640	2.0340	3.6812	10.195	9.1129	2.0340	1.5413
	2	2.0500	4.4388	7.6647	13.482	11.007	4.4388	2.7424
	3	2.2144	8.1038	11.116	18.452	14.501	8.1038	5.0423
	4	3.5598	9.4052	13.149	23.372	19.455	9.4052	8.5097
0.10	1	0.3636	2.1008	3.8405	10.323	9.1883	2.1008	1.6328
	2	2.0169	4.4927	7.6996	13.551	11.083	4.4927	2.7494
	3	2.2112	8.1231	11.192	18.484	14.542	8.1231	5.0531
	4	3.5724	9.4414	13.162	23.406	19.473	9.4414	8.5117
0.15	1	0.3629	2.2068	4.0909	10.532	9.3066	2.2068	1.7725
	2	1.9665	4.5838	7.7572	13.666	11.212	4.5838	2.7610
	3	2.2049	8.1551	11.317	18.536	14.611	8.1551	5.0715
	4	3.5903	9.5079	13.185	23.463	19.503	9.5079	8.5150
0.20	1	0.3620	2.3364	4.4152	10.819	9.4581	2.3364	1.9466
	2	1.9036	4.7125	7.8367	13.825	11.398	4.7125	2.7771
	3	2.1953	8.1996	11.488	18.609	14.708	8.1996	5.0981
	4	3.6104	9.6029	13.218	23.542	19.545	9.6029	8.5196

Table 8.12 Frequency parameters Ω for two-layered, angle-ply [45°/−45°] shallow spherical shells with different boundary conditions and length-to-radius ratios ($a/b = 1/2$, $h/a = 0.1$, $R_y = R_x$, $E_1/E_2 = 15$)

a/R_y	Mode	Boundary conditions						
		FFFC	FFCC	FCCC	CCCC	CCCF	CCFF	CFFF
0.05	1	0.3645	2.0354	3.6866	10.290	9.1764	2.0354	1.5399
	2	2.0508	4.4309	7.6754	13.543	11.072	4.4309	2.7416
	3	2.2162	8.1005	11.130	18.488	14.546	8.1005	5.0389
	4	3.5581	9.4006	13.161	23.374	19.483	9.4006	8.5105
0.10	1	0.3655	2.1093	3.8597	10.694	9.4250	2.1093	1.6277
	2	2.0196	4.4727	7.7416	13.793	11.347	4.4727	2.7464
	3	2.2188	8.1290	11.246	18.626	14.720	8.1290	5.0392
	4	3.5660	9.4297	13.211	23.415	19.582	9.4297	8.5143
0.15	1	0.3671	2.2279	4.1272	11.334	9.7832	2.2279	1.7618
	2	1.9718	4.5443	7.8480	14.198	11.816	4.5443	2.7543
	3	2.2222	8.1815	11.432	18.854	15.009	8.1815	5.0396
	4	3.5769	9.4847	13.295	23.483	19.747	9.4847	8.5199
0.20	1	0.3692	2.3752	4.4659	12.168	10.196	2.3752	1.9296
	2	1.9115	4.6421	7.9895	14.744	12.482	4.6421	2.7652
	3	2.2259	8.2561	11.680	19.168	15.409	8.2561	5.0396
	4	3.5886	9.5615	13.415	23.578	19.976	9.5615	8.5260

Table 8.13 Frequency parameters Ω for two-layered, angle-ply [45°/−45°] shallow hyperbolic paraboloidal shells with different boundary conditions and length-to-radius ratios ($a/b = 1/2$, $h/a = 0.1$, $R_y = -R_x$, $E_1/E_2 = 15$)

a/R_x	Mode	Boundary conditions						
		FFFC	FFCC	FCCC	CCCC	CCCF	CCFF	CFFF
0.05	1	0.3650	2.0546	3.6876	10.178	9.1143	2.0546	1.5421
	2	2.0492	4.4454	7.6736	13.476	11.005	4.4454	2.7422
	3	2.2213	8.1058	11.111	18.454	14.503	8.1058	5.0559
	4	3.5615	9.4232	13.155	23.370	19.460	9.4232	8.5239
0.10	1	0.3677	2.1351	3.8650	10.258	9.1955	2.1351	1.6340
	2	2.0144	4.5292	7.7353	13.531	11.070	4.5292	2.7488
	3	2.2381	8.1494	11.170	18.491	14.548	8.1494	5.1079
	4	3.5788	9.4751	13.189	23.400	19.492	9.4751	8.5679
0.15	1	0.3718	2.2438	4.1437	10.389	9.3293	2.2438	1.7680
	2	1.9620	4.6741	7.8368	13.621	11.179	4.6741	2.7593
	3	2.2633	8.2275	11.269	18.551	14.624	8.2275	5.1953
	4	3.6036	9.5554	13.245	23.450	19.546	9.5554	8.6406
0.20	1	0.3773	2.3628	4.5045	10.571	9.5130	2.3628	1.9247
	2	1.8974	4.8796	7.9764	13.746	11.330	4.8796	2.7734
	3	2.2949	8.3391	11.405	18.636	14.730	8.3391	5.3187
	4	3.6320	9.6628	13.324	23.520	19.621	9.6628	8.7404

seen that each frequency parameter Ω of the shell increases with the increments of length-to-radius ratios a/R_x or a/R_y. The figure also shows that the effects of length-to-radius ratios a/R_x and a/R_y on the frequency parameters Ω of laminated shallow shells vary with mode sequence.

Figure 8.12 shows the similar study for the shallow shell with aspect ratio of $a/b = 1$. The figure shows that the fundamental frequency parameter Ω of the shell increases with the increase of length-to-radius ratios a/R_x or a/R_y. However, for the second mode, the maximum and minimum frequency parameters occur at $a/R_x = 0.2$, $a/R_y = -0.2$ and $a/R_x = 0$, $a/R_y = 0.2$, respectively. And for the third mode, the minimum and maximum frequency parameters occur at $a/R_x = a/R_y = 0$ (plate) and $a/R_x = a/R_y = 0.2$ (spherical curvature), respectively. Figures 8.11 and 8.12 also reveal that the effects of length-to-radius ratios on the vibration characteristics of shallow shells vary with aspect ratios.

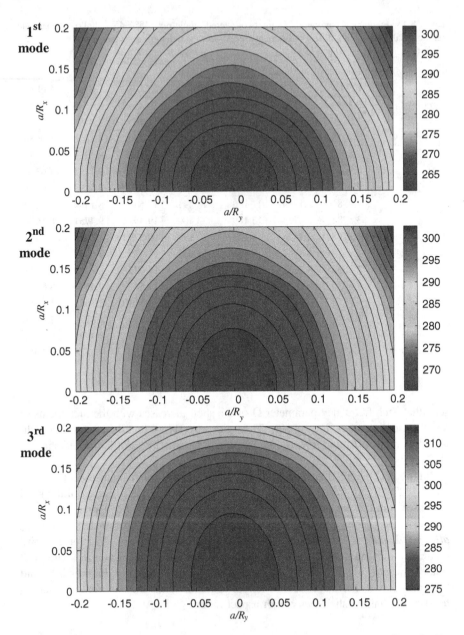

Fig. 8.11 The lowest three frequency parameters Ω versus length-to-radius ratios a/R_x and a/R_y for a $[0°/90°]$ layered shallow shell with aspect ratio of $a/b = 5$

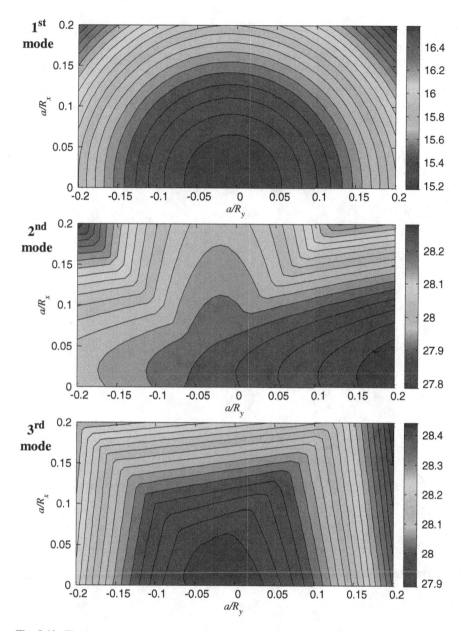

Fig. 8.12 The lowest three frequency parameters Ω versus length-to-radius ratios a/R_x and a/R_y for a $[0°/90°]$ layered shallow shell with aspect ratio of $a/b = 1$

References and Further Reading

Banerjee JR, Cheung CW, Morishima R et al (2007) Free vibration of a three-layered sandwich beam using the dynamic stiffness method and experiment. Int J Solids Struct 44:7543–7563

Bardell NS, Dunsdon JM, Langley RS (1998) Free vibration of thin, isotropic, open, conical panels. J Sound Vib 217(2):297–320

Bardell NS, Langley RS, Dunsdon JM et al (1999) An h-p finite element vibration analysis of open conical sandwich panels and conical sandwich frusta. J Sound Vib 226(2):345–377

Carrera E (1999a) Multilayered shell theories accounting for layerwise mixed description, part 1: governing equations. AIAA J 37:1107–1116

Carrera E (1999b) Multilayered shell theories accounting for layerwise mixed description, part 2: numerical evaluations. AIAA J 37:1117–1124

Carrera E (2002) Theories and finite elements for multilayered, anisotropic, composite plates and shells. Arch Comput Methods Eng 9(2):87–140

Carrera E (2003) Historical review of Zig-Zag theories for multilayered plates and shells. Appl Mech Rev 56(3):287–308

Carrera E, Brischetto S, Nali P (2011) Plates and shells for smart structures: classical and advanced theories for modeling and analysis. Wiley, New Jersey, pp 1–309

Chakraverty S (2009) Vibration of plates. CRC Press, New York, pp 1–411

Chen WQ, Lv CF, Bian ZG (2004) Free vibration analysis of generally laminated beams via state-space-based differential quadrature. Compos Struct 63:417–425

Chen YH, Jin GY, Liu ZG (2013) Free vibration analysis of circular cylindrical shell with non-uniform elastic boundary constraints. Int J Mech Sci 74:120–132

Chern YC, Chao CC (2000) Comparison of natural frequencies of laminates by 3-D theory, part II: curved panels. J Sound Vib 230(5):1009–1030

Cheng ZQ, Batra RC (2000) Deflection relationships between the homogeneous plate theory and different functionally graded plate theories. Arch Mech 52:143–158

Cheung YK, Zhou D (2001a) Free vibrations of rectangular unsymmetrically laminated composite plates with internal line supports. Comput Struct 79:1923–1932

Cheung YK, Zhou D (2001b) Vibrations analysis of rectangular symmetrically laminated composite plates with intermediate line supports. Comput Struct 79:33–41

Civalek Ö (2006) The determination of frequencies of laminated conical shells via the discrete singular convolution method. J Mech Mater Struct 1(1):163–182

Civalek Ö (2007) Numerical analysis of free vibrations of laminated composite conical and cylindrical shells: discrete singular convolution (DSC) approach. J Comput Appl Math 205:251–271

Civalek Ö (2013) Vibration analysis of laminated composite conical shells by the method of discrete singular convolution based on the shear deformation theory. Compos B Eng 45:1001–1009

Dogan A, Arslan HM (2009) Effects of curvature on free vibration characteristics of laminated composite cylindrical shallow shells. Sci Res Essay 4(4):226–238

© Science Press, Beijing and Springer-Verlag Berlin Heidelberg 2015
G. Jin et al., *Structural Vibration*, DOI 10.1007/978-3-662-46364-2

Du JT (2009) Study on modeling methods for structural vibration, enclosed sound field and their coupling system subject to general boundary conditions. Ph.D. thesis, Harbin Engineering University, Harbin, pp 1–104

Fantuzzi N, Tornabene F, Viola E (2014) Generalized differential quadrature finite element method for vibration analysis of arbitrarily shaped membranes. Int J Mech Sci 79:216–251

Fazzolari FA, Carrera E (2013) Advances in the Ritz formulation for free vibration response of doubly-curved anisotropic laminated composite shallow and deep shells. Compos Struct 101:111–128

Ferreira AJM, Roque CMC, Jorge RMN (2006) Modelling cross-ply laminated elastic shells by a higher-order theory and multiquadrics. Compos Struct 84(19):1288–1299

Ferreira AJM, Roque CMC, Jorge RMN (2007) Nature frequencies of FSDT cross-ply composite shells by multiquadrics. Compos Struct 77:296–305

Frederiksen PS (1995) Singly-layer plate theories applied to the flexural vibration of completely free thick laminates. J Sound Vib 186(5):743–759

Gautham BP, Ganesan N (1997) Free vibration characteristics of isotropic and laminated orthotropic spherical caps. J Sound Vib 204(1):17–40

Ghavanloo E, Fazelzadeh SA (2013) Free vibration analysis of orthotropic doubly curved shallow shells based on the gradient elasticity. Compos B Eng 45:1448–1457

Hajianmaleki M, Qatu MS (2012a) Static and vibration analyses of thick, generally laminated deep curved beams with different boundary conditions. Compos B Eng 43:1767–1775

Hajianmaleki M, Qatu MS (2012b) Transverse vibration analysis of generally laminated two-segment composite shafts with a lumped mass using generalized differential quadrature. J Vib Control 19(13):2013–2021

Hajianmaleki M, Qatu MS (2013) Vibrations of straight and curved composite beams: a review. Compos Struct 100:218–232

Hosseini-Hashem Sh, Ilkhani MR, Fadaee M (2013) Accurate natural frequencies and critical speeds of a rotating functionally graded moderately thick cylindrical shell. Int J Mech Sci 76:9–20

Hu XX, Sakiyama T, Matsuda H et al (2002) Vibration of twisted laminated composite conical shells. Int J Mech Sci 44:1521–1541

Ich Thinh T, Nguyen MC (2013) Dynamic stiffness matrix of continuous element for vibration of thick cross-ply laminated composite cylindrical shells. Compos Struct 98:93–102

Jafari-Talookolaei RA, Abedi M, Kargarnovin MH et al (2012) An analytical approach for the free vibration analysis of generally laminated composite beams with shear effect and rotary inertia. Int J Mech Sci 65:97–104

Jin GY, Ye TG, Chen YH et al (2013a) An exact solution for the free vibration analysis of laminated composite cylindrical shells with general elastic boundary conditions. Compos Struct 106:114–127

Jin GY, Ye TG, Ma XL et al (2013b) A unified approach for the vibration analysis of moderately thick composite laminated cylindrical shells with arbitrary boundary conditions. Int J Mech Sci 75:357–376

Jin GY, Ye TG, Jia XZ et al (2014a) A general Fourier solution for the vibration analysis of composite laminated structure elements of revolution with general elastic restraints. Compos Struct 109:150–168

Jin GY, Su Z, Shi SX et al (2014b) Three-dimensional exact solution for the free vibration of arbitrarily thick functionally graded rectangular plates with general boundary conditions. Compos Struct 108:565–577

Karami G, Malekzadeh P, Mohebpour SR (2006) DQM free vibration analysis of moderately thick symmetric laminated plates with elastically restrained edges. Compos Struct 74:115–125

Khdeir AA (1995) Dynamic response of cross-ply laminated circular cylindrical shells with various boundary conditions. Acta Mech 112:117–134

Khdeir AA, Reddy JN, Frederick D (1989) A study of bending, vibration and buckling of cross-ply circular cylindrical shells with various shell theories. Int J Eng Sci 27(1):1337–1351

Khdeir AA, Reddy JN (1997) Free and forced vibration of cross-ply laminated composite shallow arches. Int J Solids Struct 34(10):1217–1234

Kovács B (2013) Vibration analysis of layered curved arch. J Sound Vib 332:4223–4240

Kurpa L, Shmatko T, Timchenko G (2010) Free vibration analysis of laminated shallow shells with complex shape using the R-functions method. Compos Struct 93:225–233

Lam KY, Loy CT (1995a) Analysis of rotation laminated cylindrical shells by different thin shell theories. J Sound Vib 186(1):23–35

Lam KY, Loy CT (1995b) Influence of boundary conditions and fibre orientation on the natural frequencies of thin orthotropic laminated cylindrical shells. Compos Struct 31(1):21–30

Lam KY, Loy CT (1998) Influence of boundary conditions for a thin laminated rotating cylindrical shell. Compos Struct 41(3):215–228

Lam KY, Li H, Ng TY et al (2002) Generalized differential quadrature method for the free vibration of truncated conical panels. J Sound Vib 251(2):329–348

Lee JJ, Yeom CH, Lee I (2002) Vibration analysis of twisted cantilevered conical composite shells. J Sound Vib 225(5):965–982

Leissa AW (1969) Vibration of plates (NASA SP-160). US Government Printing Office, Washington, DC, pp 1–353

Leissa AW (1973) Vibration of shells (NASA SP-288). US Government Printing Office, Washington, DC, pp 1–428

Leissa AW, Chang JD (1996) Elastic deformation of thick, laminated composite shells. Compos Struct 35:153–170

Leissa AW, Qatu MS (2011) Vibrations of continuous systems. McGraw-Hill, New York, pp 1–507

Leissa AW, MacBain JC, Kielb RE (1984) Vibrations of twisted cantilever plates: summary of previous and current results. J Sound Vib 96:159–173

Li WL (2000) Free vibrations of beams with general boundary conditions. J Sound Vib 237:709–725

Li WL (2002) Comparison of Fourier sine and cosine series expansions for beams with arbitrary boundary conditions. J Sound Vib 255(1):185–194

Li WL (2004) Vibration analysis of rectangular plates with general elastic boundary supports. J Sound Vib 273:619–635

Li J, Hua H, Shen R (2008) Free vibration of laminated composite circular arches by dynamic stiffness analysis. J Reinf Plast Compos 27:851–870

Librescu L, Khdeir AA, Frederick D (1989a) A shear deformable theory of laminated composite shallow shell-type panels and their response analysis I: free vibration and buckling. Acta Mech 76:1–33

Librescu L, Khdeir AA, Frederick D (1989b) A shear deformable theory of laminated composite shallow shell-type panels and their response analysis II: static response. Acta Mech 77:1–12

Liew KM, Huang YQ, Reddy JN (2005) Vibration analysis of symmetrically laminated plates based on FSDT using the moving least squares differential quadrature method. Comput Methods Appl Mech Eng 194:4265–4278

Liew KM, Xiang Y, Kitipornchai S (1997) Vibration of laminated plates having elastic edge flexibility. J Eng Mech 123:1012–1019

Liew KM, Wang CM, Xiang Y et al (1998) Vibration of mindlin plates: programming the p-version Ritz method. Elsevier, Amsterdam, pp 1–202

Liew KM, Zhao X, Ferreira AJM (2011) A review of meshless methods for laminated and functionally graded plates and shells. Compos Struct 93:2031–2041

Lim CW, Liew KM, Kitipornchai S (1998) Vibration of cantilevered laminated composite shallow conical shells. Int J Solids Struct 35(15):1695–1707

Lin CC, Tseng CS (1998) Free vibration of polar orthotropic laminated circular and annular plates. J Sound Vib 209(5):797–810

Lin KC, Hsieh CM (2007) The closed form general solutions of 2-D curved laminated beams of variable curvatures. Compos Struct 79:606–618

Liu B, Xing YF, Qatu MS et al (2012) Exact characteristic equations for free vibrations of thin orthotropic circular cylindrical shells. Compos Struct 94:484–493

Love AEH (1944) A treatise on the mathematical theory of elasticity. Dover Pubications Inc, New York, pp 1–643

Malekzadeh P, Setoodeh AR (2009) DQM in-plane free vibration of laminated moderately thick circular deep arches. Adv Eng Softw 40:798–803

Mantari JL, Oktem AS, Soares CG (2011) Static and dynamic analysis of laminated composite and sandwich plates and shells by using a new higher order shear deformation theory. Compos Struct 94(1):37–49

Marzani A, Tornabene F, Viola E (2008) Nonconservative stability problems via generalized differential quadrature method. J Sound Vib 315:176–196

Matsunaga H (2001) Vibration and buckling of multilayered composite beams according to higher order deformation theories. J Sound Vib 246(1):47–62

Messina A, Soldatos KP (1999a) Vibration of completely free composite plates and cylindrical shell panels by a high-order theory. Int J Mech Sci 41:891–918

Messina A, Soldatos KP (1999b) Influence of edge boundary conditions on the free vibrations of cross-ply laminated circular cylindrical shell panels. J Acoust Soc Am 106(5):2608–2620

Messina A, Soldatos KP (1999c) Ritz-type dynamic analysis of cross-ply laminated circular cylinders subjected to different boundary conditions. J Sound Vib 27(4):749–768

Ng TY, Li H, Lam KY (2003) Generalized differential quadrature for free vibration of rotating composite laminated conical shells with various boundary conditions. Int J Mech Sci 45:567–587

Pagano NJ (1970) Exact solutions for rectangular bidirectional composites and sandwich plates. J Compos Mater 4:20–34

Pinto Correia IF, Mota Soares CM, Mota Soares CA et al (2003) Analysis of laminated conical shell structures using higher order models. Compos Struct 62:383–390

Price NM, Liu M, Taylor RE et al (1998) Vibrations of cylindrical pipes and open shells. J Sound Vib 218(3):361–387

Qatu MS, Leissa AW (1991) Free Vibrations of completely free doubly curved laminated composite shallow shells. J Sound Vib 151(1):9–29

Qatu MS (1992) In-plane vibration of slightly curved laminated composite beams. J Sound Vib 159(2):327–338

Qatu MS (1993) Theories and analyses of thin and moderately thick laminated composite curved beams. Int J Solids Struct 30(20):2743–2756

Qatu MS, Elsharkawy AA (1993) Vibration of laminated composite arches with deep curvature and arbitrary boundaries. Comput Struct 47(2):305–311

Qatu MS (1995a) Natural vibration of free, laminated composite triangular and trapezoidal shallow shells. Compos Struct 31:9–19

Qatu MS (1995b) Vibration studies on completely free shallow shells having triangular and trapezoidal planforms. Appl Acoust 44:215–231

Qatu MS (1996) Vibration analysis of cantilevered shallow shells with triangular trapezoidal planforms. J Sound Vib 191(2):219–231

Qatu MS (1999) Accurate equations for laminated composite deep thick shells. Int J Solids Struct 36(19):2917–2941

Qatu MS (2002a) Recent research advances in the dynamic behavior of shells: 1989–2000. Part 1: laminated composite shells. Appl Mech Rev 55(4):325–349

Qatu MS (2002b) Recent research advances in the dynamic behavior of shells: 1989–2000. Part 2: homogeneous shells. Appl Mech Rev 55(5):415–434

Qatu MS (2004) Vibration of laminated shells and plates. Elsevier, San Diego, pp 1–409

Qatu MS (2011) Effect of inplane edge constraints on natural frequencies of simply supported doubly curved shallow shells. Thin-Walled Struct 49:797–803

Qatu MS, Sullivan RW, Wang WC (2010) Recent research advances on the dynamic analysis of composite shells: 2000–2009. Compos Struct 93(1):14–31

Qu Y, Long X, Wu S et al (2013a) A unified formulation for vibration analysis of composite laminated shells of revolution including shear deformation androtary inertia. Compos Struct 98:169–191

Qu Y, Long X, Yuan G et al (2013b) A unified formulation for vibration analysis of functionally graded shells of revolution with arbitrary boundary conditions. Compos B Eng 50:381–402

Qu Y, Long X, Li H et al (2013c) A variational formulation for dynamic analysis of composite laminated beams based on a general higher-order shear deformation theory. Compos Struct 102:175–192

Qu Y, Hua H, Meng G (2013d) A domain decomposition approach for vibration analysis of isotropic and composite cylindrical shells with arbitrary boundaries. Compos Struct 95: 307–321

Rao SS (2007) Vibrations of continuous systems. Wiley, New Jersey, pp 1–720

Reddy JN (1984) Exact solutions of moderately thick laminated shells. J Eng Mech 110(5): 794–809

Reddy JN, Asce M (1984) Exact solutions of moderately thick laminated shells. J Eng Mech 110:794–809

Reddy JN, Liu CF (1985) A higher-order shear deformation theory of laminated elastic shells. Int J Eng Sci 23(3):319–330

Reddy JN (2002) Energy and principles and variational methods in applied mechanics, 2nd edn. Wiley, New Jersey, pp 1–572

Reddy JN (2003) Mechanics of laminated composite plates and shells: theory and analysis, 2nd edn. CRC Press, Florida, pp 1–831

Sai Ram KS, Sreedhar Babu T (2002) Free vibration of composite spherical shell cap with and without a cutout. Comput Struct 80:1749–1756

Schmidt R, Reddy JN (1988) A refined small strain and moderate rotation theory of elastic anisotropic shells. J Appl Mech 55:611–617

Shu C (1996) Free vibration analysis of composite laminated conical shells by generalized differential quadrature. J Sound Vib 194(4):587–604

Singh AV, Kumar V (1996) Vibration of laminated shells on quadrangular boundary. J Aerosp Eng 9:52–57

Singh BN, Yadav D, Iyengar NGR (2001) Free vibration of laminated spherical panel with random material properties. J Sound Vib 224(4):321–338

Soedel W (2004) Vibrations of shells and plates, 3rd edn. Marcel Dekker, New York, pp 1–553

Soldatos KP, Messina A (2001) The influence of boundary conditions and transverse shear on the vibration of angle-ply laminated plates, circular cylinders and cylindrical panels. Comput Methods Appl Mech Eng 190:2385–2409

Soldatos KP, Shu XP (1999) On the stress analysis of cross-ply laminated plates and shallow shell panels. Compos Struct 46:333–344

Su Z, Jin GY, Shi SX et al (2014a) A unified accurate solution for vibration analysis of arbitrary functionally graded spherical shell segments with general end restraints. Compos Struct 111:271–284

Su Z, Jin GY, Shi SX et al (2014b) A unified solution for vibration analysis of functionally graded cylindrical, conical shells and annular plates with general boundary conditions. Int J Mech Sci 80:62–80

Su Z, Jin GY, Ye TG (2014c) Free vibration analysis of moderately thick functionally graded open shells with general boundary conditions. Compos Struct 117:169–186

Swaddiwudhipong S, Tian J, Wang CM (1995) Vibration of cylindrical shells with intermidiate ring supports. J Sound Vib 187(1):69–93

Timoshenko SP (1921) On the correction for shear of the differential equation for transverse vibrations of prismatic beams. Philos Mag 6:744–746

Tolstov GP (1976) Fourier series. Prentice-Hall, New Jersey, pp 1–336

Tong L (1993) Free vibration of composite laminated conical shells. Int J Mech Sci 35(1):47–61

Tong L (1994a) Free vibration of laminated conical shells including transverse shear deformation. Int J Solids Struct 31(4):443–456

Tong L (1994b) Effect of transverse shear deformation on free vibration of orthotropic conical shells. Acta Mech 107:65–75

Toorani MH, Lakis AA (2000) General equations of anisotropic plates and shells including transverse shear deformations, rotary inertia and initial curvature effects. J Sound Vib 237 (4):561–615

Tornabene F (2009) Free vibration analysis of functionally graded conical, cylindrical and annular shell structures with a four-parameter power-law distribution. Comput Methods Appl Mech Eng 198:2911–2935

Tornabene F (2011) 2-D GDQ solution for free vibrations of anisotropic doubly-curved shells and panels of revolution. Compos Struct 93:1854–1876

Tornabene F, Viola E (2007) Vibration analysis of spherical structural elements using the GDQ method. Comput Math Appl 53:1538–1560

Tornabene F, Viola E (2008) 2-D solution for free vibrations of parabolic shells using generalized differential quadrature method. Eur J Mech A-Solids 2:1001–1125

Tornabene F, Viola E (2009a) Free vibrations of four-parameter functionally graded parabolic panels and shell of revolution. Eur J Mech A-Solids 28:991–1013

Tornabene F, Viola E (2009b) Free vibration analysis of functionally graded panels and shells of revolution. Meccanica 44:255–281

Tornabene F, Viola E, Inman DJ (2009) 2-D differential quadrature solution for vibration analysis of functionally graded conical, cylindrical and annular shell structures. J Sound Vib 328: 259–290

Tornabene F, Liverani A, Caligiana G (2012a) Laminated composite rectangular and annular plates: a GDQ solution for static analysis with a posteriori shear and normal stress recovery. Compos B Eng 43:1847–1872

Tornabene F, Liverani A, Caligiana G (2012b) Static analysis of laminated composite curved shells and panels of revolution with a posteriori shear and normal stress recovery using generalized differential quadrature method. Int J Mech Sci 61:71–87

Tornabene F, Liverani A, Caligiana G (2012c) General anisotropic doubly-curved shell theory: a differential quadrature solution for free vibrations of shells and panels of revolution with a free-form meridian. J Sound Vib 331:4848–4869

Tornabene F, Fantuzzi N, Viola E et al (2013a) Radial basis function method applied to doubly-curved laminated composite shells and panels with a general higher-order equivalent single layer theory. Compos B Eng 55:642–659

Tornabene F, Viola E, Fantuzzi N (2013b) General higher-order equivalent single layer theory for free vibrations of doubly-curved laminated composite shells and panels. Compos Struct 104:94–117

Tornabene F, Fantuzzi N, Viola E et al (2014a) Static analysis of doubly-curved anisotropic shells and panels using CUF approach, differential geometry and differential quadrature method. Compos Struct 107:675–697

Tornabene F, Fantuzzi N, Viola E et al (2014b) Winkler-Pasternak foundation effect on the static and dynamic analyses of laminated doubly-curved and degenerate shells and panels. Compos B Eng 57:269–296

Vel SS, Batra RC (1999) Analytical solutions for rectangular thick laminated plates subjected to arbitrary boundary conditions. AIAA J 37:1464–1473

Vel SS, Batra RC (2000a) The generalized plane strain deformations of thick anisotropic composite laminated plates. Int J Solids Struct 37:715–733

Vel SS, Batra RC (2000b) Closure to the generalized plane strain deformations of thick anisotropic composite laminated plates. Int J Solids Struct 38:483–489

Viola E, Tornabene F (2009) Free vibrations of three parameter functionally graded parabolic panels of revolution. Mech Res Commun 36:587–594

Viola E, Dilena M, Tornabene F (2007) Analytical and numerical results for vibration analysis of multi-stepped and multi-damaged circular arches. J Sound Vib 299:143–163

Viola E, Rossetti L, Fantuzzi N (2012) Numerical investigation of functionally graded cylindrical shells and panels using the generalized unconstrained third order theory coupled with the stress recovery. Compos Struct 94(12):3736–3758

Viola E, Tornabene F, Fantuzzi N (2013) General higher-order shear deformation theories for the free vibration analysis of completely doubly-curved laminated shells and panels. Compos Struct 95:639–666

Viola E, Rossetti L, Fantuzzi N et al (2014) Static analysis of functionally graded conical shells and panels using the generalized unconstrained third order theory coupled with the stress recovery. Compos Struct 112:44–65

Viswanathan KK, Lee JH, Aziz ZA et al (2012) Vibration analysis of cross-ply laminated truncated conical shells using a spline method. J Eng Math 76:139–156

Vlasov VZ (1951) Basic differential equations in the general theory of elastic shells (NACA-TM-1241). National Advisory Committee for Aeronautics, Washington, DC, United States, pp 1–58

Wang JTS, Lin CC (1996) Dynamic analysis of generally supported beams using Fourier series. J Sound Vib 196:285–293

Wang S (1997) Free vibration analysis of skew fiber-reinforced composite laminates based on first-order shear deformation plate theory. Comput Struct 63:525–538

Wu CP, Lee CY (2011) Differential quadrature solution for the free vibration analysis of laminated conical shells with variable stiffness. Int J Mech Sci 43:1853–1869

Wu CP, Wu CH (2000) Asymptotic differential quadrature solutions for the free vibration of laminated conical shells. Comput Mech 25:346–357

Xing YF, Liu B (2009a) New exact solutions for free vibrations of thin orthotropic rectangular plates. Compos Struct 89:567–574

Xing YF, Liu B (2009b) Exact solutions for the free in-plane vibrations of rectangular plates. Int J Mech Sci 51:246–255

Xing YF, Liu B (2009c) New exact solutions for free vibrations of rectangular thin plates by symplectic dual method. Acta Mech Sin 25:265–270

Xing YF, Liu B, Xu T (2013) Exact solutions for free vibration of circular cylindrical shells with classical boundary conditions. Int J Mech Sci 75:178–188

Xu H (2010) The Fourier spectral element method for vibration and power flow analysis of complex dynamic systems. Ph.D. thesis, Wayne State University, Detroit, Michigan, pp 1–10

Ye JQ (2003) Laminated composite plates and shells: 3D modelling. Springer, London, pp 1–273

Ye TG, Jin GY, Chen YH et al (2013) Free vibration analysis of laminated composite shallow shells with general elastic boundaries. Compos Struct 106:470–490

Ye TG, Jin GY, Su Z et al (2014a) A modified Fourier solution for vibration analysis of moderately thick laminated plates with general boundary restraints and internal line supports. Int J Mech Sci 80:29–46

Ye TG, Jin GY, Su Z et al (2014b) A unified Chebyshev-Ritz formulation for vibration analysis of composite laminated deep open shells with arbitrary boundary conditions. Arch Appl Mech 84:441–471

Ye TG, Jin GY, Chen YH et al (2014c) A unified formulation for vibration analysis of open shells with arbitrary boundary conditions. Int J Mech Sci 81:42–59

Ye TG, Jin GY, Shi SX et al (2014d) Three-dimensional free vibration analysis of thick cylindrical shells with general end conditions and resting on elastic foundations. Int J Mech Sci 84:120–137

Ye TG, Jin GY, Su Z (2014e) Three-dimensional vibration analysis of laminated functionally graded spherical shells with general boundary conditions. Compos Struct 116:571–588

Zenkour AM (1998) Vibration of axisymmetric shear deformable cross-ply laminated cylindrical shells—a variational approach. Int J Eng Sci 36:219–231

Zenkour AM, Fares ME (2001) Bending, buckling and free vibration of nonhomogeneous composite laminated cylindrical shells using a refined first order theory. Compos B Eng 32:237–247

Zhang X, Li WL (2009) Vibrations of rectangular plates with arbitrary non-uniform elastic edge restraints. J Sound Vib 326:221–234

Zhang L, Xiang Y (2006) Vibration of open circular cylindrical shells with intermediate ring supports. Int J Solids Struct 43:3705–3722

Zhang XM (2001) Vibration analysis of cross-ply laminated composite cylindrical shells using the wave propagation approach. Appl Acoust 62(11):1221–1228

Zhang XM (2002) Parametric analysis of frequency of rotating laminated composite cylindrical shells with the wave propagation approach. Comput. Methods Appl Mech. Eng. 191:2029–2043

Zhao X, Liew KM, Ng TY (2003) Vibration analysis of laminated composite cylindrical panels via a meshfree approach. Int J Solids Struct 40(1):161–180

Zhou D (1995) Natural frequencies of elastically restrained rectangular plates using a set of static beam functions in the Rayleigh-Ritz method. Comput Struct 57:731–735

Zhou D (1996) Natural frequencies of rectangular plates using a set of static beam functions in the Rayleigh-Ritz method. J Sound Vib 189:81–87

Zhou D (2001) Free vibration of multi-span Timoshenko beams using static Timoshenko beam functions. J Sound Vib 241:725–734

Printed in the United States
By Bookmasters